THE SERENGETI'S GREAT MIGRATION

"WHETHER SEEN AT EYE LEVEL OR FROM THE AIR, THE GREAT FLOWING BIOMASS OF NEAR TWO MILLION WILDEBEEST, ZEBRA AND GAZELLE ACROSS THE HIGH PLAINS OF THE SERENGETI IN MYSTERIOUS 'MIGRATION' TOWARDS THE DISTANT RAINS, IS AMONG THE VERY LAST VAST WILDLIFE SPECTACLES ON EARTH, BEYOND COMPARISON WITH ANYTHING WE KNOW EXCEPT THE CALVACADES OF BISON ON THE GREAT PLAINS OF NORTH AMERICA AND THE CANOPIES OF PASSENGER PIGEONS WHOSE MANY MILLIONS WERE STILL DARKENING THE SKIES ONLY 30 YEARS BEFORE THE LAST SPECIMEN OF THE MOST NUMEROUS BIRD SPECIES EVER TO EXIST ON EARTH DIED IN A ZOO IN 1913.

ONLY BY CHANCE, TWO DECADES EARLIER, WAS THE SAME GROTESQUE AND OUTRAGEOUS FATE NOT VISITED ON THE BISON WHICH HAS SURVIVED AS A MORE OR LESS CAPTIVE CREATURE OF LARGE GRASSY MENAGERIES SUCH AS PARKS. UNTIL MANKIND PROFOUNDLY UNDERSTANDS THAT ALL LIFE IS ONE, THE FUTURE OF THE WILDEBEEST AND ITS TRAVELLING COMPANIONS WILL REMAIN PRECARIOUS, INCREDIBLE AS THAT MAY SEEM TO ANY OF US FORTUNATE ENOUGH TO HAVE TAKEN THIS BEAUTIFUL BOOK INTO OUR HANDS AND BEHELD ITS PLENTY."

PETER MATTHIESSEN

THE SERENGETI'S

ABBEVILLE PRESS

NEW YORK | LONDON | PARIS

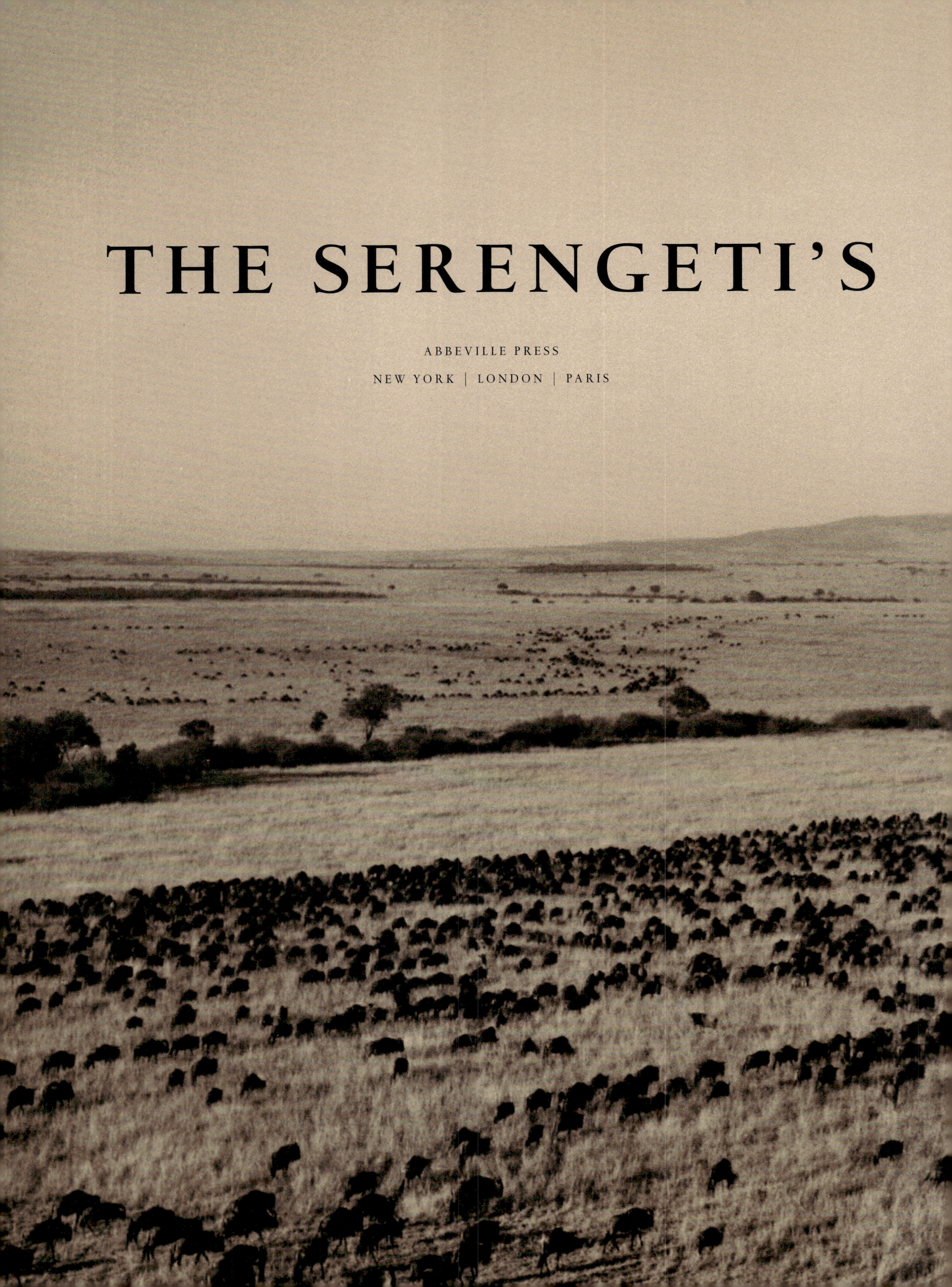

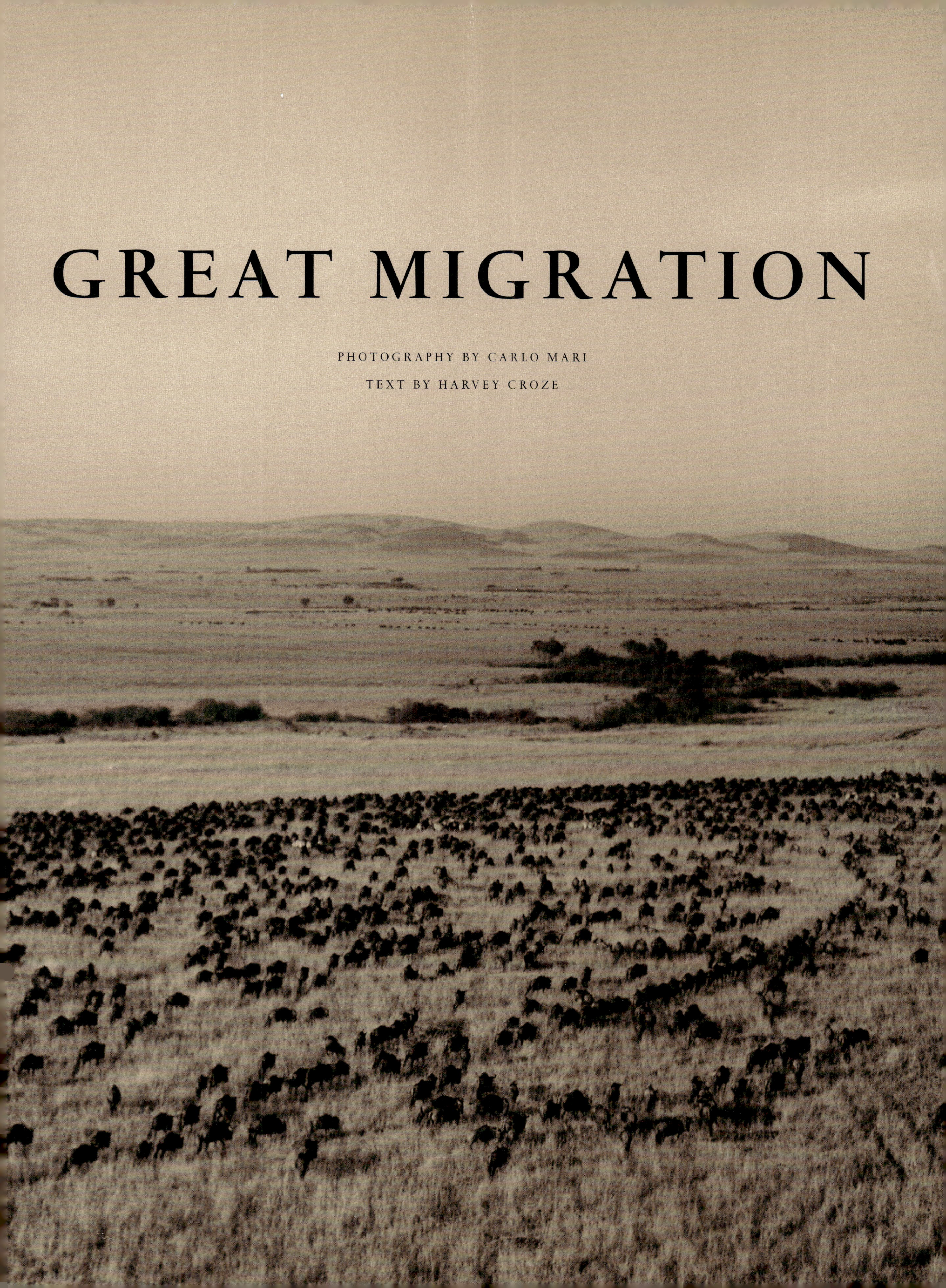

GREAT MIGRATION

PHOTOGRAPHY BY CARLO MARI

TEXT BY HARVEY CROZE

To Isabel Brazas Croze and Arlen White Linn
Harvey Croze

Designed by Olivier Assouline

First published in the United States of America by Abbeville Press in 2000

First published in Great Britain by The Harvill Press in 1999

 The text of this book was set in Spectrum. Printed and bound in Italy by EBS, Verona.

Inquiries should be addressed to
Abbeville Publishing Group,
22 Cortlandt Street,
New York, N.Y. 10007

First edition
10 9 8 7 6 5 4 3 2 1

Library in Congress Cataloging-in-Publication Data
Mari, Carlo.
Serengeti's great migration/photographs by Carlo Mari ; text by Harvey Croze.
p. cm.
ISBN 0-7892-0669-2 (alk. paper)
1. Ungulata--Migration--Tanzania--Serengeti Plain. 2. Ungulata--Predators of--Tanzania--Serengeti Plain. 3. Ungulata--Predators of--Tanzania--Serengeti Plain--Pictorial works. 4. Ungulata--Predators of--Tanzania--Serengeti Plain--Pictorial works. I. Croze, Harvey. II. Title.

QL737.U4 M37 2000
599.6"1568'0967827--dc21

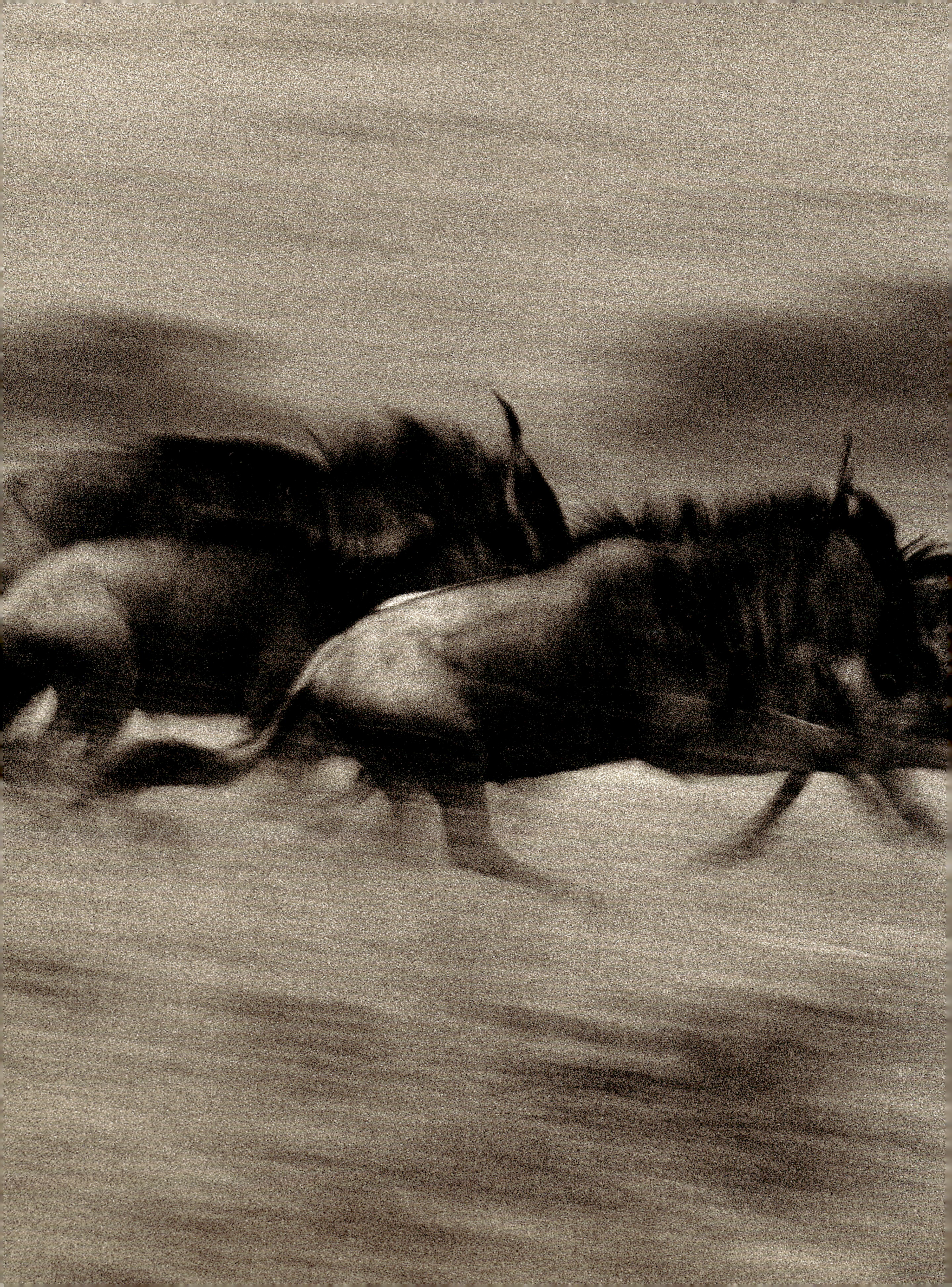

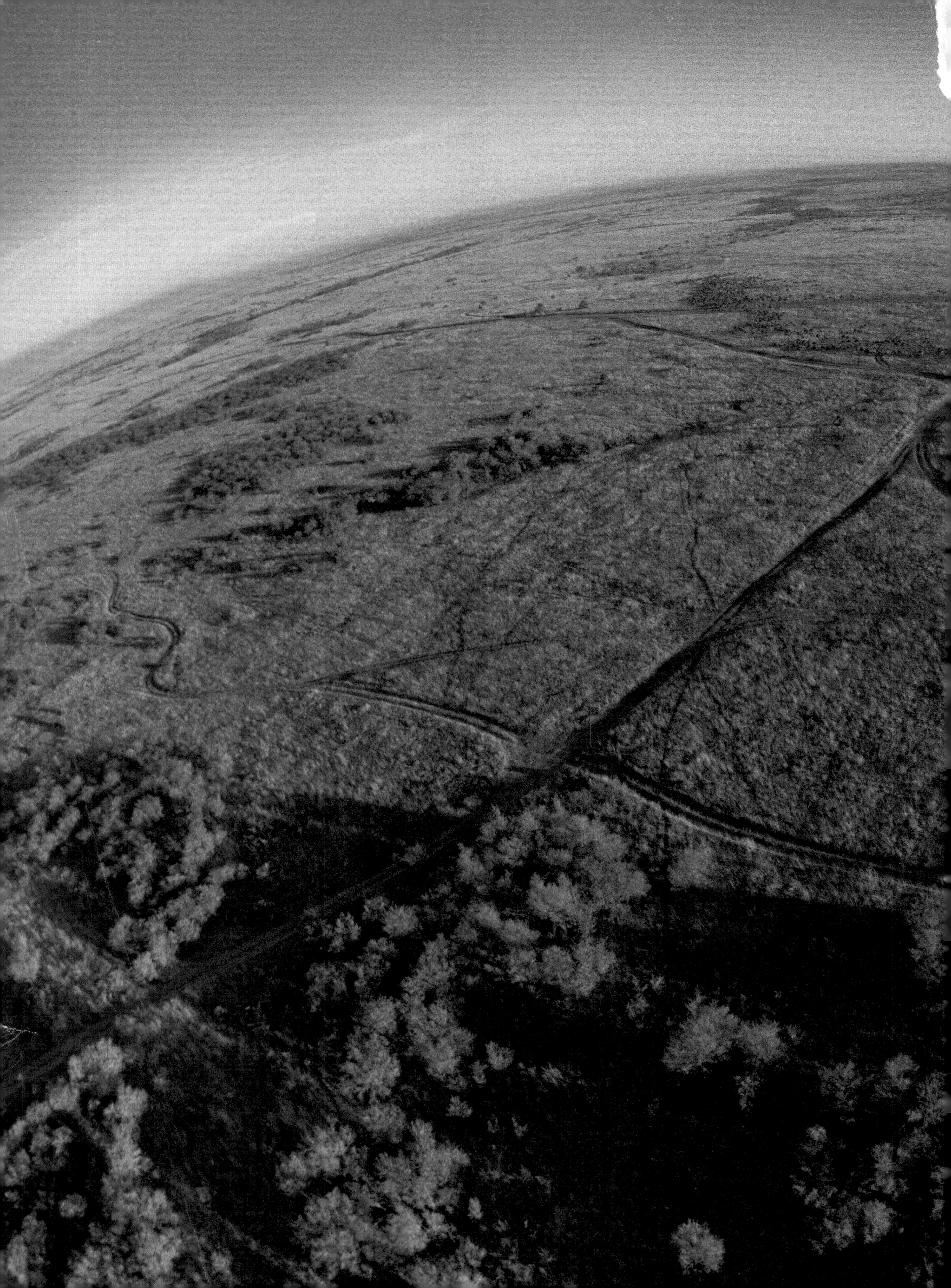

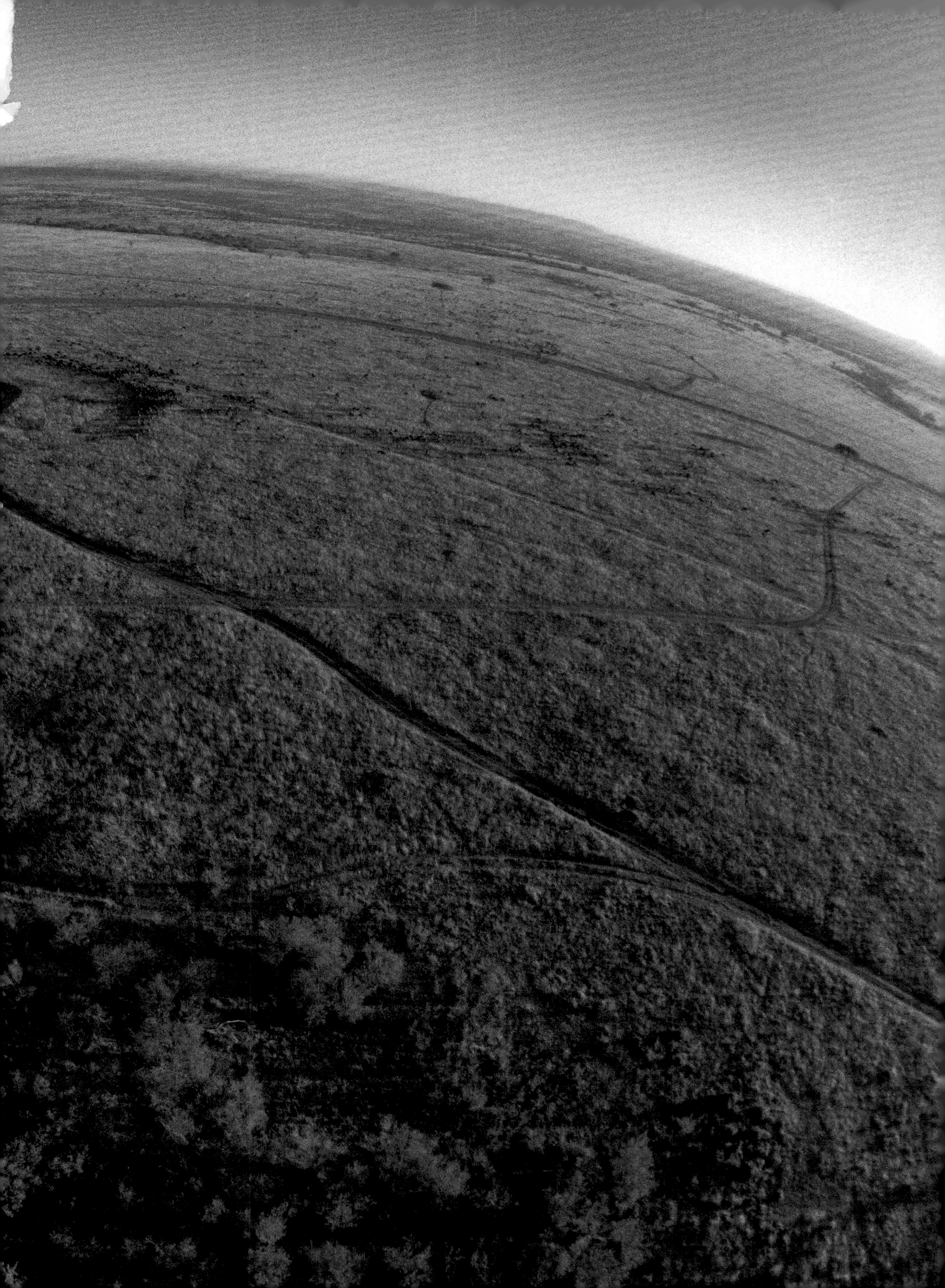

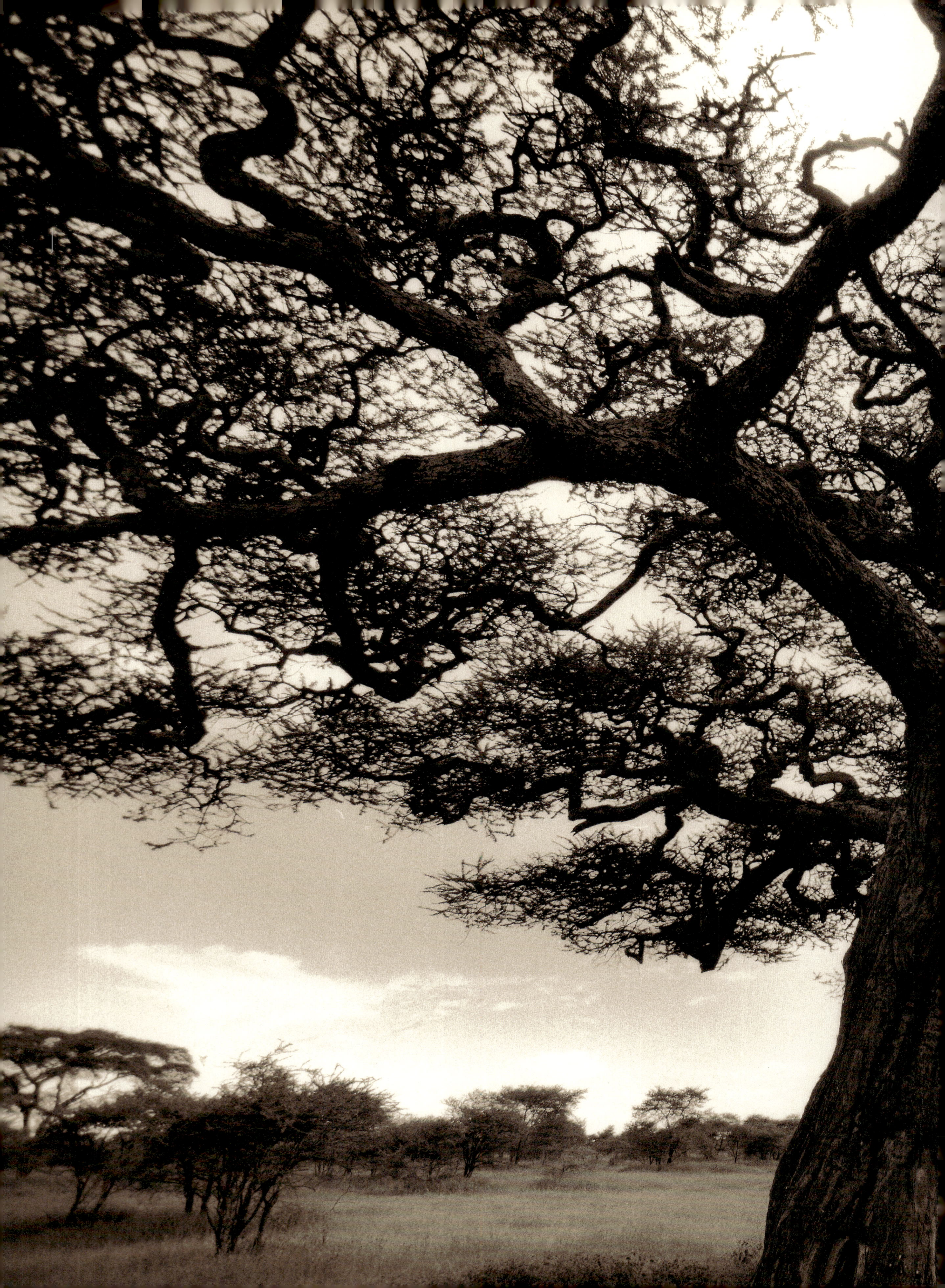

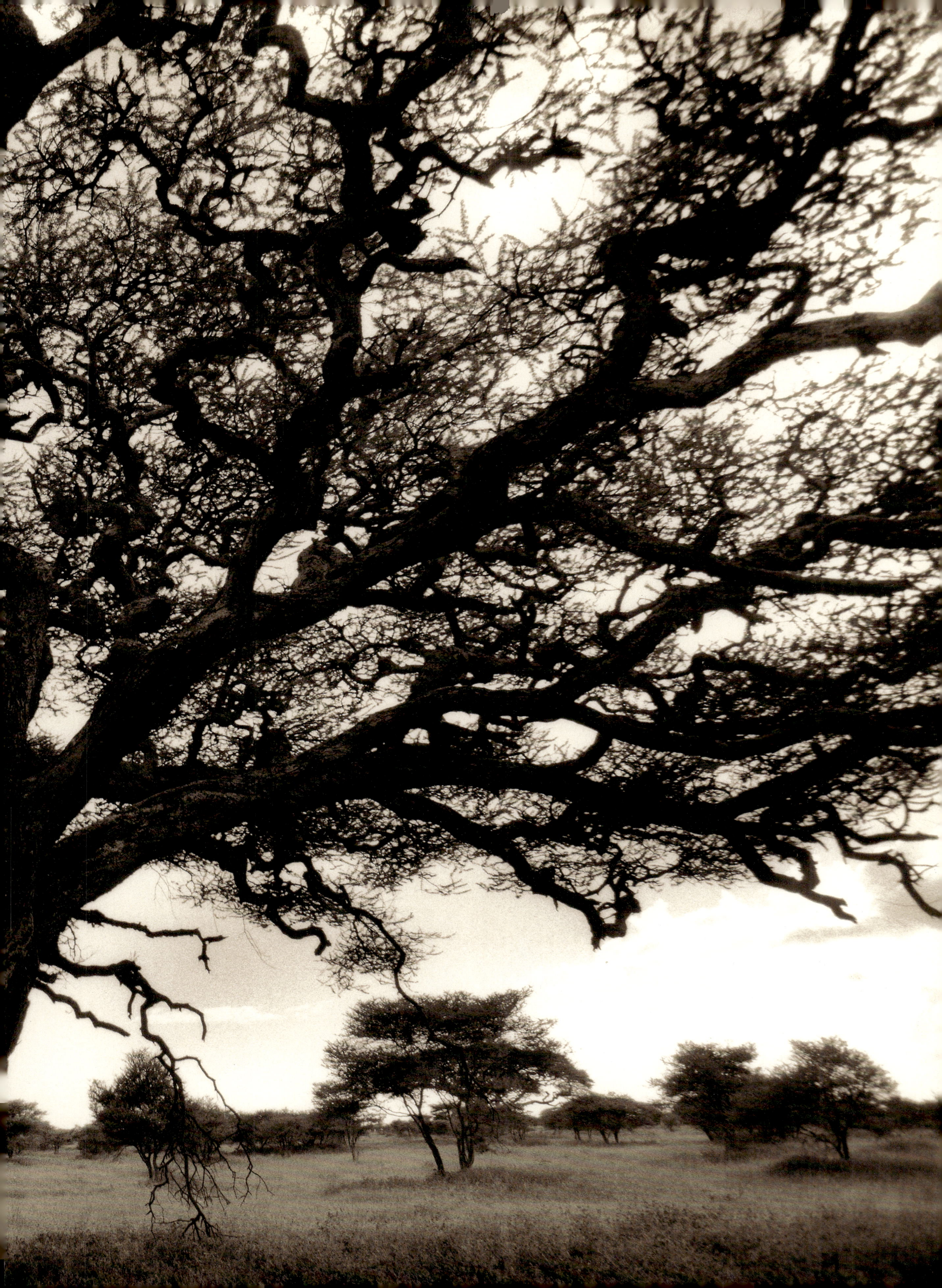

"DRAMATIS BESTIALIS"

MAIN MIGRATORS	
WILDEBEEST = 1,300,000	
THOMSON'S GAZELLE = 360,000	
ZEBRA = 191,000	1,863,000
ELAND = 12,000	
RESIDENTS	
TOPI = 95,000	
IMPALA = 76,000	
BUFFALO = 46,000	
GRANT'S GAZELLE = 26,000	
KONGONI = 14,000	276,000
GIRAFFE = 9,000	
WARTHOG = 6,000	
WATERBUCK = 2,000	
ELEPHANT = 2,000	
GRAND TOTAL	2,139,000

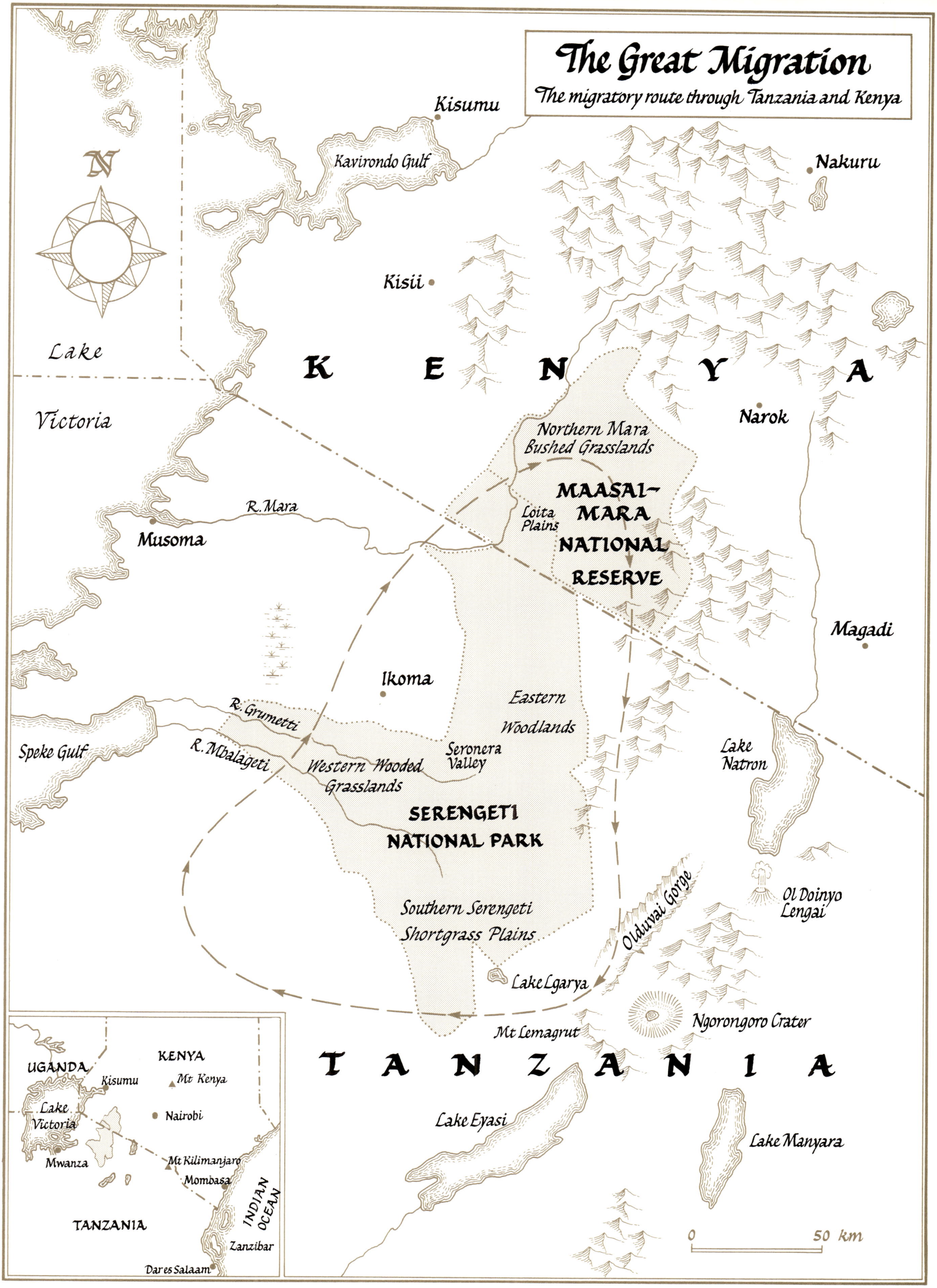
The Great Migration
The migratory route through Tanzania and Kenya
Kisumu
Kavirondo Gulf
Nakuru
N
Kisii
Lake
Victoria
K E N Y A
Narok
Northern Mara
Bushed Grasslands
MAASAI-
MARA
NATIONAL
RESERVE
Loita
Plains
R. Mara
Musoma
Magadi
Ikoma
Eastern
Woodlands
R. Grumetti
Speke Gulf
R. Mbalageti
Seronera
Valley
Western Wooded
Grasslands
Lake
Natron
SERENGETI
NATIONAL PARK
Olduvai Gorge
Ol Doinyo
Lengai
Southern Serengeti
Shortgrass Plains
Lake Lgarya
Ngorongoro Crater
Mt Lemagrut
T A N Z A N I A
Lake Eyasi
Lake Manyara
0
50 km
UGANDA
KENYA
Kisumu
Mt Kenya
Lake
Victoria
Nairobi
Mwanza
Mt Kilimanjaro
Mombasa
INDIAN
OCEAN
TANZANIA
Zanzibar
Dar es Salaam

WILDEBEEST

{*Connochaetes taurinus*}, n., pl. ~e, a.k.a GNU

A LARGE, BEARDED ANTELOPE OF THE ACACIA SAVANNA AND SHORTGRASS PLAINS. DEEP-CHESTED AND SHORT-NECKED, HIGH SHOULDERS AND THIN LEGS... HEAD: BROAD, BLUNT MUZZLE. HORNS: COWLIKE, SMOOTH, WITH ENLARGED BOSS, WIDER AND MUCH THICKER IN MALE... COAT: SHORT AND GLOSSY WITH VERTICAL STRIPS OF LONGER HAIR; A BEARD AND LAX OR UPSTANDING MANE. COLOUR: VARIES WITH SUBSPECIES, GENDER (MALES DARKER), SEASON, AND INDIVIDUALITY; RANGING FROM SLATE GREY TO DARK BROWN; LIGHTER ABOVE/DARKER BELOW; BLACK FACE, MANE, TAIL, AND STRIPES, BEARD BLACK (EXCEPT WHITE-BEARDED SPECIES)[1].

SMALLER THAN AN AFRICAN BUFFALO, BIGGER THAN A GAZELLE, THE GNU IS THE MOST NUMEROUS MEMBER OF THE IMPORTANT SUB-FAMILY OF THE LONG-FACED ALCELAPHINE ANTELOPES THAT INCLUDES IMPALA, TOPI AND BONTEBOK, HARTEBEESTE (Kongoni) AND THE RARE HUNTER'S HARTEBEESTE. A "SOCIAL GRAZER" WHO, LIKE A SOCIAL DRINKER, CONGREGATES IN RESPONSE TO LOCAL DISTRIBUTION OF KEY RESOURCES, IN THE GNU'S CASE, SHORTGRASS PASTURES AND WATER. BIOLOGISTS WITH A TAXONOMIC BENT ARE FINDING SUBTLE COLORATION AND MORPHOLOGICAL

(1) Estes, R. D. (1993). *The Safari Companion: A Guide to Watching African Mammals*. White River Junction, Vermont, Chelsea Green.

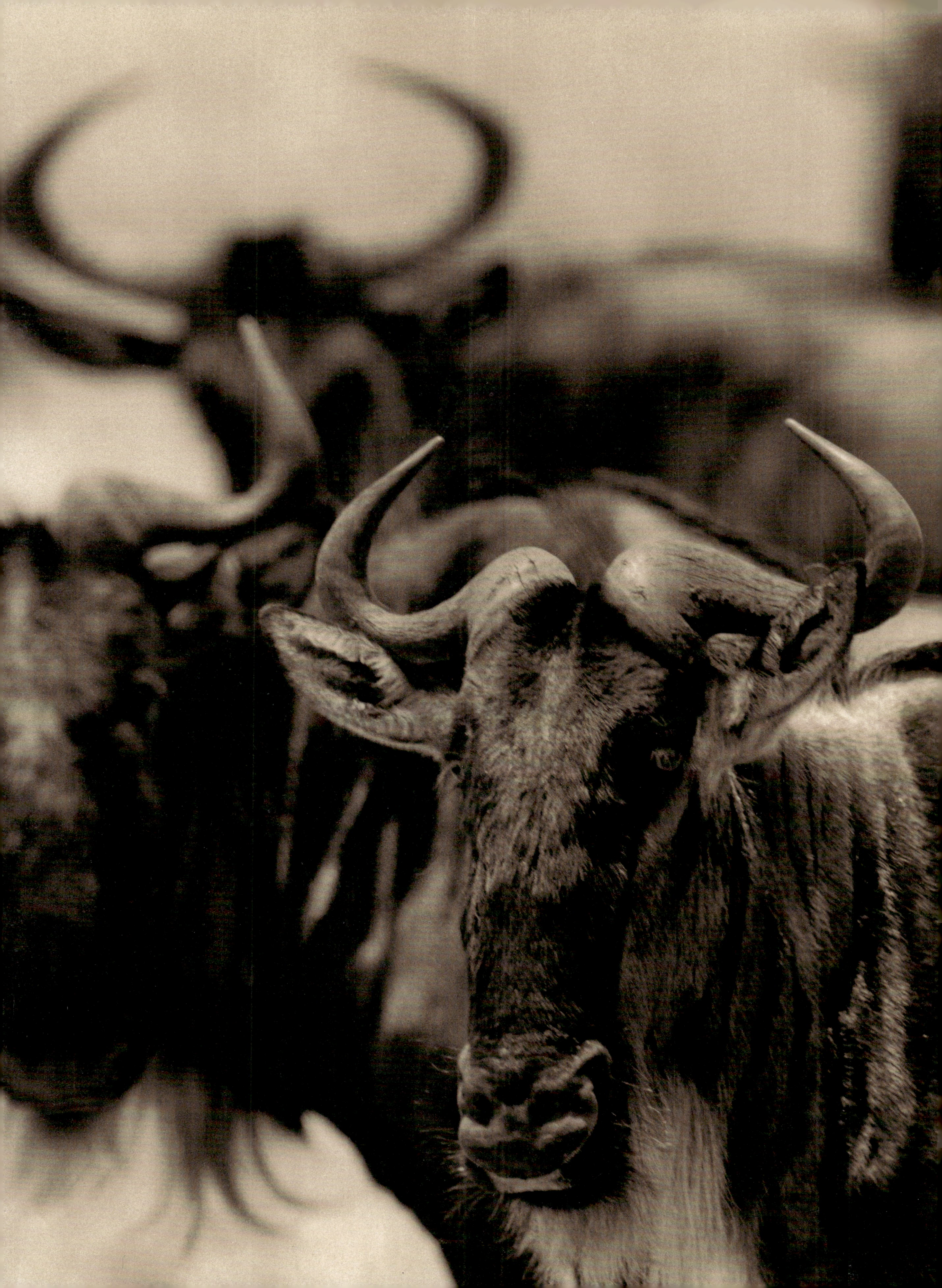

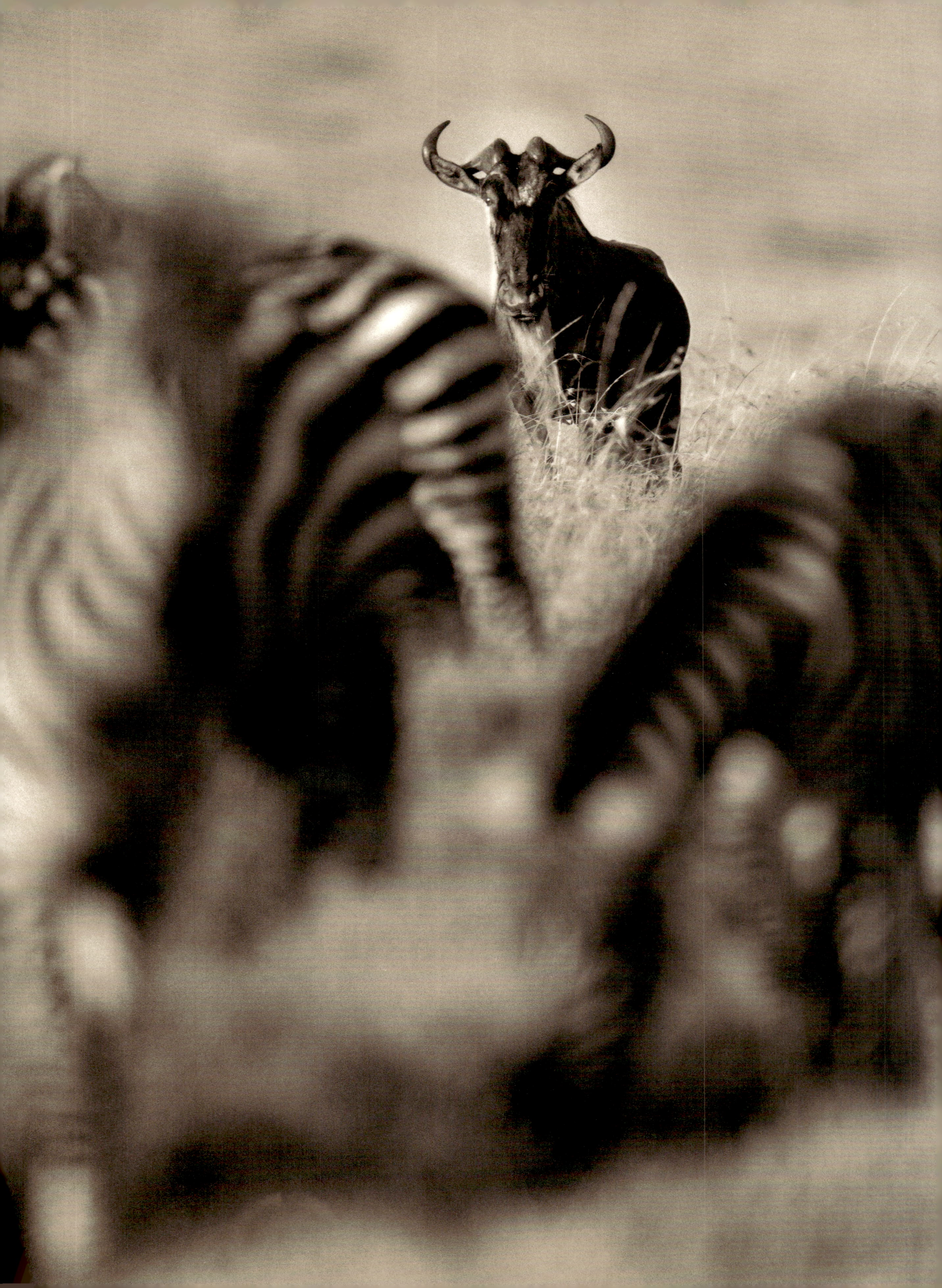

DIFFERENCES BETWEEN SPECIES POPULATIONS AND GIVING NAMES TO SUBSPECIES OR RACES. THE GNU HAS AT LEAST SIX TO ITS CREDIT. OUR HERO OF THE SERENGETI MIGRATION IN KENYA AND TANZANIA WEST OF THE GREGORY RIFT IS THE WESTERN WHITE-BEARDED WILDEBEEST, *Connochaetes taurinus mearnsi*. THE POPULATION OF RELATIVES IN CENTRAL AND SOUTHERN AFRICA IS DISTINGUISHED WITH TWO ENGLISH NAMES, THE BLUE WILDEBEEST OR BRINDLED GNU (*C. t. taurinus*), EXCEPT FOR THOSE THAT LIVE IN SOUTHERN TANZANIA AND MOZAMBIQUE AS FAR AS THE ZAMBEZI RIVER, WHICH ARE, OF COURSE, JOHNSTON'S OR THE NYASSA WILDEBEEST (*C. t. johnstoni*) AND THE LOT CONFINED TO THE LUANGWA VALLEY IN ZAMBIA, NAMELY, COOKSON'S WILDEBEEST (*C. t. cooksoni*). IT IS LIKELY THAT IF YOU WOULD CONFINE ALL SIX SUBSPECIES TO ONE BIG PASTURE, THEY WOULD ALL HAPPILY INTERBREED AND BECOME AGAIN ONE ARCHETYPAL GNU. BUT LET US INTRODUCE THE BEAST:

A dumpy, thick-necked, long-faced antelope with horns that flare out sideways and then upwards (rather like a cow's). The flat, rather square nasal plate (with hair-lined, flap-edged nostrils) is bounded by an even broader, grass-nibbling mouth. Dense tufts of hair on the long, convex muzzle conceal very active pre-orbital glands and help to diffuse scent to them. The muzzle is black in all subspecies, as is the shaggy mane and tail. The body colours of the four or five different subspecies varies from dark grey-brown to slate blue to pale greyish fawn, with variable degrees of brindling. This strongly textured streaking of the neck, shoulders and flanks mimics the visual effect of a long, lank mane very closely. The neck and chin are bearded in long, black, brown, cream or white hair. The short legs are brown or ochre, with pedal glands between the large true hooves (there are prominent false lateral hoooves)[2].

GRATEFUL WE SHOULD ALL BE FOR THE BREVITY OF THE ONOMATOPOEIC WORD "GNU", WHICH ECHOES THE SOUND THE BEAST MAKES DAY AFTER DAY, NIGHT AFTER NIGHT, OVER AND OVER DURING THE ENDLESS CYCLE OF THE MIGRATION. THE REPEATED REFERENCE TO THE ANIMALS BEING "COW-LIKE" IS PARTICULARLY APT GIVEN THEIR DEMEANOUR, FROM THE MILDLY FOOLISH AS THEY MOVE FORWARD TO INVESTIGATE YOU TO THE TRUE "MAD COW" OF THE PLAINS AS THEY LEAP ABOUT DURING THE RUT. THEY ARE ABOUT THE SIZE OF THE ECONOMICAL AND LEAN MAASAI ZEBU CATTLE, WITH MALES UP TO TWENTY TWO INCHES AT THE SHOULDER AT A MAXIMUM WEIGHT OF AROUND 600 POUNDS. SEASONED WILDEBEESTE-WATCHERS CONCLUDE THAT GNUS HAVE BEEN DESIGNED BY A COMMITTEE AND ASSEMBLED WITH SPARE PARTS.

(2) Kingdon, J. (1997). *The Kingdon Field Guide to African Mammals*. London, Academic Press.

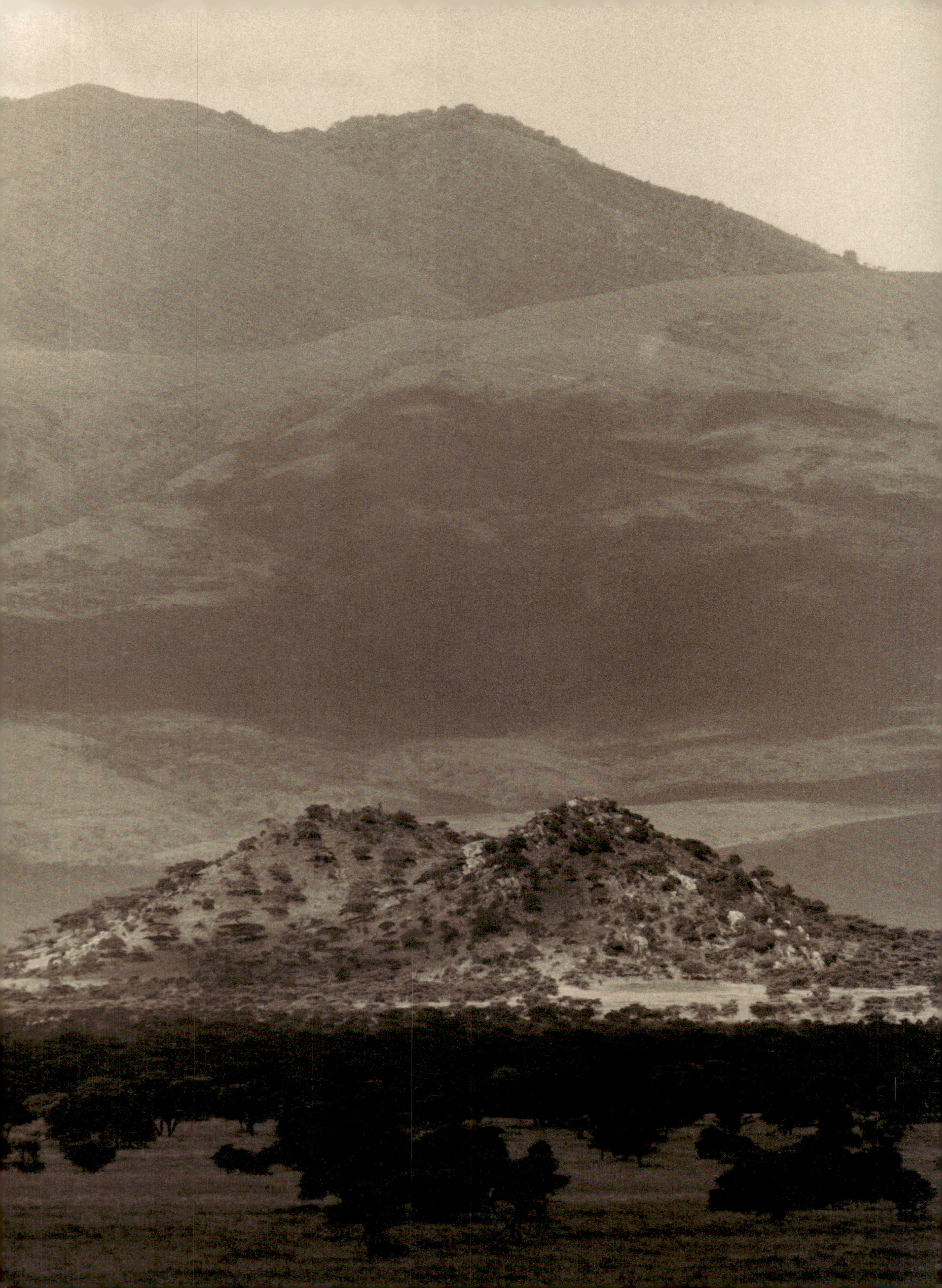

FOREWORD

Richard D. Estes

I WRITE THIS AS I JOURNEY TO TANZANIA TO OBSERVE THE RUT OF THE SERENGETI WILDEBEEST. THIS ANNUAL EVENT, COMING AFTER THE LONG RAINS WHEN THE WILDEBEESTE MIGRATE FROM THE SERENGETI PLAINS INTO THE WOODLAND ZONE, IS ONE OF THE PLANET'S PREMIER ANIMAL HAPPENINGS. IN THE SPACE OF THREE WEEKS, HALF A MILLION COWS ARE BRED, EACH DURING AN OESTRUS LASTING ONLY ONE DAY. THE NOISE AND CONFUSION ARE INDESCRIBABLE, AS BREEDING BULLS GALLOP INTO THE RIVER OF PASSING WILDEBEESTE, ROUNDING UP THE COWS. THE BASSO CALLING OF THOUSANDS OF MALES, LIKE A CHORUS OF GIANT BULLFROGS, SHATTERS THE AIR, WHILE THE CLASH OF HORNS AS COMPETING BULLS RAM ONE ANOTHER HEAD-ON PUNCTUATES THE DIN.

LOOKING AT CARLO MARI'S MAGNIFICENT IMAGES HAS PRIMED ME FOR THE GREAT MIGRATION, BRINGING BACK MEMORIES OF THE YEARS I HAVE SPENT STUDYING WILDEBEEST BEHAVIOUR AND ECOLOGY. HE HAS CAPTURED THE SPIRIT OF THE SERENGETI AND MAASAI MARA, AND INCLUDED THE WHOLE CAST OF CHARACTERS: FIRST AND FOREMOST THE MASS MOVEMENTS OF THE HERBIVORES, WITH THE WILDEBEEST AS THE KEYSTONE SPECIES, AND ALSO THE PREDATORS AND SCAVENGERS THAT DISPOSE OF THE OLD, THE UNFIT, AND THE UNLUCKY.

MUCH HAS CHANGED SINCE I FIRST EXPERIENCED THE GREAT MIGRATION IN 1963. THE REALM OF AFRICAN WILDLIFE HAS BEEN PROGRESSIVELY REDUCED AS THE HUMAN POPULATION DOUBLED TO OVER 500 MILLION AND CONTINUES TO EXPAND, THE SERENGETI IS THE LAST INTACT PLAINS ECOSYSTEM AND THE ABUNDANCE OF MIGRATORY WILDEBEESTE, GAZELLE AND ZEBRA REMAINS AT AN HISTORIC HIGH THAT IS UNEQUALLED ELSEWHERE IN AFRICA. THE PARK IS FAR BETTER PROTECTED FROM POACHERS THAN IT WAS IN THE 1970s AND 80s, WHEN RHINOS WERE NEARLY ELIMINATED AND AN ELEPHANT POPULATION ESTIMATED AT 3,000 WAS REDUCED TO ABOUT 300 HIGHLY STRESSED ANIMALS. BUT A MILLION PEOPLE ARE NOW CROWDED BETWEEN LAKE VICTORIA AND THE WESTERN BORDER OF THE PARK WITH A GROWING DEMAND FOR LAND AND MEAT. THE SERENGETI IS A WORLD HERITAGE SITE AND INTERNATIONAL BIOSPHERE RESERVE AND THE TANZANIAN GOVERNMENT WILL NEED THE SUPPORT OF THE DEVELOPED NATIONS TO PRESERVE THIS AND THE OTHER NATURAL AREAS FOR FUTURE GENERATIONS.

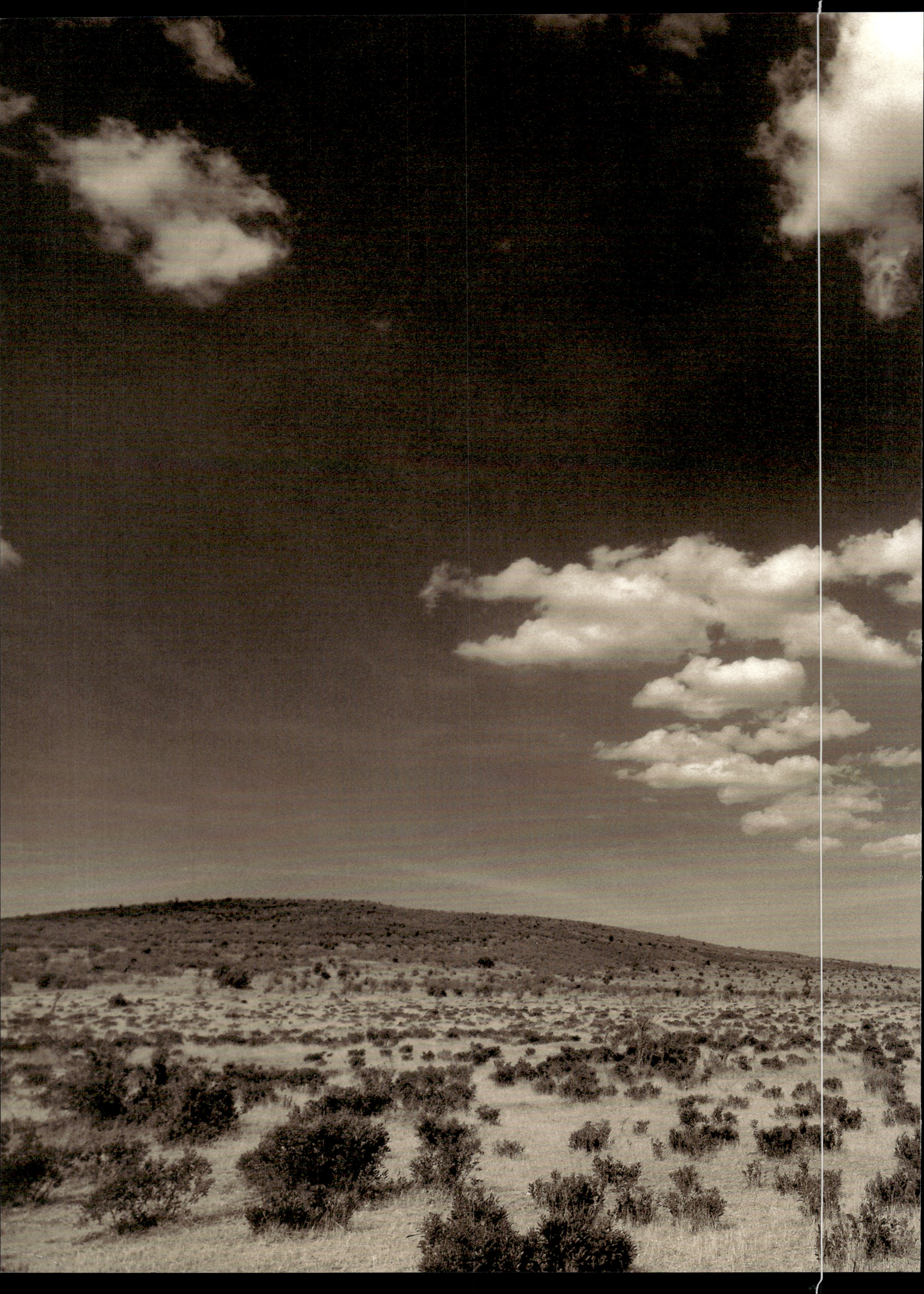

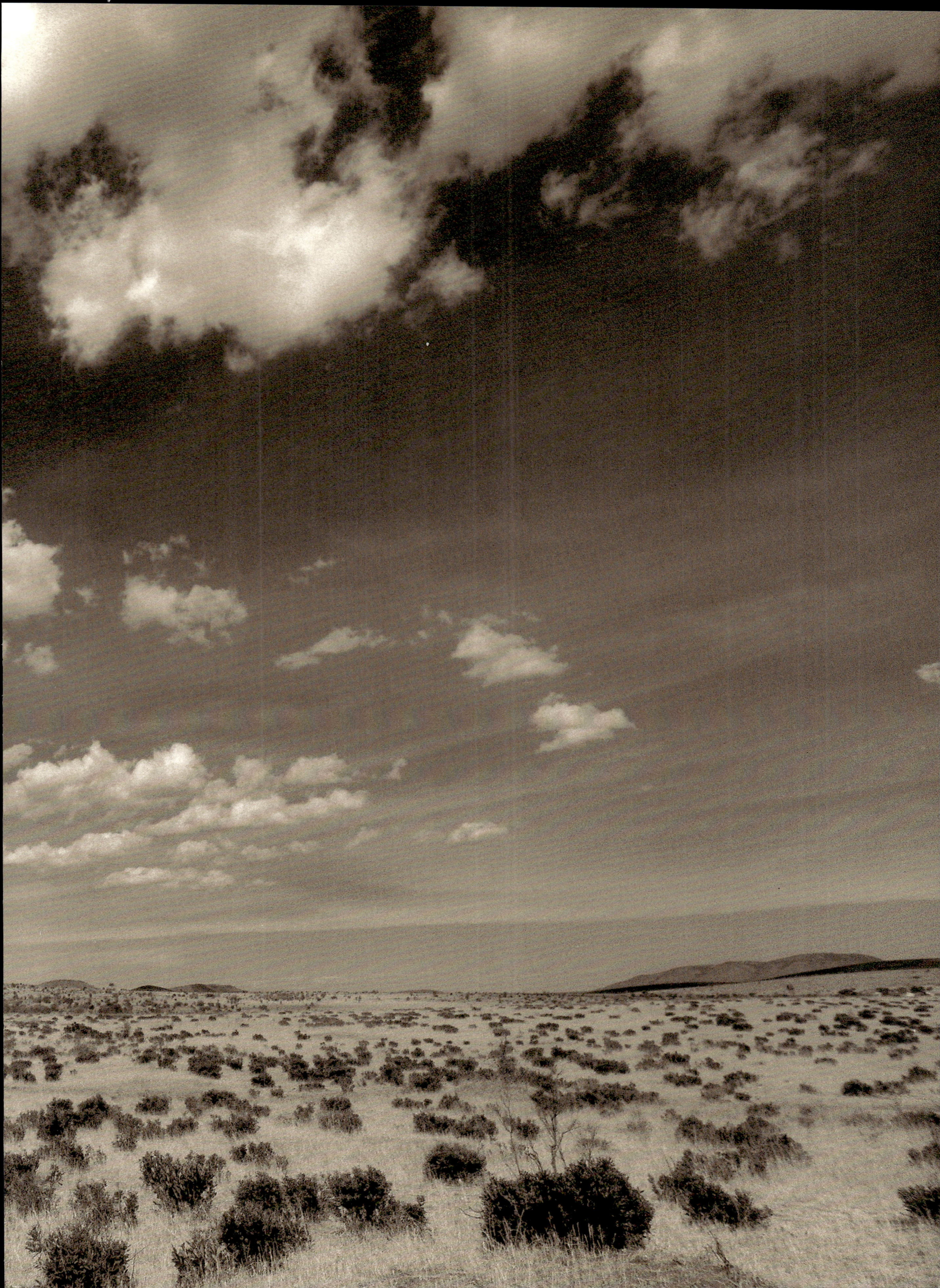

THE GNUS' LONG MARCH EACH YEAR ROUND THE SERENGETI-MAASAI MARA ECOSYSTEM IS DRIVEN FROM WITHIN, GUIDED FROM WITHOUT. THEIR GENES ARE PROGRAMMED TO MOVE, THE AVAILABILITY

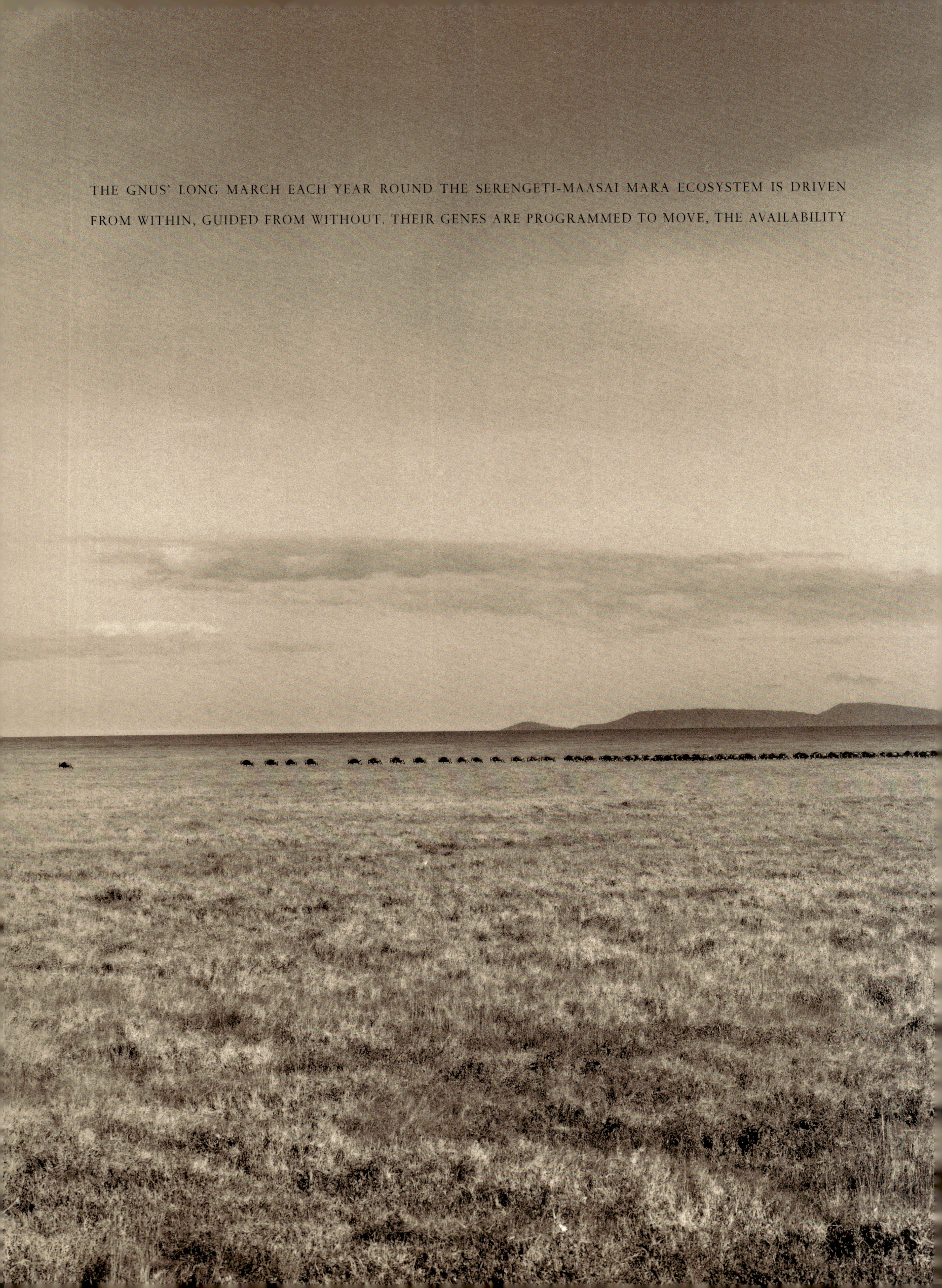

OF FOOD TELLS THEM WHERE TO MOVE. THE ECOSYSTEM'S SEASONAL OFFERINGS CHART THE COURSE FOR THE HERDS, AWAY FROM DANGEROUS SHOALS OF PREDATORS, TOWARDS THE WELL-WATERED UNEATEN PASTURES.

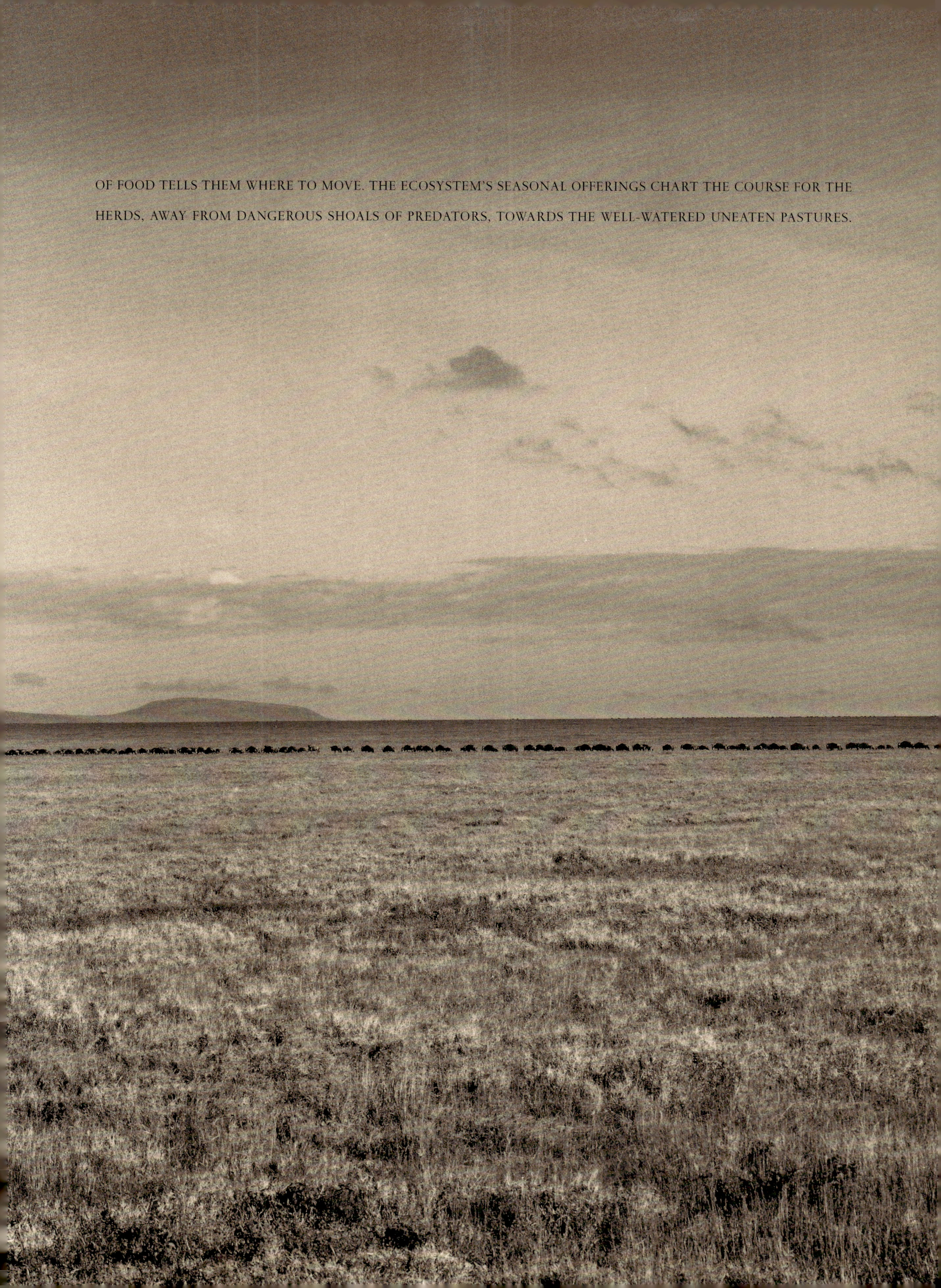

A 300-MILE ROUND TRIP
FROM THE SOUTHERN SERENGETI TO THE MARA IN THE NORTH

EVERY CIRCLE IS AS OLD AS THE HILLS

YOU CAN START ANYWHERE TO FOLLOW A CIRCULAR ROUTE. TAKE A MAP OF NORTHERN TANZANIA AND YOU SEE THE SERENGETI NATIONAL PARK TUCKED UP AGAINST THE KENYAN BORDER WHERE THE MARA RIVER CROSSES IT. THE RIVER RUNS TO LAKE VICTORIA, STEAMY AND SHIMMERING IN THE WEST. TO THE EAST A BROKEN RIDGE OF HILLS CLIMBS GRADUALLY SOUTHWARDS TO THE GOL MOUNTAINS AND OL-DOINYO LENGAI, THE LAST ACTIVE VOLCANO IN EAST AFRICA. TO THE SOUTH, ON THE LOWER SLOPES OF NGORONGORO CRATER AND LEMAGRUT MOUNTAIN NEAR LAETOLI ANTHROPOLOGISTS FOUND FOSSILISED FOOTPRINTS, EVIDENCE THAT OUR ANTECEDENTS WERE WALKING UPRIGHT SOME 3.6 MILLION YEARS AGO. BEGIN YOUR CIRCLE AROUND THERE, NEAR OLDUVAI GORGE. TRACE A LONG SWEEPING ARC, FIRST WEST NEARLY TO THE LAKE, THEN NORTH THROUGH THE IKORONGO CONTROLLED AREA PAST THE OLD GERMAN FORT IKOMA, ON UP INTO THE MASAI MARA NATIONAL RESERVE, THEN BACK SOUTH INSIDE THE HILLS BEARING NAMES SUCH AS KUKA, LOBO, LONGOSSA AND GRUMACHEN, BACK TO WHERE WE STARTED. THE ELONGATED CIRCLE ON YOUR MAP IS MILLIONS OF YEARS OLD. YOUR FINGER HAS MOVED THE EQUIVALENT OF 300 MILES. IT IS THE PATH OF THE SERENGETI WILDEBEEST MIGRATION. FOR DISTANCE AND NUMBERS, IT IS THE GREATEST ANIMAL SHOW LEFT ON EARTH.

IS A PICTURE WORTH A THOUSAND WORDS? OF COURSE. PICTURES LIKE PARAGRAPHS TELL LITTLE TALES. IN THE SAME WAY WE DIP INTO A BOOK, READ AND APPRECIATE A FINELY CRAFTED PARAGRAPH, IN A COLLECTION OF PHOTOGRAPHS WE PAUSE AND GAZE AT A PARTICULARLY FINE IMAGE WITH A POWERFUL COMPOSITION AND A VISUAL WEB OF SUBCONSCIOUS LINKS. OUR APPRECIATION OF THE PARAGRAPH OR THE PHOTOGRAPH WHETS OUR APPETITE FOR MORE. WE CONTINUE TO DIP OR GAZE UNTIL THE FLOOD OF WORDS OR IMAGES LEADS OUR MINDS TO WONDER ABOUT THE PROCESSES THAT LINK THE PIECES. WE ARE SEARCHING FOR THE STORY: FOR A TALE WELL TOLD.

THIS IS MORE THAN A PICTURE BOOK LOOKING INTO THE LIFE AND TIMES OF THE GNU. IT GLIMPSES A CONTINUUM OF LIFE, A JOURNEY WITHOUT END, A PERPETUAL CIRCLE OF PASSAGE, BIRTH AND DEATH, ENCOMPASSING ALL THAT LIVING THINGS DO, WHICH ALTHOUGH ENDLESSLY REPETITIVE, LIKE LIFE ITSELF, IS ENDLESSLY CAPTIVATING. THE CENTREPIECE IS THE MIGRATION, A ROLLICKING PRODUCTION THAT OFFERS BOTH COMEDY AND TRAGEDY, WITH CHORUSES AND CAVORTING BY A CAST OF THOUSANDS.

THE MAIN PLAYERS ARE NEARLY TWO MILLION MIGRANT HERBIVORES – WILDEBEESTE, ZEBRA, THOMSON'S GAZELLE AND ELAND, ALL A PART OF THE BIG CIRCLE. THEY TREK ANNUALLY THROUGH THE HOME RANGES OF SOME 300,000 FELLOW HERBIVORES, LOCAL RESIDENTS LIKE TOPI, BUFFALO, GIRAFFE, KONGONI (HARTEBEESTE), GRANT'S GAZELLE, WATERBUCK AND WARTHOGS. THE ANTAGONISTS COMPRISE THE GUILD OF PREDATORS – LIONS, HYENAS, LEOPARDS, CHEETAHS, HUNTING DOGS AND JACKALS – THAT WORRY THE MIGRANTS AND RESIDENTS MORE OR LESS THROUGHOUT AND FORM THE HEART OF THE DRAMA. COMIC RELIEF IS COMPLIMENTS OF THE UGLY SCAVENGERS, THE LEERING HYENA, AND, LIKE MANY COMPLEX LEAD ROLES, THE FOOLISH GNU ITSELF.

THE SCRIPT IS A *COMMEDIA DELL'ARTE* IN WHICH THE ROLES ARE STEREOTYPED, EVENTS AND OUTCOMES PREDETERMINED, BUT WITH CONSIDERABLE LATITUDE FOR IMPROVISATION CONTAINED ONLY BY THE FUNDAMENTAL DRIVE TO STAY ALIVE LONG ENOUGH TO REPRODUCE. ALTHOUGH THE EXACT ROUTE OF THE MIGRATION AROUND THE ECOSYSTEM VARIES FROM YEAR TO YEAR, THE CYCLE ALWAYS RUNS ITS COURSE. AT THE RISK OF EXHAUSTING THE METAPHOR, THE SERENGETI ECOSYSTEM PRESENTS US WITH A REVOLVING STAGE, NEARLY 25,000 SQUARE MILES IN AREA, THE SIZE OF SWITZERLAND, WITH FOUR MAJOR HABITATS AND FOUR SEASONS CONVENIENTLY MAPPABLE TO THE CARDINAL POINTS OF THE COMPASS. THE MAP AND THE SEASONS OFFER A STRUCTURE[1]: FIRST FOLLOWING THE MIGRATION THROUGH THE WESTERN WOODED-GRASSLANDS DURING THE LONG DRY SEASON, NORTH TO THE MASAI MARA'S OPEN PLAINS AND BUSHLANDS FOR THE SHORT RAINS, BACK THROUGH THE EASTERN WOODLANDS DURING THE SHORT DRY SEASON, AND ON TO THE SHORTGRASS PLAINS TO THE SOUTH IN TIME FOR THE LONG RAINS AND THE BIG BIRTH. WE COULD JUMP ON TO THE STAGE AT ANY POINT. FOR THE SAKE OF ARGUMENT, WE WILL FOLLOW IN THE FOOTSTEPS OF OUR ANCESTORS AND BEGIN IN THE SOUTHWEST, UNDER THE SHADOW OF LEMAGRUT, WHERE THE SHORTGRASS PLAINS GIVE WAY TO THE WESTERN WOODED-GRASSLANDS.

I

WESTERN WOODED-GRASSLANDS

Long Dry Season, June–September

WE PICK UP THE ANNUAL CLOCKWISE JOURNEY OF THE WILDEBEEST AS IT PROCEEDS INTO THE WESTERN WOODED-GRASSLANDS OF THE SERENGETI NATIONAL PARK. THE VAST OPEN PLAINS OF RED OAT GRASS (*THEMEDA TRIANDRIA*) WAVE LIKE PROVERBIAL FIELDS OF GRAIN UNDER SCATTERED ACACIA TREES UP TO THE EDGES OF THE ITONJO HILLS AND THE NYARABORO PLATEAU. THE PRODUCTIVITY OF THESE WOODED GRASSLANDS HAS FOR CENTURIES MADE THEM IRRESISTIBLE TO MASAI HERDS AND HERDERS FROM THE NGORONGORO CRATER HIGHLANDS. THE BOUNDARY OF THE PLAINS AND THE WOODED-GRASSLAND BELT IS NOT ONLY A SUBTLE EAST–WEST MOISTURE AND SOIL GRADIENT, BUT ALSO A NORTH–SOUTH LINE OF *KOPYES*, APTLY NAMED *INSELBERGS*, "ISLAND MOUNTS" BY THE 19TH-CENTURY EXPLORERS OF "GERMAN EAST AFRICA". PILES OF ROUNDED GRANITIC BOULDERS, SOME AS BIG AS HOUSES, JUT ABOVE THE PLAIN AND PROVIDE CATCHMENT AND MICRO-HABITATS FOR A HOST OF WATER-LOVING, SHADE-SEEKING PLANTS AND ANIMALS. THEY ALSO PROVIDED GATHERING PLACES FOR THE MASAI BEFORE THE SERENGETI NATIONAL PARK WAS GAZETTED IN 1951.

IN EARLIER TIMES, THE HERDERS BIDED THEIR TIME WAITING FOR THE RAINS, WHILST THE CATTLE GRAZED ON GREEN FRINGES AT THE FOOT OF THE *KOPYES* WHERE SOME WATER HAD ACCUMULATED AS IT DOES ON THE EDGE OF A TARMAC ROAD. TO A RATTLING OF PEBBLES, THEY PLAYED THE ANCIENT BOARD GAME *KHALA* OR *MBAO* IN DOUBLE LINES OF SEVEN PITS THAT WE CAN STILL FIND GOUGED INTO THE FLAT ROCKFACE. THE MIDDLE EASTERN GAME MUST HAVE COME TO AFRICA WITH THE DHOW TRADE FROM THE ARABIAN PENINSULA, MAKING ITS WAY FROM THE COAST VIA IVORY TRADE ROUTES. WHEN NOT GAMING OR IDLING, THE *IL-MURRAN* (MASAI WARRIOR HERDERS) USED OCHRE, ASH AND CHARCOAL TO RENDER ROCK PORTRAITS OF DOMINANT SPECIES LIKE ELEPHANTS AND WILDEBEESTE, TO REPEAT

THE GEOMETRIC MOTIFS OF THEIR BUFFALO-HIDE WARRIOR SHIELDS, OR TO LAMPOON THE COLONIALS WHO WOULD EVENTUALLY EXPEL THEM FROM THEIR ARCADIA. UNMISTAKABLE ARE THE CHALKY WHITE CARICATURES WITH BIG HATS AND HANDS IDLE ON THEIR HIPS.

THE WESTERN WOODED-GRASSLANDS OF THE SERENGETI ECOSYSTEM ARE A WONDERFUL MOSAIC OF HILLS AND VALLEYS, PLATEAUS AND FLOODPLAINS ALONG THE MBALANGETI AND GRUMETI RIVERS THAT DRAIN WESTWARDS INTO LAKE VICTORIA. GRASS THERE IS IN PLENTY AND MOISTURE ACCUMULATED IN THE VALLEY BOTTOM SOILS, AUGMENTED WITH THE OCCASIONAL "OUT OF SEASON" RAIN SHOWER FROM THE CONGO BASIN AND LAKE WEATHER SYSTEM. IT IS ENOUGH TO SUSTAIN THE HERDS AS THEY THREAD THEIR WAY THROUGH THE VALLEYS, IN SOME YEARS ALMOST TO THE SHORES OF THE GREAT LAKE. SWINGING BACK NORTH AND EAST, THEY PAUSE FOR SOME DAYS OR WEEKS ON THE FLOODPLAINS UNTIL THEIR RELENTLESS "PREDATION" ON THE GRASS, COMBINED WITH THAT OF THE LOCAL RESIDENT TOPI AND KONGONI, THINS THE PASTURES AND THEIR STOMACHS TELL THEM IT IS TIME TO MOVE ON.

GRASS |

WITHIN HOURS OF THE FIRST RAINFALL THAT BREAKS THE DROUGHT, THE GRASS BEGINS TO SPROUT. SOME CLAIM THAT AS THE RAINS PICK UP, THEY CAN LITERALLY SEE THE GRASS GROW. CERTAINLY GROWTH RATES OF A FEW INCHES IN A FEW HOURS HAVE BEEN MEASURED. GRASS IS ONE OF THE MOST SUCCESSFUL TERRESTRIAL LIFE FORMS. A WALK VIRTUALLY ANYWHERE ON EARTH WILL TREAD OVER GRASS. THE SERENGETI LION RESEARCHERS JANETTE HANBY AND DAVID BYGOTT LOOKED BEYOND THEIR BIG CATS, AND SAW THAT[2]:

"Serengeti grasses are special, like the soils in which they grow. There are many different kinds, each with a skill for survival. Some sleep through the dry season as fleshy roots, others as seeds, but all are awakened by the rains and begin a frantic race to cover as much ground as possible. Some can do this by growing thick and tall, choking their competitors; others send long runners in all directions and yet others spring up quickly from seeds and colonise bare patches. There are specialists which can cling to loose sand, or suck moisture from the soda-saturated soil beside waterholes. A few grasses gain an advantage by being unpalatable. One tastes terribly of turpentine, others have tough fibres or crystal teeth to lacerate grazing mouths.
There is even a grass that thrives on being eaten... For many thousands of years animals had come to the Serengeti and chewed a living off the grasses. The mat-grass had exploited this situation. When nibbled flat by grazers this kind of grass spreads out and grows quickly. The grazers clip down any other surrounding grasses as well so the spreading mat has room to grow. If for some reason the grazers don't come, the neighbouring grasses can grow tall and begin to shade and crowd out the flat grass and it loses its vigour. So the spreading grass actually needs to be eaten to thrive."

GRASS SPECIES ARE "INCREASE STRATEGISTS" THAT MAKE A LIVING BY GROWING, REPRODUCING AND DYING BACK IN ONE SHORT SEASON. GREAT EMPHASIS IS PERFORCE PLACED ON SEEDS. THE PLANTS THEMSELVES ARE JUST A FEW THIN LEAVES, ONE OR TWO STEMS AND A SEED HEAD THAT WEIGHS AS MUCH AS THE REST OF THE PLANT. CLEARLY THESE ARE ORGANISMS WHOSE SUCCESS LIES IN THEIR ABILITY TO FLOURISH WHEN CONDITIONS ARE RIGHT. SINCE THE CORRECT CONDITIONS FOR GROWTH AND REPRODUCTION MAY BE LIMITED TO A FEW SHORT WEEKS OF RAIN IN THE SERENGETI, THE GRASSES THERE HAVE EVOLVED TO REPRODUCE AS QUICKLY AS POSSIBLE.

ORGANISMS THAT PRODUCE AN ABUNDANCE OF SEED ARE TENACIOUS COLONISERS, ESPECIALLY WHEN THE SEEDS ARE EQUIPPED TO ENHANCE DISPERSAL. SOME HAVE HOOKS TO ATTACH TO THE HAIR OF A PASSING ANIMAL, OTHERS HAVE AN EDIBLE AND TOUGH SEED COATING TO ATTRACT HERBIVORES ON THE ONE HAND AND SURVIVE THEIR DIGESTIVE JUICES ON THE OTHER. STILL OTHERS SPORT FEATHERY DEVICES TO CATCH AND FLOAT ON THE WIND. IF CONDITIONS ARE NOT FAVOURABLE FOR FLOWERING, THE PLANTS PROPAGATE BY VEGETATIVE MEANS, LITERALLY CREEPING OVER THE BARE SURFACE AND PUTTING DOWN ROOTS WHERE THERE IS A BIT OF SOIL AND A TOUCH OF MOISTURE. THE ROOTS MAY REACH DOWN SEVERAL METRES AND CONTRIBUTE TO THE BREAKDOWN OF SOIL. GIVEN ITS SIMPLE STRUCTURE AND RAPID GROWTH, GRASS IS ONE OF THE FASTEST MEANS TO PRESENT THE ELEMENTS TO THE EATERS. GRASS IS TO SOIL AS WILDEBEESTE ARE TO GRASS. THE PRIMARY PRODUCTION OF THE AFRICAN GRASSLANDS IS PROLIFIC, PERHAPS GREATER THAN THAT OF ANY OTHER ECOSYSTEM ON EARTH, EVEN FORESTS. DURING THE RAINS EACH SQUARE YARD OF GRASSLAND CAN PRODUCE ALMOST TWO POUNDS OF EDIBLE MATERIAL EVERY MONTH – SOME 2,000 POUNDS FOR EVERY 250 ACRES. THIS RAPID CONVERSION OF MATERIALS INTO AN EASILY AVAILABLE AND EDIBLE FORM CREATES THE OPPORTUNITY FOR NUMEROUS HERBIVORES TO EXIST, AND THE VERY GRAZING OF THOSE ANIMALS STIMULATES THE GRASS SWARD TO PRODUCE EVEN MORE THAN IT WOULD WITHOUT ANIMAL MOWING.

CHOICE | LOOKING OUT OVER THE SERENGETI GRASSLANDS, YOU MAY WELL WONDER WHY WILDEBEESTE ARE NOT EVERYWHERE, WANDERING RANDOMLY THROUGH THE GRASSLANDS EATING WHATEVER THEY BUMP INTO. A PROCESS OF SELECTION IS TAKING PLACE. THE WILDEBEEST, DESPITE ITS RATHER LIMITED MENTAL PROWESS IS MAKING A CHOICE: THE CHOICE OF BEING HERE OR GOING OVER THERE, A CHOICE THAT BEGINS JUST BEYOND THE WILDEBEEST'S NOSE. ALTHOUGH GRASS IS ABUNDANT IN SPACE, IT IS SPORADICALLY AVAILABLE OVER TIME. EVERY SEASON – WINTER OR DRY – GRASS EFFECTIVELY DISAPPEARS. IN THE NORTH, IT GETS COVERED WITH SNOW OR JUST STOPS GROWING IN THE COLD. IN AFRICA, IF IT DID NOT GET EATEN UP, IT WOULD DRY UP. CONSIDER THE GNU'S VIEW OF ITS SERENGETI WORLD, AS DESCRIBED BY RESEARCHERS WHO BROKE THE STANDARD SCIENTIFIC REPORTING MOULD BY SQUEEZING THEMSELVES INTO THE BEAST'S MIND[3]:

"It spends much of its time with head bent down, walking along, taking bite after bite after bite, continually hearing the cacophony of the surrounding herd, and, we presume, reassured and protected from the Serengeti's predators by that dissonance. It bites and swallows, bites and swallows, until its rumen signals 'full'. Then it lies down, regurgitates and remasticates the previously swallowed forage, swallows it again, and thus accelerates the extraction of nutrients as the forage is processed by the complex of microbes and chemicals in its digestive tract. From time to time come a few days when it and its companions cannot fill their rumens fast enough to meet their needs, so the herd moves to another location. As the weeks proceed, the herd moves again and again – less during the wet season when foraging on the Serengeti short-grass plains, incessantly during the dry season when forage quality and quantity are both meager...

As it walks along feeding, taking several bites per minute, our wildebeest, moving through visually homogeneous grasslands, in fact encounters a constantly varying environment... It encounters different individual grass plants at different stages of growth, different genotypes of the same grass species, different grass species, a mixture of grasses, forbs, shrubs, and, off the Serengeti plains, tree seedlings, saplings, and adults. As it moves from area to area, it encounters grasslands that are genetically short, medium, and tall. All may be ungrazed, lightly grazed, moderately grazed, or heavily grazed. Some landscape regions it moves through are flat to gently undulating, others are highly dissected by ravines, gorges, and steep hills."

IT MAY JUST BE ENDLESS PLAINS OF GRASS TO US, BUT TO THE GNU, THE SERENGETI ECOSYSTEM IS A VERITABLE SMORGASBORD. TOWARDS THE END OF THE "SHORT" DRY SEASON AROUND MARCH, IT IS THE LACTATING FEMALES THAT FIRST START TO FEEL THE PINCH AS THE SHORTGRASS SERENGETI PLAINS DELIVER LESS AND LESS NUTRITION. MOREOVER, WATER HOLES ARE FAST BECOMING FEW, SHALLOW AND FOUL, AND GNUS NEED TO DRINK EVERY TWO OR THREE DAYS. ALTHOUGH THE ANIMALS ARE IN TOP CONDITION AFTER WEEKS OF FEEDING OFF THE FERTILE VOLCANIC SHORTGRASS PASTURES, THE FEMALES' RESTLESSNESS BEGINS TO COALESCE INTO MILLING GROUPS THAT ARE HALF PUSHED, HALF DRAWN TO THE WESTERN HILLS.

HOW DO THE MIGRATING GNUS KNOW WHICH WAY TO GO? THE QUESTION MAY SOUND TRIVIAL, BUT IT HAS AT LEAST TWO ANSWERS, ONE BUILT-IN, THE OTHER EXTERNAL. THE "INTERNAL KNOWLEDGE" IS ETCHED INTO THEIR VERY CELLS, ENCODED IN THE DNA OF THEIR GENES BY HUNDREDS OF THOUSANDS OF YEARS OF NATURAL SELECTION. ON AVERAGE, GNUS THAT WENT THE "WRONG WAY", THAT IS, WHERE THERE WERE NO REGENERATING SWARDS OF GREEN GRASS AND FREE WATER TO DRINK, DIED BEFORE LEAVING PROGENY WITH THE SAME UNFORTUNATE PROPENSITY. ON THE OBVERSE SIDE OF THE NATURAL SELECTION COIN ARE THOSE GNUS WHO WENT THE "RIGHT WAY". OVER THE MILLENNIA, THEY FOUND ON AVERAGE SUFFICIENT FOOD AND WATER, AND CONSEQUENTLY SURVIVED LONG ENOUGH TO PRODUCE MORE GNUS WITH THE SAME USEFUL TRAITS: "IT'S GETTING DRY, MY NEW CALF IS DRAINING MY RESERVES, I'M GETTING HUNGRY, I MIGHT JUST MOVE A LITTLE FURTHER..."

OLD AFRICAN HANDS ARE CONVINCED THAT THE WILDEBEESTE ALSO REACT POSITIVELY TO DISTANT CONVECTIVE THUNDER STORMS. ALTHOUGH WE KNOW OF NO EXPERIMENTAL EVIDENCE SHOWING CONCLUSIVELY THAT, AT THE SIGHT OF DISTANT DARKENING CLOUDBANKS AND FLASHES OF LIGHTNING, THE RUMBLE OF THUNDER, AND THE OZONE-RICH SMELL OF NEWLY WETTED SOIL WAFTING ON THE FRESHENING WIND, THE MILLING HERDS FOCUS AND SET OFF IN ONE PARTICULAR DIRECTION. IT WOULD BE SURPRISING IF EVEN THE GNU COULD OVERLOOK SUCH BLATANT PORTENTS OF CHANGE.

THE RUT

SOCIETY AFFECTS US ALL: IN ONE WAY OR ANOTHER WE ARE ALL HERD ANIMALS. THE "ENFORCED COLLECTIVISM" OF THE GNUS IS BROUGHT ABOUT BY THE ECOLOGICAL IMPERATIVE OF BEING PRESENT WHERE AND WHEN THE PASTURES GROW. THE PROXIMITY OF LIKE-MINDED BEASTS, THEIR SMELL, THEIR SOUNDS, THE CONSTANT SIGHT OF THEM – LIKE BEING IN A CROWDED MIXED SHOWER ROOM – INEVITABLY TRIGGERS HORMONAL CHANGES IN

BOTH SEXES. GONADS GROW, HORMONES FLOW, A COMMUNAL STATE OF EXCITEMENT ENSUES AND THE ANIMALS COME INTO "BREEDING MODE". THIS TIME IS KNOWN AS THE "RUT", A TERM THAT HAS ITS ORIGINS IN THE LATIN WORD *RUGIRE*, TO ROAR, OF WHICH THERE IS MUCH DURING THE RUT.

ANYONE WHO HAS SPRAYED AGAINST GARDEN PESTS, CHASED A RABBIT FROM THE LETTUCE PATCH OR A NEIGHBOUR'S COW FROM THE WRONG SIDE OF THE FENCE, OR, FAR MORE TRAGICALLY, FOUGHT TO FEND OFF INVASION KNOWS THE IMPERATIVE OF TERRITORIALITY. THE TWO MOST IMPORTANT THINGS IN A MAMMAL'S LIFE (ARGUABLY INCLUDING OUR OWN) ARE FOOD AND A MATE. SPACE IS REQUIRED TO ENJOY EITHER, AND THE LAND (OR LAKE BOTTOM OR CORAL REEF) THAT IS NECESSARY TO PROVIDE SUFFICIENT RESOURCES FOR OUR MATE(S) AND OURSELVES IS CALLED A "TERRITORY". IN MOST SPECIES, THE TERRITORY IS A FIXED AND CONSTANTLY GUARDED AND FOUGHT-OVER PATCH OF REAL ESTATE. WITH THE MIGRATORY WILDEBEESTE, THE PATCHES MOVE ALONG WITH THE TIDE OF THE HERDS. THE THREAD-LIKE PATTERNS OF STREAMING WILDEBEESTE MOVING TO THE NORTHWEST SETTLE FOR A TIME INTO CLUMPS OF FEMALES GRAZING AND BEING HERDED BY FRENZIED MALES WHO SET UP TEMPORARY TERRITORIES, AND THEN RUSH AROUND KEEPING THE FEMALES IN AND OTHER MALES OUT.

YET EVEN IN THE GREATEST MIGRATION ON EARTH, NOT EVERYONE GOES ALONG WITH THE CROWD. TO A GREATER OR LESSER DEGREE MOST MIGRATIONS ARE ACTUALLY "PARTIAL MIGRATIONS", WITH A PROPORTION OF THE POPULATION STAYING PUT SOMEWHERE ALONG THE ROUTE. THUS, THERE ARE RESIDENT POPULATIONS, SUCH AS THE 10,000 TO 20,000 GNUS THAT SEEM NOT TO LEAVE THE LOITA PLAINS IN THE KENYAN MARA WHEN THE HUNDREDS OF THOUSANDS OF ANIMALS IN THE MAIN BODY OF THE MIGRATION RETREAT SOUTH. THERE ARE ALSO APPARENT RESIDENTS IN THE FLOODPLAINS OF THE WESTERN CORRIDOR OF THE SERENGETI AND OVER 15,000 IN THE NGORONGORO CRATER THAT ALMOST CERTAINLY INTERACT AT THE EDGES OF THEIR RANGE WITH THE MIGRATION. AND, IN THE UNNATURALLY QUIET WAKE OF THE MIGRATION, THERE ALWAYS SEEM TO BE SOME TERRITORIAL MALES WHO REMAIN BEHIND. THEY MOPE AROUND, RATHER FORLORN, STILL SPOILING FOR A FIGHT AND HOPING FOR A MATE. IT IS DOUBTFUL IF THEY WILL SURVIVE THEIR CONSPICUOUS SOLITUDE TO MEET THE MIGRATION ON ITS NEXT ROUND. THE COMPLEX PATTERNS OF ANIMAL MIGRATION ARE TESTIMONY TO THE CONTINUOUS BALANCING ACT OF WEIGHING THE COSTS AND BENEFITS OF ONE SURVIVAL STRATEGY AGAINST ANOTHER. TO GO OR NOT TO GO, THAT IS THE GNU'S QUESTION.

MALE GNUS INVEST A GOOD DEAL OF EFFORT IN DECIDING WHO WILL GET CLOSE ENOUGH TO RECEPTIVE FEMALES TO MATE. MALE UNGULATES SUCH AS WILDEBEESTE AND GRANT'S GAZELLE SEEM TO DO LITTLE ELSE EXCEPT FIGHT. THEY FIGHT TO ESTABLISH A HIERARCHY, THEY FIGHT TO MAINTAIN A TERRITORY, THEY FIGHT OVER FEMALES. THE SCREENING PROCESS OF AGGRESSIVE MALE CONTESTS HELPS ENSURE THAT ONLY THE STRONGEST, MOST TENACIOUS, QUICKEST AND MOST CUNNING WILL PASS ON THEIR TRAITS AND ONLY THE FITTEST GENES WILL MAKE IT THROUGH

THE GAUNTLET OF MALE RIVALRY. GNU FIGHTS APPEAR TO BE AN EARNEST AND ARTLESS SERIES OF HEAD-BASHES AND HORN TWISTINGS. DESPITE THE APPARENT LACK OF FINESSE IN THE ENCOUNTERS, THERE ARE SUBTLE PLOYS. IF ONE ANIMAL IS LOSING GROUND, HE MAY SUDDENLY UTTER A WARNING SNORT OF THE KIND USED AFTER SIGHTING A PREDATOR; THIS DISTRACTS THE OPPONENT LONG ENOUGH FOR ESCAPE OR A LOW BLOW. A ROUND OF FIGHTING IN FULL FRENZY MAY SUDDENLY BREAK OFF, AND EACH ANIMAL BEGIN GRAZING STRENUOUSLY ALMOST NOSE TO NOSE. THIS "DISPLACEMENT GRAZING" MAY SERVE AS A SIGNAL OF CHALLENGE, LIKE TWO HUMAN *MACHOS* DISPLAYING IN A BAR, EYEBALL TO EYEBALL BELTING BACK SHOTS OF WHISKY.

DESPITE THE ENERGY SPENT IN FIGHTING, THE BATTLES ARE RARELY TO THE DEATH. THE PROPENSITY TO KILL – EXCEPT UNDER EXTREME LIFE OR DEATH SITUATIONS – IS HARDLY A TRAIT WORTH PASSING ON. THE FIGHTS IN AN HYPOTHETICAL POPULATION OF BLOODTHIRSTY WILDEBEESTS WOULD IN TIME SPIRAL OUT OF CONTROL AND NUMBERS DWINDLE THROUGH SELF-DESTRUCTION. THOU SHALT NOT KILL EXCEPT UNDER SPECIAL CIRCUMSTANCES. KILLING TO PROTECT YOUR MATE OR OFFSPRING IS IN TURN A CASE OF SELF-DEFENCE IN ORDER TO PROPAGATE YOUR GENES AND ALLOW YOUR OFFSPRING TO CARRY THEM INTO THE FUTURE. THE SYSTEMATIC KILLING OF MALE RIVALS WOULD EVENTUALLY PUT YOUR FUTURE GENE BEARERS AT GREATER RISK THAN THE PROPENSITY TO POSTURE, FIGHT TO A POINT AND THEN BREAK OFF BEFORE THINGS GET OUT OF HAND.

YET, REGARDLESS OF THE LEAPS, HEADBUTTS AND CHASES BETWEEN RIVAL MALES, AT THE END OF THE DAY IT IS THE FEMALE WHO CHOOSES. SHE IS HELPED ALONG IN THIS REGARD, SEDUCED BY SENSUOUS WILDEBEEST FOREPLAY. WHEN THE MOON IS FULL – THE LUNAR CYCLE SEEMS TO COINCIDE WITH THE MATING PEAK – WATCH FOR HIS COMMANDING "ROCKING CANTER" AS HE APPROACHES A RIVAL OR POTENTIAL MATE; MARVEL AT THE RIPPLING SPLENDOUR OF HIS SEXY "LOWSTRETCH"; THRILL TO THE PENETRATING TONES OF HIS "FAST CALLING", A HORMONE-HEAVY REFRAIN THAT HAS BEEN LIKENED FAVOURABLY TO A "CHORUS OF GIANT FROGS". ADD TO THIS THE NAUGHTINESS OF "URINATION ON DEMAND" AND THE BLATANT SPLENDOUR OF HIS MANHOOD AS HE REARS UP IN FRONT OF HER IN THE "COPULATORY DISPLAY"... HOW CAN SHE RESIST? YET, THOUGH MALES MAY REPEL OTHER SUITORS, DASH MADLY ABOUT TO CORRAL AND CORNER A FEMALE ON HEAT, UNLESS SHE DECIDES TO STOP, HOLD FAST, HUNKER DOWN A BIT AND UNTUCK HER TAIL, CONSUMMATION IS NEARLY IMPOSSIBLE. SHOULD SHE BE FORCED AGAINST HER WILL – RAPED IN FACT – THERE IS SOME EVIDENCE THAT FEMALE ANTELOPES ARE ABLE TO EXPEL AN UNWANTED LOAD OF SEMEN – "SPERM-DUMPING" – THE ULTIMATE GESTURE OF REJECTION. BUT CHOOSE THEY DO: MOST OF THE MATING OF THE SERENGETI GNU POPULATION OCCURS DURING THE NORTHWEST LEG OF THE MIGRATION AND NEARLY 90% OF THE ADULT FEMALES – SOMETHING IN THE REGION OF 400,000 ANIMALS – GET PREGNANT.

2

NORTHERN MARA BUSHED GRASSLAND

Short Rains, October–November

THE MASAI MARA, TODAY A GREAT EXPANSE OF BUSHED GRASSLAND, STRETCHES NORTHWARD BEYOND THE TANZANIA–KENYA BORDER AND FETCHES UP AGAINST THE ISINYA PLATEAU AND ITONJA HILLS TO THE NORTH, THEN MERGES WITH THE LOITA PLAINS TO THE NORTHEAST WHERE IT ROLLS INTO THE FOOTHILLS OF THE MAU NAROK AND THE LOITA FOREST. OPEN, ROLLING, A LONG VIEW AT EVERY POINT, AN ENDLESS AREA TO ROAM. IF YOU ARE LUCKY ENOUGH TO BE THERE BEFORE FOLLOWING THE MIGRATION, THE BIG CATS OR A HERD OF ELEPHANTS, TAKE A MOMENT TO EXPERIENCE A "360". THERE ARE FEW OTHER PLACES IN THE WORLD WHERE YOU CAN STAND ON A SLIGHT RISE, TURN A FULL CIRCLE AND LET YOUR EYE SWEEP THE FAR HORIZON AT EVERY CARDINAL COMPASS POINT. THE SKY WILL INVARIABLY PROVIDE YOU WITH AN OVERARCHING FIRMAMENT WITH CLOUDSCAPES OF DIZZYING VARIETY.

BUT IT WAS NOT ALWAYS SO. THE CONSTANCY OF AFRICA IS ITS CONSTANT CHANGING. ONLY A QUARTER OF A CENTURY AGO, "THE MARA" (AS THE COMBINATION OF THE MASAI MARA NATIONAL RESERVE AND THE LOITA PLAINS ARE KNOWN) WAS RATHER MORE WOODED GRASSLAND WITH HILL THICKETS THAN THE OPEN GRASSLANDS WE SEE TODAY, ESPECIALLY ON THE SERENGETI SIDE. THE HILLTOPS WERE CAPPED WITH DENSE THICKETS, THE PLAINS MORE PUNCTUATED WITH ACACIAS. THE SERENGETI GNUS HARDLY BOTHERED TO CROSS INTO THE MARA AT ALL. TWO OF AFRICA'S MOST EFFECTIVE HABITAT MODIFIERS, FIRE AND ELEPHANTS, HAVE COMBINED OVER THE YEARS TO CHANGE THE FACE OF THE MARA. ANNUAL GRASS FIRES – USUALLY LIT BY MASAI HERDERS TO ENCOURAGE A QUICK REGROWTH OF FRESH GREEN GRASS, OR BY CARELESS TOURISTS – TAKE A HEAVY TOLL ON REGENERATING TREE SEEDLINGS ON THE LOWER SLOPES AND PERNICIOUSLY ERODE AWAY THE EDGE OF THICKETS ON THE RISES, LEAVING ONLY ONE OR TWO FIRE-RESISTANT SPECIES. ELEPHANTS HAVE ALWAYS BASHED THE ODD ADULT TREE OR PLOUGHED THEIR WAY THROUGH THE THICKETS, BUT THEIR IMPACT ON THE ADULT PART OF THE WOODY PLANT POPULATION SEEMS TO BE LESS IMPORTANT THAN THEIR "PREDATION" ON TREE SEEDLINGS THAT HAVE ESCAPED FIRE. ON AVERAGE, IN THE LONG-TERM CYCLE FROM WOODLAND TO GRASSLANDS, FIRE PERTURBS, ELEPHANTS MAINTAIN.

IF ENOUGH OF THE TREE SEEDLINGS "GET AWAY", THE WOODLANDS WILL EVENTUALLY COME BACK. THE GNUS PLAY AN IMPORTANT ROLE HERE. IN THE WAKE OF THE MIGRATORY HERDS, THERE IS VIRTUALLY NO GRASS LEFT ABOVE GROUND, ONLY THE ODD UNPALATABLE SHRUB AND THE TREE SEEDLINGS (OR MINIATURE TREES THAT ARE TRYING TO GROW BACK FROM AN OLD ROOTSTOCK). NO GRASS MEANS NO FIRE THAT DRY SEASON. OVER THE NEXT YEAR, THE LITTLE TREES GAIN ALMOST TWO FEET AND BECOME A BIT MORE ROBUST. IF BY CHANCE THE HERDS PASS OVER THE SAME GROUND IN THE NEXT YEAR, THE FIRE HAZARD IS REMOVED ONCE AGAIN AND THE LITTLE TREES GROW SLIGHTLY BIGGER. ELEPHANTS ARE NOT IMMORTAL: THEIR POPULATIONS WAX AND WANE, PUTTING MORE OR LESS PRESSURE ON THE HABITAT. IF THEY ALSO LET UP FOR A FEW SEASONS ON THE REGENERATING TREES, IT IS POSSIBLE TO REVERT TO A WOODLAND PHASE. IT MAY TAKE YEARS, BUT WHO IS IN A HURRY EXCEPT FOR US HUMANS? SO FOR NOW, AND FOR THE YEARS TO COME, THE MARA WILL BE AS WE SEE IT: A ROLLING GRASSY OPEN INVITATION TO MIGRATORY HERDS.

RAIN | BY THE END OF THE SHORT BUT FEROCIOUS DRY SEASON, MORE TIME IS SPENT WATCHING THE HORIZON THAN THE GAME BOARD. EVERY FEBRUARY, THE ANNUAL 23.5-DEGREE TILT OF THE EARTH'S AXIS SHIFTS THE "METEOROLOGICAL EQUATOR" A BIT TO THE SOUTH OF THE GEOGRAPHICAL EQUATOR OVER THE AFRICAN CONTINENT. THE GREAT PERENNIAL HEMISPHERIC WIND EDDIES CAUSED BY THE SPINNING OF THE EARTH ARE PULLED INTO THAT "INTER-TROPICAL CONVERGENCE ZONE", THE SO-CALLED *ITCZ*. THE REAL POSITION OF THE ITCZ IN ANY PARTICULAR YEAR WITH RESPECT TO THE EARTH'S MIDRIFF, DETERMINES ANNUAL SHIFTS AND STRENGTHS OF THE TRADE WINDS: IN EASTERN AFRICA, ROUND ABOUT LATE MARCH, THE MONSOON SWINGS FROM THE NORTHEAST TO THE SOUTHEAST. ALMOST OVERNIGHT, INSTEAD OF DELIVERING THE BONE-DRY AIR FROM THE RUB-AL-KHALI IN THE EMPTY HEART OF THE ARABIAN PENINSULA, THE KUSI BRINGS SOFT, MOIST WINDS FROM THE SOUTHERN INDIAN OCEAN. AT THE SAME TIME, THE INCREASED TEMPERATURES FROM THE SUN BEATING DIRECTLY OVER THE CONGO BASIN START TO BUILD UP CONVECTIVE STORM SYSTEMS FAR TO THE WEST OF THE SERENGETI.

THE DIFFERENCE BETWEEN THE WET AND DRY SEASONS IN THE SERENGETI IS OFTEN THE DIFFERENCE BETWEEN AN INCIPIENT DUSTBOWL AND A LUSH MEADOW. RAINFALL IN THE WOODED GRASSLANDS IS NOT ONLY LOW, IT IS ALSO TOO UNPREDICTABLE TO PRODUCE THE CONSERVATIVE, SLOW-GROWING VEGETATION FOUND IN TEMPERATE REGIONS. 70 YEARS IN EVERY 100 WILL RECEIVE LESS THAN THIRTY INCHES, AND THE RAIN THAT FALLS IS NOT EVENLY DISTRIBUTED FOR THE CONVENIENCE OF PLANT REPRODUCTIVE CYCLES. IN THE SERENGETI-MARA ECOSYSTEM, MOST RAIN CRASHES DOWN IN A COUPLE OF MONTHS OF THE WET SEASON: SOME SIX WEEKS OF "LONG RAINS" AND A FORTNIGHT OF "SHORT RAINS". FURTHERMORE, IT FALLS TOO HARD AND RUNS OFF TOO QUICKLY FOR THE SOIL, AND HENCE THE PLANTS, TO ABSORB IT ALL. THE RAIN THAT DOES SOAK IN IS JUST ENOUGH TO SUSTAIN THOSE FORMS OF

LIFE ADAPTED TO SURVIVE THROUGH THE EIGHT MONTHS OF THE DRY SEASON. THE PULSE OF ALTERNATING WET AND DRY SEASONS IS THE MAIN MOVER OF THE FLYWHEEL OF THE ANNUAL WILDEBEEST CYCLE. IF ALL GOES WELL, THEN, AND THERE IS NO EL NIÑO, NO "SOUTHERN OSCILLATION" OF THE WARM WATERS OF THE PACIFIC, NO ANOMALOUS TROPICAL STORMS FERMENTING OUT OF THE WESTERN INDIAN OCEAN, NO HEIGHTENED SUNSPOT ACTIVITY, AND NO – WE KNOW NOT EXACTLY WHAT ELSE – THEN THE SERENGETI WILL HAVE A NORMAL YEAR: THE "SHORT" RAINS WILL BEGIN IN LATE OCTOBER TURNING THE MASAI MARA INTO A HUGE GRASSY FEAST FOR THE GNUS.

DISEASE & PARASITES | AMONG THE MOST IMPORTANT PERTURBATIONS THAT ALTER AND RE-SHAPE AN ECOSYSTEM, ALONG WITH THINGS LIKE RAIN AND FIRE, ARE THE LOWLY WORMS AND GERMS. PARASITIC ORGANISMS – VIRUSES, BACTERIA, WORMS AND FLUKES – TAKE A HUGE TOLL ON POPULATIONS, FAR GREATER THAN LARGE PREDATORS, AND PUT ENORMOUS DEMANDS ON HIGHER ORGANISMS TO TAKE PREVENTATIVE AND AVOIDANCE MEASURES. IF THERE IS NO TIME OR MECHANISM TO ADAPT, THE VICTIMS SIMPLY SUCCUMB. AT THE END OF THE LAST CENTURY, 95% OF THE SERENGETI WILDEBEESTE AND BUFFALO ALONG WITH LARGE NUMBERS OF OTHER HERBIVORES DIED FROM A RINDERPEST EPIDEMIC THAT WAS PASSED ON TO THEM BY MAASAI CATTLE. THE VIRUS WAS BROUGHT TO AFRICA FROM ITS ASIAN-EUROPEAN ORIGIN IN IMPORTED CATTLE EITHER DURING THE ITALIAN INVASION OF ETHIOPIA IN 1889 OR WITH BLACK SEA CATTLE AS EARLY AS 1884 ACCOMPANYING THE RELIEF OF GENERAL GORDON IN KHARTOUM. THE TOLL ON THE MASAI WAS WORSE THAN WAR: TWO-THIRDS OF THEM DIED. OTHER TRIBES FARED LITTLE BETTER IN A MAELSTROM OF BIOLOGICAL CATASTROPHE: FAMINE, SMALLPOX, MAN-EATING LIONS... THE DISEASE CONTINUED TO RUN ITS COURSE FOR BOTH PEOPLE AND ANIMALS, PERIODICALLY COMING AND GOING. IMPROVED VETERINARY SERVICES IN THE 1950s LED TO IMMUNISATION OF CATTLE AND A WEAKENING OF THE VIRUS'S CHOKEHOLD ON BOTH DOMESTIC AND WILD ANIMALS UNTIL IN THE EARLY 1960s NO FURTHER INFECTIONS WERE DETECTED. SINCE THEN, THE WILDEBEESTE HAVE RECOVERED THEIR NUMBERS TO NEARLY 1.5 MILLION.

THE EFFECT OF ERADICATION OF RINDERPEST SURGED THROUGH THE ECOSYSTEM. CONSIDER THE FOLLOWING WEB OF CAUSE AND EFFECT. AFTER THE VIRUS IS CONQUERED, WILDEBEESTE INCREASE; ON THE PLAINS THEY EAT MORE GRASS; LESS GRASS ALLOWS FOR MORE HERBS AND FORBS; MORE HERBS MEANS MORE FOOD FOR BROWSERS LIKE IMPALA; MORE IMPALA PROVIDE MEALS FOR CHEETAHS, SO CHEETAHS INCREASE... MEANWHILE, IN THE WOODLANDS, THE WILDEBEESTE ALSO EAT MORE GRASS; LESS GRASS FEEDS FEWER BUFFALO, SO BUFFALO DECREASE; LESS GRASS ALSO FUELS FEWER FIRES; SMALL TREES INCREASE; GIRAFFE NUMBERS GROW; LARGE TREES BECOME FEWER... THERE ARE THREE THINGS TO BE LEARNED FROM THIS: ECOSYSTEMS ARE COMPLEX; PERTURBATIONS IN ONE PART OF THE FOOD CHAIN AFFECT ALL THE OTHERS; THERE IS NO SUCH THING AS THE CLIMAX HABITAT TYPE.

WILDEBEESTE APPEAR TO US RATHER SILLY AT THE BEST OF TIMES, BUT WHEN WE SEE ONE ALONE ON THE PLAINS, TURNING ENDLESSLY IN TIGHT CIRCLES, IT SEEMS IT REALLY HAS TAKEN LEAVE OF ITS SENSES. INDEED IT HAS, DRIVEN TO DISTRACTION BY A PARASITE. A PARTICULARLY UNPLEASANT SPECIES OF BOTT FLY DEPOSITS ITS LIVE LARVAE ON GRASS STEMS, NOT TO EAT THE GRASS, BUT TO AWAIT THE MOIST WARM MUZZLE OF A GRAZING GNU. THE LARVA RESPONDS BY CRAWLING QUICKLY UP THE UNSUSPECTING GNU'S NOSE. IT BURROWS UP INTO THE COMPLEX NASAL TISSUE TO FEED AND EVENTUALLY PUPATES. THE FLY WILL HATCH WITHIN A FEW DAYS, AND WRIGGLE ABOUT UNTIL THE GNU SNEEZES AT THE IRRITATION AND EXPELS THE ADULT FLY INTO THE WORLD ON ITS MAIDEN FLIGHT TO BEGIN THE CYCLE AGAIN. BUT THINGS OCCASIONALLY GO BADLY WRONG. THE USUAL FLY NESTING SITE IS NOT TOO FAR FROM THE BRAIN, AND SOMETIMES THE LARVA BURROWS TOO FAR, INTO THE BRAIN. TISSUE IS TISSUE TO THIS NOXIOUS CREATURE, AND IT MUNCHES ITS WAY INTO WHAT MODEST BRAIN THE POOR GNU HAS. THE RESULT IS A PERMANENT, PATHETIC AND FATAL DISORIENTATION, A SAD CIRCLE DANCE OF DEATH THAT ENDS WHEN THE GNU FINALLY STUMBLES AND IS SET UPON AND DISPATCHED BY A PASSING PREDATOR. SINCE THERE IS A 50:50 CHANCE OF THE INVADING LARVA CRAWLING UP EITHER NOSTRIL, THEY SEEM TO INVADE THE BRAIN IN DIFFERENT AREAS: SOME OF THE DOOMED GNUS TURN ONLY LEFT, THE OTHERS ONLY RIGHT.

ALTHOUGH THE NUMBERS ARE NOT YET IN, IT IS LIKELY THAT MORE GNUS SUCCUMB TO PARASITES THAN TO PREDATORS. MOST PARASITES, HOWEVER, ARE NOT AS DEVIOUS AS THE BOTT FLY; THIS LOWLY NEMATODE WORM LIVES IN THE RUMINANT GUT DIRECTLY COMPETING FOR FOOD WITH THE HOST. IT IS SURPRISING BUT A TYPICAL OCCURENCE AT THE ONSET OF THE RAINS IN THE RANGELAND CATTLE COUNTRY OF EASTERN AFRICA. AFTER THE FIRST WEEK OR SO THE CATTLE LOOK, IF ANYTHING, WORSE THAN THEY DID AT THE END OF THE DRY SEASON. THE REASON IS THAT THE FIRST FLUSH OF GREEN GRASS DOES NOT GET FAR ENOUGH IN THEIR GUTS TO PUT ANY MEAT OR FAT ON THEIR BONES – IT GOES DIRECTLY INTO THE WAITING WORMS WHO, BEING SMALLER, CAN METABOLISE IT FASTER. THE CATTLE, AND PRESUMABLY THE GNUS, ACTUALLY GET LESS NUTRITION UNTIL THE PARASITE POPULATION LEVELS OUT AND THE MAMMALS TAKE CONTROL (FOR A WHILE) OF THEIR GUTS AGAIN.

PREDATORS | ALTHOUGH GNUS HAVE STAMINA AND WILL PUT UP A VIGOROUS THOUGH FUTILE FIGHT IF PUT TO BAY, IN GENERAL THEY ARE RUN DOWN AND REGULARLY KILLED BY HYENAS AND LIONS – IN THAT ORDER. THEY WILL LOSE WEAKENED OR VERY YOUNG ANIMALS TO WILD DOGS, LEOPARDS, AND CHEETAHS. REAL SUCCESS, HOWEVER, IN GNU-CATCHING REQUIRES COOPERATIVE HUNTING. CALVES, OF COURSE, ATTRACT A LARGER GROUP OF INTERESTED PARTIES, INCLUDING ALL OF THE BIG PREDATORS, PLUS LESSER ONES SUCH AS JACKALS, RATELS AND THE OCCASIONAL EAGLE. FOR THE WILDEBEESTE, THERE IS CERTAINLY MORE SAFETY IN NUMBERS FOR DISTRACTION THAN

IN SPEED FOR FLIGHT. IN FULL FLIGHT, GNU MOTHERS WITH CALVES AT HOOF HAVE BEEN OBSERVED JOCKEYING THEIR POSITION RELATIVE TO THE THREAT IN ORDER TO PUT THE CALF ON THE FAR SIDE.

LIFE-THREATENING DANGER NOTWITHSTANDING, A GROUP OF GNUS WILL BEHAVE TOWARDS A CONSPICUOUSLY PASSING PREDATOR IN MUCH THE SAME WAY A HERD OF COWS RESPOND TO YOU WHEN YOU STOP NEAR THEIR FIELD FOR A PICNIC. THEY PEER RATHER STUPIDLY, BOB THEIR HEADS UP AND DOWN (PROBABLY TO BRING AS MANY HIGH-RESOLUTION RETINAL CELLS AS POSSIBLE INTO PLAY TO IDENTIFY THE INTRUSION), AND MAY EVEN TAKE A FEW TENTATIVE STEPS FORWARD FOR A BETTER VIEW. ONLY WHEN THE PREDATOR SHOWS OBVIOUS INTEREST WILL THEY TURN AND BOLT. AS THE HERDS MOVE ALONG, THEY ARE EXPOSED TO A CHANGING ARRAY OF FELINE THREATS. CHEETAH LIVE ONLY IN OPEN GRASSLAND AND, BEING RELATIVELY FEW IN NUMBER, POSE NO GREAT THREAT. LIONS, ON THE OTHER HAND, INHABIT BOTH OPEN AND WOODED HABITATS, WHILST LEOPARDS ARE MAINLY A THREAT ALONG THE ROCKY AND HILLY PORTIONS OF THE MIGRATION ROUTE IN THE WOODLANDS.

IMAGINE AN INVETERATE LOVER OF ICE-CREAM APPROACHING AN ICE-CREAM VAN. THAT IS HOW A LION MUST FEEL WHEN IT COMES OVER A RIDGE TO SET EYES ON THE FIRST ARRIVING WAVES OF THE ANNUAL WILDEBEEST MIGRATION. SO MUCH FOOD, AND IT IS BEING DELIVERED, LIKE THE ICE-CREAM IN THE VAN COMING DOWN THE STREET. IF ONLY IT WERE THERE ALL THE TIME. BUT, LIKE THE GNUS, IT ONLY PASSES THROUGH. THIS IS WHY FOUR OUT OF FIVE OF THE SERENGETI LIONS ARE TERRITORIAL WITH THE FEMALES DEFENDING SOME 124 SQUARE MILES AGAINST INVADERS. THE TERRITORIES HAVE TO BE LARGE IN ORDER TO CONTAIN ENOUGH FOOD ON AVERAGE OVER THE YEAR FOR A PRIDE THAT CAN COMPRISE AS MANY AS 16 ADULT FEMALES. TYPICALLY, THERE ARE SOME HALF DOZEN IN A PRIDE, FEWER MALES, PLUS THE YOUNG OF THE YEAR – THERE COULD BE OVER A TON OF LION TO MAINTAIN.

WHEN THE MIGRATION IS IN FULL SWING OVER ANY PARTICULAR TERRITORY, THE LIONS PRACTICALLY BUMP INTO THE GNUS. MEALS ARE FREQUENT AND BELLIES ARE MOSTLY FULL. THE LION STRATEGY OF COOPERATIVE HUNTING IS PRACTISED TO A LESSER DEGREE – ONE OR MORE LIONESSES STALK THE GNUS UNTIL THEY TURN AND RUN INTO THE CLUTCHES OF THE THIRD LIONESS HIDING IN LONG GRASS OR BEHIND A TERMITE MOUND ON THE OTHER SIDE. WE INEVITABLY THINK OF THE MERCILESS PREDATOR TEARING OUT THE THROAT OF ITS HAPLESS PREY. THE TRUTH IS AS MERCILESS BUT DIFFERS IN DETAIL. IN FACT LIONS TYPICALLY SECURE THEIR MEAL BY A COMBINATION OF HUGE CAT CLAWS TO NAIL THE AMBUSHED GNU, WEIGHT TO ANCHOR IT TO THE SPOT, CLAWS AGAIN TO WORK ALONG THE BODY TO THE FRONT END, AND THEN MASSIVE JAWS TO CLAMP OVER THE UNFORTUNATE GNU'S MUZZLE. THE DOOMED GNU IS UNABLE TO BREATHE AND THE LION'S WEIGHT AND STRENGTH COMBINE TO SUFFOCATE IT, OFTEN CAUSING IT TO CHOKE ON ITS OWN BLOOD. AS HORRIBLE AS SUCH A DEMISE MAY SEEM, IT IS LIKELY THAT SHORTLY AFTER THE LION HAS STRUCK, THE GNU ENTERS A CATATONIC STATE OF SHOCK. ALTHOUGH ITS DEEP NERVOUS SYSTEM KEEPS THE ANIMAL STRUGGLING WILDLY, IN FACT IT IS PROBABLY ALREADY UNCONSCIOUS OF FEAR, PAIN AND THE END OF ITS

MIGRATION AND LIFE. THE EPISODE IS QUICKLY BROUGHT TO A CLOSE AS THE REST OF THE PRIDE JOIN THE CAPTOR TO TUCK INTO THE GNU FROM EVERY SIDE. THERE IS MUCH JOSTLING AND SNARLING. THE LARGEST FEMALES GET THEIR SHARE FIRST, THE YOUNGER CATS AND THE CUBS WHEN AND AS THEY ARE ABLE. IF THEY ARE ALL LUCKY, THEY CAN FINISH MOST OF THE GNU IN A FEW HOURS BEFORE A MANED MALE IN THE NEIGHBOURHOOD HAS A CHANCE TO DETECT THE EVENT AND RUSH IN TO STEAL THE KILL. THE MALE LION HAS NO QUALMS IN THIS REGARD AND WILL PURLOIN HIS SUPPER FROM THE FEMALES WHENEVER HE CAN. HIS CAPACITY FOR MEAT IS PRODIGIOUS: THE RENOWNED NATURALIST GEORGE SCHALLER ONCE OBSERVED A MALE CONSUME SIXTY-SIX POUNDS IN FIVE AND A HALF HOURS[4]. OF COURSE, PREDATORS LIKE LIONS AND HYENAS, ALSO HAVE THE OTHER MIGRATORY SPECIES (ZEBRA, GAZELLE, ELAND) AVAILABLE ON THE MENU, AS WELL AS THE RESIDENTS (TOPI, KONGONI, IMPALA, ETC.). BUT THE PASSING OF THE MIGRATION FROM THE PLAINS CAUSES A 20-FOLD DROP IN WHAT'S ON THE PREDATORS' SHELVES. THUS LION TERRITORIALITY IS FLEXIBLE: AS THE MIGRATION MOVES ON, THE LIONS MUST RANGE FURTHER AFIELD THAN THEIR AVERAGE SIX TO EIGHT MILES A DAY. THIS RESULTS IN MORE AGONISTIC ENCOUNTERS WITH NEIGHBOURING PRIDES, AND, MORE IMPORTANTLY, IT DRAINS BODY RESOURCES, PARTICULARLY OF YOUNG CUBS. OF THE PLAYFUL CUBS WE SEE TODAY TRYING TO SNATCH SCRAPS AT A KILL, ONLY ONE OR TWO WILL MAKE IT THROUGH TO ADULTHOOD.

DO NOT JUMP TO CONCLUSIONS WHEN YOU SEE A HYENA SKULKING OFF WITH A SNATCHED BONE. FOR YEARS WE ASSUMED THAT HYENAS MUST BE SCAVENGERS, GIVEN THEIR UNSAVOURY HABIT OF LITERALLY DIVING INTO DEAD CARCASSES FOR A SNACK. BUT DESPITE THE FACT THAT HYENAS WILL EAT ALMOST ANYTHING – FROM ANTELOPES TO SAFARI BOOTS – ON AVERAGE THEY ARE MORE HUNTERS THAN SCAVENGERS. THE NOBEL LAUREATE, NIKO TINBERGEN, IN HIS FOREWORD TO HANS KRUUK'S CLASSIC WORK, THE SPOTTED HYENA[5], OBSERVED THAT:

"... popular images of animals are often amazingly wrong, and without discarding anything that experienced hunters told him, [Kruuk] quite rightly started by simply trusting the evidence of his own eyes... how right he was. By taking the trouble to study the hyena wherever and whenever he could create an opportunity (for instance by observing the hyenas throughout the night when they are most active), he made a surprising discovery: the spotted hyena is a truly formidable predator, no less formidable or less dreaded than the wolf, lion, or wild dog... they resemble wolves hunting deer and moose."

THUS A GROUP OF HYENAS (OFTEN RELATED) – KNOWN AS A CLAN – SHARE AND GUARD A HUNTING TERRITORY. WORKING TOGETHER USUALLY AT NIGHT, A GROUP OF SIX TO TWENTY HYENAS BEGIN A CASUAL, LEISURELY WALK TOWARDS A GROUP OF MILLING GNUS. THE GNUS, WHO HAVE BEEN IGNORING THE HYENAS TILL THAT MOMENT, SUDDENLY SEEM TO UNDERSTAND THAT THE PREDATORS MEAN BUSINESS AND THEY TURN AND START TO MOVE AWAY. ONE OR MORE OF THE HYENAS BREAKS INTO A RUN: THE COMICAL ROCKING GAIT THAT SEEMS DESIGNED TO DISPLAY INDIFFERENCE BUT WHICH IN FACT IS A DEADLY SIGNAL TO THOSE WHO RECOGNISE IT. THEY INCREASE THE PACE AND FAN OUT BEHIND THE RETREATING GNUS WHO ARE NOW STARTING TO GALLOP. THE GNUS JOSTLE AMONGST THEMSELVES, NOT TOTALLY IN PANIC BUT MORE STRATEGICALLY TO TRY TO SHIELD THE YOUNG ANIMALS ON THE

SIDES AWAY FROM THE CLOSING HYENAS. INEVITABLY, AFTER SOME 330 YARDS, A MOTHER AND CALF LAG BEHIND. THE COOPERATING CLAN MEMBERS INSTANTLY ALTER COURSE AND BEGIN TO CUT IT OFF FROM ALL SIDES. A SHORT SPRINT SETS THE LEAD HYENA CLAMPING ITS STEEL-TRAP JAWS ON TO THE UNFORTUNATE CALF'S RUMP. THE REST OF THE CLAN CLOSES IN QUICKLY. THE CALF'S MOTHER MAY TURN AND PUT UP A SHORT FIGHT. SINGLE HYENAS HAVE BEEN SENT FLYING BY A DESPERATE HEADBUTT. SINGLE HYENA HUNTS FAIL THREE TIMES AS OFTEN AS GROUP EFFORTS. THUS, AS MORE JAWS LATCH ON TO THE VICTIM, MOTHERLY LOVE DISSIPATES IN A FLASH, THE FEMALE WHEELS AND SPRINTS OFF AFTER THE RETREATING HERD TO LIVE TO BREED ANOTHER DAY.

GNUS THAT BEHAVE STRANGELY THROUGH SICKNESS OR STARVATION ARE ALSO ON THE HIT LIST FOR THE COOPERATING CLAN. AN EXPERIMENT IN NGORONGORO CRATER TURNED UP AN INTERESTING, IF UNFORTUNATE RESULT, NAMELY, THAT IT MAY BE PRECISELY THE ODD ONES OUT THAT THE HYENAS WILL FOCUS ON. A WELL-MEANING RESEARCHER HAD PAINTED THE HORNS OF A NUMBER OF GNUS A NICE GLOSSY WHITE IN ORDER TO TRACK THEIR MOVEMENTS MORE EASILY. DURING THE FOLLOWING MONTHS, ALMOST ALL OF THE EXPERIMENTAL SUBJECTS WERE KILLED IN GREAT DISPROPORTION TO THE REST OF THE POPULATION, SUGGESTING THAT THEIR STRIKING APPEARANCE MADE THEM MORE CONSPICUOUS IN THE EYES OF THE LOCAL HYENAS AS WELL AS IN THE PURE WHITE LIGHT OF SCIENCE. THE HYENA'S AWESOME TEETH AND WELL-MUSCLED JAWS CAN CRUNCH UP AND EAT AN ENTIRE WILDEBEEST, LEAVING ONLY THE HORNS. ITS DIGESTIVE SYSTEM PROCESSES NEARLY ALL FORMS OF PROTEIN SO THAT THE DROPPINGS ARE WHITE, ALMOST PURE CALCIUM MIXED WITH A BIT OF HAIR. FOR ONE SO SCRUFFY, THE HYENA LEAVES THE ECOSYSTEM NEATER THAN HE FINDS IT.

CHEETAHS ARE AS PERNICKETY AS THEY LOOK. THEIR SIZE AND WEIGHT – A BIT LESS THAN A HYENA AND ONLY A QUARTER OF A FEMALE LION – LIMITS THE RANGE OF PREY THEY CAN SUCCESSFULLY ATTACK AND BRING DOWN. SINCE THEY CHOSE NOT TO JOIN FORCES LIKE THE HYENAS, THEY ARE LIMITED TO OUTRUNNING SMALL OPEN-GRASSLAND PREY, MAINLY THOMSON'S GAZELLE. AFTER CAREFULLY SELECTING A PRIME THOMMIE (CONTRARY TO CONVENTIONAL WISDOM, THEY DO NOT WASTE THEIR TIME WITH SUB-STANDARD FARE), THEY SIMPLY OUT-MANOEUVRE IT WITH LEGENDARY SPEED AND AGILITY[6]. A QUICK STRANGULATION AND A HASTY MEAL OF RUMP STEAK – LEAVE THE ENTRAILS AND BONY BITS TO THE SCAVENGERS – AND THEY ARE OFF BEFORE THE BULLYING LION OR LOUTISH HYENA CAN ARRIVE TO PINCH THE KILL. OCCASIONALLY A MATING PAIR OF CHEETAHS OR NEWLY INDEPENDENT SIBLINGS WILL GANG UP TO BRING DOWN A GNU, BUT EVEN THEN THEY CAN ONLY HANDLE YOUNG CALVES NO OLDER THAN FOUR MONTHS. CHEETAHS ADORN THE GNU'S MIGRATION CYCLE BUT THEY DO NOT MUCH AFFECT IT.

DISMISSED AS "VERMIN" AND SHOT BY COLONIAL PARK WARDENS, THE AFRICAN WILD DOG IS IN FACT AN ENGAGING, HIGHLY SOCIAL AND COLOURFUL BEAST – THE LATIN NAME, *LYCAON PICTUS*, MEANS "PAINTED WOLF". HALF THE SIZE OF A HYENA, WILD DOGS ARE BY ALL ACCOUNTS THE MOST EFFICIENT PREDATORS ON THE PLAINS, BUT IT TAKES A

HUGE COOPERATIVE EFFORT FOR A GROUP OF FORTY-FOUR-POUND DOGS TO PULL DOWN AND DISPATCH A 265-POUND GNU, AND THEY RARELY SUCCEED IN CATCHING ANYTHING LARGER THAN A YEARLING. JONATHAN SCOTT WATCHED A GROUP OF KNOWN INDIVIDUALS IN THE MARA CHASE TWICE INTO GROUPS OF WILDEBEESTE WITHOUT LOCATING A TARGET. ONE OF THEM

"had even initiated a half-hearted attack on a herd of bachelors, charging towards them and lunging at their ankles. It was a test of nerves for both predator and prey: approach too close and the dog risked being injured by a sweep of those curved horns; run and the wildebeest might provoke an all-out attack by the whole pack. But the bulls were not easily intimidated into fleeing from the dogs. They stood their ground, snorting and grunting, shaking their massive horned heads, forming a black wall that barred the free passage of the pack. The bulls edged forward, bolder by the second, driving [the dog] in front of them, forcing him to turn and lope away. After running a few paces, the old male stopped and faced the herd again, lowering his head and chasing them back. Perhaps it was all just a game, a sign of frustration on the part of the hungry dog. But for the wildebeest it ensured that this particular herd would not provide an easy meal for the wild dogs.[7]"

THE HISTORICAL ANIMOSITY SPRUNG FROM HUMAN REVULSION AT THE DOGS' MERCILESS AND RATHER BLOODY TECHNIQUES OF DISPATCHING TIMOROUS ANTELOPES BY TEARING THEM TO PIECES. BUT THE PREY'S DEMISE IS SWIFT AND SURE, AND IN ANY EVENT IT IS NEITHER RELEVANT NOR INTERESTING TO IMPOSE OUR VALUE JUDGEMENTS ON THE NATURAL WORLD. "CRUELTY" AND "COMPASSION" HAVE NO MEANING TO PLANTS AND ANIMALS, AND "ETHICS" DESCRIBES NO NATURAL PROCESS. WILD DOGS AND OTHER PREDATORS KILL OUT OF NECESSITY, AS INDEED WE USED TO. IT IS A DEEP AND BITTER IRONY THAT HUMANS, THE MOST SYSTEMATIC, EFFICIENT, IRRATIONAL AND UNETHICAL KILLERS OF ALL TIME HAVE THE EFFRONTERY TO PASS JUDGEMENT ON THE SURVIVAL STRATEGY OF GROUPS OF SMALL CANINES MAKING A LIVING IN THE SERENGETI ECOSYSTEM.

WE COULD JUST AS WELL TURN THE TABLES AND WAX ANTHROPOMORPHIC IN FAVOUR OF THE DOGS. WE COULD OBSERVE, FOR EXAMPLE, THAT A WILD DOG KILL IS REMARKABLE FOR THE POLITE MUTUAL RESPECT THAT EXISTS AMONG THE PACK MEMBERS, AND FOR THEIR ALMOST GENTLE DEFERENCE TO ONE ANOTHER. THERE IS LITTLE SQUABBLING OR FIGHTING. THEY DO THEIR KILLING AS A TIGHTLY COORDINATED UNIT, WORKING WITH AN OBVIOUS TEAM SPIRIT. ONCE THE MESSY BUT NECESSARY BUSINESS OF KILLING IS OVER AND DONE WITH, THE DOGS SHARE THE MEAL IN A HIERARCHICAL ORDER, EACH CALMLY WAITING ITS TURN, PERHAPS IN THE KNOWLEDGE THAT IF THERE IS NOT ENOUGH TO GO ROUND THIS TIME, THEN THEY WILL KILL AGAIN UNTIL EACH HAS HAD ITS FILL. THAT'S NICE, ISN'T IT? ALMOST HUMAN PERHAPS?

SCAVENGERS

IF YOU SHOULD HAPPEN TO TAKE A NAP IN THE OPEN ON THE SERENGETI PLAINS, YOU MAY BE SUDDENLY AWAKENED BY A FLAP OF WINGS AND THE THUD OF A TWENTY-TWO-POUND BIRD LANDING NOT FAR FROM YOU. VULTURES ARE OPPORTUNISTS WHO USE HIGH-LEVEL AERIAL SURVEILLANCE TO KEEP CONSTANTLY ON THE

WATCH, WAITING FOR SOMEONE TO DIE. THEY RANGE OVER THOUSANDS OF SQUARE MILES, AND MAY COMMUTE HUNDREDS OF MILES A DAY FROM THEIR CLIFF NEST SIGHTS IN THE GOL MOUNTAINS TO WHEREVER IN THE ECOSYSTEM THE MIGRATION MAY BE. THEY SOAR ALONG LIKE MENACING SPY PLANES, WITH BARELY A WING STROKE, CAREFULLY SENSING AND USING THERMAL UP-DRAUGHTS AND AIR RISING FROM HILL SLOPES.

THIS LOW-BUDGET TRAVELLING IS EFFICIENT AND ALLOWS THEM TO MAINTAIN LARGE, STRONG BODIES THAT ARE NECESSARY FOR COMPETITION AT A CARCASS. AT FIRST SIGHT THE VULTURES AROUND A CARCASS SEEM TO BE IN A FIERCE FREE-FOR-ALL OVER FOOD. BUT LOOK AGAIN. THERE IS AN ORDER. FIGHTS ARE USUALLY BETWEEN MEMBERS OF THE SAME SPECIES. DIFFERENT SPECIES ARRIVE ON THE KILL AT DIFFERENT TIMES IN A PREDICTABLE ORDER AND THEY UTILISE DIFFERENT PARTS OF THE CARCASS WITH SPECIAL EQUIPMENT. THE SHAPE OF THEIR BILL GIVES CLUES TO WHETHER THEY ARE "RIPPERS", SUCH AS THE LARGE LAPPET-FACED AND WHITE-HEADED, "GULPERS", LIKE THE MEDIUM-SIZED WHITE-BACKED AND RUPPELL'S GRIFFON OR "SCRAPPERS", LIKE THE SMALL HOODED AND EGYPTIAN VULTURES. EACH SPECIES OCCUPIES A DISTINCT NICHE, WHICH ALLOWS THEM TO CO-EXIST, MORE OR LESS PEACEFULLY. IT WOULD BE KINDER TO CONSIDER VULTURES AS PREDATORS WHO HAPPEN TO PREY ON ANIMALS THAT ARE ALREADY DEAD. IN THE SERENGETI, AS IN ALL ECOSYSTEMS, EVENTUALLY EVERYONE BECOMES VULTURE "PREY". WILD DOGS DEAL WITH GAZELLE AND GNUS IN VERY DIFFERENT WAYS. BUT ONCE DEAD, WE ALL BECOME JUST ONE KIND OF VULTURE FOOD.

RIVER CROSSINGS

THE GNU'S MIGRATION ROUTE IS CUT AGAIN AND AGAIN WITH THE DRAINAGE LINES OF THREE MAJOR RIVERS THAT RUN INTO LAKE VICTORIA: FROM NORTH TO SOUTH, THE MARA, THE GRUMETI AND THE MBALANGETI. THESE LARGE PERENNIAL RIVERS ARE A MAJOR OBSTACLE TO THE ESSENTIALLY TERRESTRIAL GNUS AND THEIR CO-MIGRANTS. MOREOVER, THEIR TRIBUTARIES, WHICH DRAIN SURROUNDING HILLS AND PLAINS, CAN BECOME VIOLENT TORRENTS A FEW HOURS AFTER A SHORT RAINFALL.

THE RIVERS ARE CLEARLY OBJECTS OF TERROR FOR THE GNUS. THE CHURNING WATER IS ENOUGH IN ITSELF, BUT INVARIABLY THERE ARE ALSO GLOOMY THICKETS ALONG THE BANKS THAT THE GNUS SEEM TO INSTINCTIVELY TRY TO AVOID. AS THE NUMBERS BUILD UP IN FRONT WITH THE LEAD ANIMALS (USUALLY MALES) HESITATING AND DITHERING AT THE BANK, THE CROWDS ARRIVING FROM THE REAR PUT GREATER AND GREATER PRESSURE ON THOSE AT THE WATER'S EDGE UNTIL THEY ARE LITERALLY PUSHED OVER THE EDGE. ONCE IN, SHEER SURVIVAL TAKES OVER AND THEY SWIM LIKE MAD THINGS FOR THE OPPOSITE SHORE. YOUNG ANIMALS ARE OFTEN OVERWHELMED BY WAVES OR FLAILING ADULTS AND HELPLESSLY GO UNDER. ADULTS, TOO, ARE SWEPT AWAY IF THE FLOW IS FIERCE, EITHER TO DROWN OR GET TRAPPED IN BANK-SIDE TANGLES WHERE THEY ARE AWAITED BY THE CROCODILES. SUCH DROWNINGS OCCUR TWO OR THREE TIMES A YEAR, WITH UP TO 500 ANIMALS PERISHING AT A TIME.

THE PENCHANT FOR CROSSING WATER BODIES AGAINST ALL ODDS SEEMS TO BE BUILT INTO THE WILDEBEEST GENOME. ALAN ROOT[8] FILMED A BIZARRE EPISODE IN 1973 IN WHICH THE AVANT-GARDE OF THE MIGRATION MOVING WEST OFF THE SHORTGRASS PLAINS FOUND ITSELF CONFRONTED WITH THE MIGHTY LAKE LAGARJA, A SEASONAL POND TWO YARDS DEEP AND A COUPLE OF MILES ACROSS NESTLED IN THE SOUTHWEST BORDER OF THE SERENGETI NATIONAL PARK. RATHER THAN SKIRT THE EDGE OF THE LAKE, WHICH IN THE MIND OF THE GNUS WOULD HAVE LED THEM OFF THE EDGE OF THE KNOWN WORLD, THEY BRAVELY PLOUGHED ACROSS. THE LAKE WAS SHALLOW ENOUGH TO ALLOW THE ADULTS TO HALF WALK, HALF SWIM THEIR WAY THROUGH, BUT THE CALVES ONLY A FEW WEEKS OLD COULD NOT KEEP UP. WHEN THE ADULTS GOT TO THE FAR SIDE, MOST OF THE MOTHERS HAD LOST THEIR CALVES. THEY MILLED AROUND A BIT MOOING, THEN BEING GOOD MOTHERS ALBEIT OF LITTLE BRAIN, THEY RETRACED THEIR FOOTSTEPS BACK INTO THE MUDDY WATER TO LOOK FOR THE YOUNGSTERS. A FEW WERE LUCKY, BUT MOST REACHED THE OPPOSITE SHORE WITHOUT THEIR OFFSPRING. NATURALLY, THEY TURNED BACK TO THE LAKE TO CARRY ON THE SEARCH OR THE MIGRATION OR WHATEVER ELSE OCCURRED TO THEM. SOON LAKE LAGARJA WAS ACHURN WITH WILDEBEEST ADULTS PLYING BACK AND FORTH, YOUNGSTERS BLEATING, FLOUNDERING AND EVENTUALLY SINKING BENEATH THE MUDDY WAVES. THIS GROTESQUE PANTOMIME OF MOTHERLY LOVE WENT ON FOR SEVEN DAYS UNTIL THE MOMENTUM OF THE HERDS ARRIVING FROM THE SOUTHEAST FORCED ALL STRAGGLERS ON BEYOND THE MAD DRYING SHORES THAT WERE STREWN WITH THE CARCASSES OF MUCH OF THAT YEAR'S CALF PRODUCTION. A GOOD YEAR FOR THE SCAVENGERS, A BAD ONE FOR THE GNUS.

THE CROCODILE IS A SURVIVOR OF JURASSIC PARK THAT FINDS ITS NICHE AS A TOP CARNIVORE IN THE AQUATIC FOODCHAIN: IT SPENDS MOST OF ITS LIFE IN THE WATER EATING FISH. IT CAN, HOWEVER, MOVE QUICKLY ON LAND – ALMOST AT A GALLOP – TO SNAP UP ANIMALS THAT LINGER TOO LONG AT THE WATER'S EDGE. RIVER CROCS IN PARTICULAR GET CONSIDERABLE NOURISHMENT FROM TERRESTRIAL ANIMALS. BUT BESIDES SHORT DASHES, NESTING AND BASKING, THE CROCODILE MAKES ITS LIVING IN THE WATER. THE JAWS OF THE CROCODILE IN WHICH THE DROWNED GNU END UP NEVER EVOLVED TO DO MUCH MORE THAN CLAMP. THE PRIMITIVE TEETH CAN ONLY PIN DOWN; THEY NEITHER SLICE NOR GRIND. THE TOOTHED TRAP THUS SNAPS SHUT OVER ANY CONVENIENT PART OF AN ANIMAL, DEAD OR ALIVE. THE CROCODILE HOLDS FAST, UNDERWATER IF IT NEEDS TO DROWN THE PREY. IT HAS EVOLVED THE PHYSIOLOGICAL MEANS TO HOLD ITS BREATH FOR MORE THAN AN HOUR. WITH THE PREY HELD TIGHT, IT THEN SPINS ITS BODY VIOLENTLY AROUND WITH A CORKSCREW SNAP OF ITS MUSCULAR TAIL. CHUNKS OF MEAT, BONE AND SINEW ARE THUS TWISTED OFF LIKE PIECES OF TOFFEE AND SWALLOWED WHOLE. CROCODILES HAVE THE PATIENCE OF THE PRIMITIVE. THEY FLOAT FOR HOURS WITH JUST THEIR EYES AND NOSTRILS BREAKING THE SURFACE, WATCHING, WAITING. AS THE NOISY MILLING GNUS SCRAMBLE DOWN THE RIVERBANK, THEY DON'T NOTICE THE "FLOATING LOG", THAT EDGES SLOWLY NEARER TO GREET THEIR CROSSING WITH "GENTLY SMILING JAWS".

3

EASTERN WOODLANDS

Short Dry Season, December–February

IN THE MARA, AROUND CHRISTMAS TIME, THE MIGRATION IS QUICKLY USING UP THE GRASS GROWTH ENGENDERED BY THE SHORT RAINS. ALTHOUGH THE "SHORT DRY SEASON" IS BRIEF, IT IS INTENSE AS THE ANNUAL TILT OF THE EARTH ON ITS ACCESS PRESENTS THIS PART OF THE WORLD TO THE FULL FACE OF THE SUN. VEGETATION THAT IS NOT EATEN QUICKLY DESICCATES, AND THE PRESSURE IS ONCE AGAIN ON THE HERDS TO SEEK FODDER AND WATER. THE EASTERN WOODLANDS OF THE SERENGETI WITH THEIR SCATTERED ACACIA AND COMMIPHORA TREES AND NUMEROUS SMALL WATERCOURSES AND VALLEY BOTTOMS, ARE AN EFFECTIVE RESERVE OF GRASS TO SUSTAIN THE GNUS AS THEY TRAVEL BACK SOUTH TO THE SHORTGRASS PLAINS.

IN PRIMEVAL TIMES A FREAK LIGHTNING STRIKE OR A SPARK FROM A VOLCANO IGNITED GRASS FIRES. TODAY IT IS MORE LIKELY TO BE A MAN WITH A MATCH, DOING WHAT HIS ANCESTORS HAVE DONE FOR THE LAST TWO MILLION YEARS: LIGHTING FIRES FOR MANAGEMENT OR FUN. SINCE MORE THAN 75% OF AFRICAN GRASSLANDS GO UP IN SMOKE EVERY YEAR, FIRE HAS BECOME AN INTEGRAL MODIFYING FACTOR OF GRASSLAND ECOLOGY. THE MASAI LIGHT GRASS FIRES TO ENCOURAGE THE GREEN GRASS SHOOTS THAT APPEAR IMMEDIATELY AFTER A BURN WITH EVEN THE LIGHTEST SHOWER. SOME TREE SPECIES ADAPT TO BURNING BY DEVELOPING A THICK FIREPROOF BARK, OTHERS "RETREAT" UNDERGROUND. A STUNTED BUT LIVING TWO-FOOT-HIGH ACACIA IN A REGULARLY BURNED AREA MIGHT HAVE A ROOTSTOCK SIX INCHES THICK – EVIDENCE THAT THE PART OF THE PLANT ABOVE GROUND HAS BEEN BURNED BACK REGULARLY FOR A DECADE OR MORE. SOME GRASSLANDS ARE MAINTAINED BY FIRE: OTHERS MODIFIED AS FIRE-RESISTANT SHRUBS CREEP IN. THE EFFECTS ARE VARIED, DEPENDING ON LOCAL CONDITIONS.

WE SHOULD NOT THINK OF OUR HABITAT DESCRIPTIONS AS FIXED AND FOREVER. MOST OF THE SERENGETI ECOSYSTEM (EXCEPTING THE SOUTHERN SHORTGRASS PLAINS AND THE WESTERN FLOODPLAINS) IS IN REALITY A SHIFTING PATCHWORK OF GRASSLAND AND WOODLAND. THE SIZE AND LOCATION OF THE PATCHES CHANGES CONTINUALLY, DEPENDING ON THE SHORT-TERM (RELATIVE TO A HUMAN LIFE SPAN, THAT IS) CHANGES IN IMPACT OF ANIMALS, FIRE AND RAINFALL. WE OFTEN ARE LED TO BELIEVE THAT ALL VEGETATION UNDERGOES SUCCESSIVE STAGES OF GROWTH

UNTIL A CLIMAX FORM IS REACHED. BUT THIS IS NOT SO IN AFRICA'S WOODED GRASSLANDS. IN MANY AREAS TREES ARE DYING AND WOODLANDS SEEM TO BE DISAPPEARING. IS THIS AN ECOLOGICAL DISASTER? PERHAPS NOT, SINCE THE WOODED GRASSLAND IS NOT A CLIMAX FORM, NOT AN END PRODUCT IN ITSELF, BUT A STAGE IN A CENTURIES-LONG CYCLE. IN A WOODLAND, SOME OF THE TREES START TO DIE FROM OLD AGE OR ARE PUSHED OVER BY ELEPHANTS. THEIR DEATH ALLOWS SEEDLINGS TO SPROUT THAT WOULD OTHERWISE HAVE WITHERED IN THE SHADE OF THE ADULT. MORE ROBUST SPECIES OF GRASS CAN ALSO NOW THRIVE. INCREASED GRASS COVER, TOGETHER WITH A PARTICULARLY LONG DRY SEASON, INCREASES THE FIRE HAZARD. FIRES RETARD THE REGENERATION OF THE ACACIA WOODLAND BY BURNING BACK YOUNG TREES MOVING US FURTHER INTO A GRASSLAND PHASE. IF THERE IS A STRING OF GOOD WET SEASONS, THERE IS MORE FOOD ON OFFER FOR GRAZERS LIKE GNUS AND GAZELLE. THEIR POPULATIONS GROW, EAT MORE GRASS, AND THEREBY REDUCE POTENTIAL FUEL FOR A FIRE. WITH LESS FIRE, YOUNG TREES BEGIN TO GROW BEYOND THE FIRE-CRITICAL HEIGHT OF TEN FEET. GRASSLAND BEGINS TO REVERT TO WOODLAND. AND SO THE CYCLE GOES ON FOR A HUNDRED YEARS OR MORE.

MAN & THE WILDEBEEST | NOT EVERYONE WILL EAT A GNU. THE MASAI, FOR EXAMPLE, WON'T TOUCH THEM WITH A SPEAR POINT. THEY WOULD, HOWEVER, TUCK INTO AN ELAND, PRESUMABLY BECAUSE TO THEIR TASTE AN ELAND IS MORE LIKE A COW AND THEREFORE WORTHY OF A WARRIOR'S PALATE. WILDEBEEST MEAT, HARVESTED ON RANCHES WITH SPECIAL DISPENSATION, IS AVAILABLE OVER SOME NAIROBI BUTCHER COUNTERS AND ON THE MENU OF THE CARNIVORE RESTAURANT, SO ONE CAN DISCOVER FOR ONESELF THAT THE MAASAI HAVE A POINT. HOWEVER, MEAT IS MEAT TO A POOR MAN, AND IT IS ESTIMATED THAT THE LOCAL CITIZENS, PARTICULARLY ALONG THE NORTHWESTERN BORDERS OF THE SERENGETI NATIONAL PARK REMOVE SOME 135,000 MIGRATORY HERBIVORES (MAINLY WILDEBEESTE, BUT ALSO THOMSON'S GAZELLE AND ZEBRA) ALONG WITH AN ESTIMATED 75,000 RESIDENT HERBIVORES. THE ANIMALS ARE HUNTED PRIMARILY WITH THE TRADITIONAL METHOD OF PASSIVE SNARING. A STAKED-DOWN WIRE LOOP IS CONCEALED AROUND THE EDGES OF A NATURAL BREAK IN A THICKET OR IN THE NARROW GAP AT THE END OF A MAN-MADE FUNNEL OF CUT THORN TREES. HUNDREDS OF THESE SNARES ARE LEFT DEPLOYED WHERE THE MIGRATION IS LIKELY TO PASS. ALTERNATIVELY, THE ANIMALS ARE ACTIVELY DRIVEN TO THE SNARE LINE BY BEATERS. PITFALL TRAPS ARE USED NEAR RIVER-CROSSINGS. FIREARMS SUCH AS ANTIQUE OR HOMEMADE MUZZLE-LOADING RIFLES ARE OCCASIONALLY DEPLOYED, AND MORE RECENTLY THE UBIQUITOUS AK-47 THAT IS FLOODING THE AFRICAN UNDERWORLD FROM NEIGHBOURING WAR ZONES IS BECOMING THE WEAPON OF CHOICE. SPEARS AND POISONED ARROWS ARE FREQUENTLY CARRIED BY HUNTERS, BUT SEEM TO THESE DAYS TO BE USED MAINLY TO DISPATCH ANIMALS STRUGGLING IN SNARES OR PITS. MANY ANIMALS WREST THEMSELVES FREE OF

THE SNARE'S STAKE AND TOTTER OFF TO DIE SLOWLY FROM INFECTION OR GANGRENE. IN SUM, OVER A MILLION PEOPLE LIVING IN A THIRTY-ONE-MILE "CATCHMENT" ZONE TO THE WEST OF THE PARK AND GAME RESERVES SATISFY PERHAPS AS MUCH AS A QUARTER OF THEIR REQUIREMENTS FOR PROTEIN AND OTHER ANIMAL PRODUCTS BY TAPPING ILLEGALLY INTO THE MIGRATION. ALTHOUGH THE WILDEBEEST POPULATION HAS BEEN STEADILY GROWING, SO HAS THE HUMAN POPULATION, AND IT IS LIKELY THAT UNCONTROLLED OFFTAKE IS UNSUSTAINABLE.

THE GREAT SERENGETI MIGRATION CANNOT BE PROTECTED BY POLICE ACTION ALONE. IN THE REMOTE BORDER AREAS OF THE ECOSYSTEM ANTI-POACHING EFFORTS ARE BOTH INEFFECTIVE AND ALIENATING. THEY COULD BE IMPROVED BY SUBSTANTIAL ADDITIONAL INVESTMENT IN ROADS, VEHICLES AND TRAINED AND MOTIVATED PERSONNEL. THAT WILL IMPROVE THE ARREST RATE BUT NOT HELP THE ALIENATION. LOCAL COMMUNITIES NEED TO BE ENGAGED RATHER THAN REPELLED, AND PROVIDED WITH OPPORTUNITIES FOR ALTERNATIVE OR AUGMENTED FORMS OF LIVELIHOOD, SUCH AS IMPROVED INFRASTRUCTURE AND MARKETING FOR DOMESTIC STOCK. IT HAS BEEN DEMONSTRATED ELSEWHERE THAT A COMBINATION OF ENLIGHTENED CONSERVATION EDUCATION AND ACCESS TO EQUITABLE SHARING OF THE BENEFITS THAT ACCRUE FROM TOURISM CAN TURN POACHERS INTO CONSERVATIONISTS, OR AT LEAST INTO RATIONAL MANAGERS OF A COMMON RESOURCE. THERE MAY BE ANOTHER REASON APART FROM TOUGHNESS AND TASTE WHY THE MASAI ARE NOT OVERLY FOND OF GNUS: THEY EAT THE SAME GRASS AS THEIR CATTLE DO. IT SEEMS, HOWEVER, HISTORICALLY, THERE HAS BEEN LITTLE OVERT ANIMOSITY AND EVEN TODAY, IN THE SHORTGRASS PLAINS BETWEEN THE SERENGETI NATIONAL PARK AND THE NGORONGORO CONSERVATION AREA THERE SEEMS TO BE ENOUGH TO GO AROUND. WHEN THERE IS NOT, THE NOMADS, BOTH HUMAN AND BOVINE, MIGRATE TO GREENER PASTURES.

IN THE NORTHERN MARA, IT IS A DIFFERENT MATTER. THE GNUS PUSH UP INTO AREAS WHERE THERE ARE MORE CATTLE THAN IN THE PAST, HERDED BY MASAI WHO ARE TENDING TO REMAIN SEDENTARY ON THEIR GROUP RANCHES IN ORDER TO TAKE ADVANTAGE OF INFRASTRUCTURE DEVELOPMENT AND THE POSSIBILITIES, ALBEIT POORLY MANAGED, OF REVENUE-SHARING THE TOURIST EXPENDITURE AT GATES AND LODGES. BUT EVEN MORE SIGNIFICANTLY, THE LOITA IS IN THE MARGIN OF THE RAINFALL ZONE WHERE WHEAT HYBRIDS CAN START TO GROW PRODUCTIVELY. THIS SETS THE STAGE FOR DIRECT COMPETITION BETWEEN MAN AND BEAST FOR LAND, AND GIVEN IMPERFECTIONS IN POLICY IMPLEMENTATION, CONFLICT IS GROWING. THE CONSERVATIONIST BERATES THE MASAI FOR DARING TO JEOPARDISE THE LAST GREAT MIGRATION ON EARTH. THE MASAI RETORTS, "I NEED TO GROW WHEAT, RAISE GRADE CATTLE TO FEED AND EDUCATE MY FAMILY ON MY LAND. IF YOU WANT TO ME PROTECT THE GREAT GNU MIGRATION FOR THE BENEFIT OF THE COUNTRY AND THE WORLD, YOU SHOULD PAY ME!" THERE IS A POINT TO BE CONSIDERED. IT HAS BEEN ESTIMATED THAT THE MARA GROUP RANCHES ARE EFFECTIVELY SUBSIDISING THE MIGRATION TO THE BENEFIT OF THE WORLD COMMUNITY AT AN OPPORTUNITY COST TO THEMSELVES FOR NOT GROWING WHEAT OR RAISING BETTER CATTLE TO THE TUNE OF MILLIONS OF DOLLARS[9]. IS THE WORLD WILLING TO PAY FOR THE GNU?

4

SOUTHERN SERENGETI SHORTGRASS PLAINS

Long Rains, March–April

THROUGHOUT THE SERENGETI ECOSYSTEM THERE ARE "BUSHED GRASSLANDS" AND THERE ARE "WOODED GRASSLANDS". IN THE MARA, THERE ARE "OPEN GRASSLANDS" KEPT OPEN BY THE COMBINED WORK OF FIRE AND ANIMALS, AND IN THE WESTERN CORRIDOR, THERE ARE GRASSLANDS MAINTAINED BY ANNUAL FLOODING OF LARGE RIVERS. IN ALL OF THESE HABITATS, IF YOU REMOVED THE MAIN MODIFYING FACTORS OF FIRE, FLOOD OR FEEDING, THEIR LONGGRASS PLAINS WOULD SUCCUMB IN TIME TO THE INEVITABLE ENCROACHMENT FROM THE EDGES OF THICK WOODLAND OR BUSHLAND. BUT THE SHORTGRASS PLAINS OF THE SOUTHEASTERN SERENGETI, THOSE ARE THE "REAL" GRASSLANDS THAT WILL NOT CHANGE FOR A THOUSAND GENERATIONS.

THE CHARACTER AND EXTENT OF THE SHORTGRASS WAS STAMPED ON ONE-THIRD OF THE SERENGETI WHEN THE PLEISTOCENE VOLCANOES OF THE NGORONGORO HIGHLANDS BLEW OUT VAST QUANTITIES OF ASH JUST OVER A MILLION YEARS AGO. THE HEAVY ASH SETTLED QUICKLY TO THE GROUND AND PRODUCED THE INGREDIENTS OF AN ALKALINE-RICH SOIL THAT PRODUCES SHORT SWEET GRASS WHEN IT RAINS. THE IRRESISTIBLE MAGNET OF VAST QUANTITIES OF NUTRITIOUS FODDER, JUST WHEN IT IS NEEDED MOST AS ALL OF THE YOUNG OF THE YEAR ARE BORN, IS WHAT PULLS THE GNUS BACK TO THE SHORTGRASS PLAINS AND WHAT KEEPS THE FLYWHEEL OF THE MIGRATION TURNING. THEY SPEND A PLEASANT, IF SOMETIMES FRANTIC, TIME DURING THE LONG RAINS FEEDING ON NUTRITIOUS GRASSES THAT ARE RICH IN MINERALS (CALCIUM, ESPECIALLY) AND STORING UP THE ENERGY THEY WILL NEED FOR LACTATION, MATING AND MARCHING. ALTHOUGH THE ADULT FEMALES IN PARTICULAR ARE DRAWN INTO THE ROUNDABOUT OF THE MIGRATION BY THEIR STOMACHS, THE CONSTANT TREKKING HAS THE ANCILLARY BENEFIT THAT IT ALLOWS THEM SIMPLY TO OUTMARCH LARGE NUMBERS OF PREDATORS.

AT A CASUAL GLANCE, IT MAY APPEAR THAT THE GNUS AND THEIR FELLOW MIGRATORS, ZEBRA AND THOMSON'S GAZELLE ARE EATING THE SAME THING. YET EVERY FARMER KNOWS THAT THE HORSES IN A FIELD EAT DIFFERENT PLANTS FROM THE COWS. YOU ARE WITNESSING PART OF A "GRAZING SEQUENCE" IN WHICH A RANGE OF HERBIVORES SHARE THE PASTURE IN A MUTUALLY BENEFICIAL WAY. A MATURE STAND OF COARSE GRASS IS UNPALATABLE TO THE

GNUS AND PHYSICALLY UNAVAILABLE TO THE LITTLE GAZELLE BECAUSE OF ITS STRUCTURE. MOREOVER, THEY ARE PUT OFF BY POTENTIALLY DANGEROUS STANDS OF TALL GRASS THAT COULD CONCEAL PREDATORS, SO THEY "LET" LARGER ANIMALS PREPARE THE WAY. VERY LARGE HERBIVORES LIKE BUFFALO AND ELEPHANTS ARE ABLE TO PLOUGH FEARLESSLY INTO CHEST-HIGH GRASS AND HARVEST WHAT THEY NEED. ONCE THEY HAVE TRAMPLED AND EATEN THE HEAVIER MATERIAL AND THERE IS SOME RE-GROWTH OF FINER GRASS, THE ZEBRA MOVE IN. ZEBRA ARE ALSO SO-CALLED "COARSE-FEEDERS" WHO ARE ABLE TO BITE OFF AND DIGEST STEMMIER GRASS. THEY ARE THEN FOLLOWED BY WILDEBEESTE WHO PREFER A BIT MORE LEAF AND A LOT LESS STEM.

THE ZEBRA AND THE WILDEBEESTE REFINE THE STRUCTURE OF THE PASTURE EVEN FURTHER AND MULCH UNEATEN MATERIAL INTO THE SOIL STIMULATING THE GROWTH OF MORE SHOOTS. THE GRASSLAND IS "READY" FOR THOSE WITH MORE DELICATE MOUTHS AND REFINED TASTES, SUCH AS GAZELLE AND KONGONI. DURING THE GROWING SEASON, THE FIRST-WAVE GRAZING OF WILDEBEESTE ACTUALLY INCREASES THE AMOUNT OF GRASS LEFT FOR THOMSON'S GAZELLE: USE BEGETS GROWTH. THE THOMMIES SELECTIVELY NIP OFF LEAFY PARTS OF THE GRASSES AND PLUCK THE FORBS LEFT BY THE LARGER HERBIVORES. THE SEQUENCE DOES NOT END THERE. IF THERE IS WATER NEARBY, WE MAY FIND HIPPOS DOING THEIR SHARE OF MOWING AND EGYPTIAN GEESE NIBBLING ON THE SHORTEST SWARDS. FINALLY, TERMITES CLEAN UP THE DEBRIS, AND BIRDS CHASE INSECTS STIRRED UP AT THE HERBIVORES' HOOVES. THE RESULT IS A HEALTHY "LAWN", OFTEN A MOSAIC OF CLOSE-CROPPED PATCHES OF FAVOURED GRASS SPECIES THAT HAVE ACTUALLY BEEN ENCOURAGED TO PRODUCE MORE BY THE ATTENTION OF THE GNU AND ITS FRIENDS OF THE HERBIVORE GUILD.

IT IS NOT ALWAYS BROTHERS AND SISTERS WORKING TOGETHER. IMAGINE THE IMPACT ON THE LOCAL RESIDENT HEBIVORES AS NEARLY TWO MILLION MIGRANT ANIMALS INVADE THEIR PASTURES AND MAKE THEMSELVES AT HOME IN THE LARDER. THE BROWSERS AND LESS NUMEROUS GRAZERS, LIKE GIRAFFE, ELEPHANTS, ORIBI AND SUCHLIKE ARE PROBABLY LITTLE AFFECTED (EXCEPT BY THE LONG-TERM CHANGES IN THE WOODLAND–GRASSLAND BALANCE). BUT THERE IS LITTLE DOUBT THAT GRAZERS LIKE THE TOPI AND BUFFALO THAT ARE RESIDENT ON THE WOODED GRASSLANDS OR FLOODPLAINS OF THE WESTERN CORRIDOR SUFFER FOOD SHORTAGES IF IT HAPPENS TO BE A YEAR THEY ARE VISITED BY THE MIGRATION. SUCH VISITS, TOGETHER WITH THE NORMAL VICISSITUDES OF RAIN-FED GRASS GROWTH, ARE THE REASON WHY THESE SPECIES, TOO, MOVE ABOUT LOCALLY. THE AMAZING THING IS THAT THESE TWO MILLION ODD ANIMALS APPEAR TO MAINTAIN A SELF-REGULATORY BALANCE: THERE ARE NO MASS STARVATIONS AND APPARENTLY NO UNFAVOURABLE CHANGES IN THE GRASSLAND PASTURE TO BARE SOIL AND UNPALATABLE SPECIES. THE FACT THAT THE POST-RINDERPEST GROWTH OF GNUS IN THE SERENGETI ECOSYSTEM SEEMED TO BE LEVELLING OFF AT SOMETHING UNDER 1.5 MILLION IN THE EARLY 1990S, SUGGESTS THAT A BALANCE BETWEEN FOOD SUPPLY AND DEMAND IS BEING STRUCK. SADLY, THE LEVEL OF INVESTMENT IN RESEARCH HAS ALSO DIMINISHED OVER THE YEARS AND THEREFORE SO HAS THE COMPREHENSIVE MONITORING OF THE SERENGETI ECOSYSTEM.

THE BIG BIRTH

SINCE, AT THE BEGINNING OF THIS TALE, WE SAW THAT ALL OF THE GNU MATING AND CONCEIVING WAS TAKING PLACE "ON THE ROAD" AS THE MIGRATION MOVED INTO THE WESTERN WOODED GRASSLANDS IN MAY/JUNE, WE CAN PREDICT THAT MOST BIRTHS WILL TAKE PLACE THE NEXT JANUARY/FEBRUARY AFTER AN EIGHT-AND-A-HALF-MONTH GESTATION. IN FACT IT IS SO DESIGNED. WITHIN THE SPACE OF THREE WEEKS EACH YEAR NEARLY 400,000 FEMALE GNUS DROP THEIR CALVES. THE PREDATORS ARE WAITING TO TUCK INTO AN INCREDIBLE FEAST OF WILDEBEEST VEAL. DESPITE THE FACT THAT THE NEWLY BORN GNUS ARE ON THEIR FEET WITHIN TWO OR THREE MINUTES OF BEING DROPPED AND RUNNING FULL TILT ALONGSIDE THEIR DAMS AFTER 15 MINUTES, THEY ARE EXCEEDINGLY VULNERABLE TO ALL THE PREDATORS LARGER THAN A BAT-EARED FOX. THEIR CONSPICUOUS TAN COLORATION DISAPPEARS AFTER ABOUT THREE MONTHS, BY THE TIME THE HERDS ARE LEAVING THE SHORTGRASS PLAINS. BUT WHILST THE ADULTS STOCK UP ON GRASS, THE CALVES ARE PRIME TARGETS.

SURPRISINGLY, THE OVERALL EFFECT OF THIS SYNCHRONISED CALVING IS TO REDUCE MORTALITY FROM PREDATION BY GLUTTING THE MARKET. THE PREDATORS DO INDEED EAT A LOT OF CALVES. BUT THEY CAN ONLY EAT SO MANY AT A TIME. MOST OF THE TIME THE PREDATORS ARE LOLLING AROUND WITH DISTENDED BELLIES WISHING FOR A CHANGE OF DIET. THE NET EFFECT IS THAT THEY EAT A SMALLER PROPORTION OF THE ANNUAL GNU PRODUCTION THAN IF CALVES WERE AVAILABLE ALL YEAR ROUND. AS THE GNUS MILL ABOUT FEEDING, AND PREDATORS DASH INTO GROUPS TO PICK OFF A MEAL, THERE SEEMS TO BE TOTAL CONFUSION OF MOTHERS AND YOUNG. YET GNU MOTHERS ARE INCREDIBLY ADEPT AT RECOGNISING THEIR CALVES, INITIALLY BY SMELL, BUT LATER BY THE QUALITY OF THEIR BLEATING. IT SEEMS THAT THE YOUNG MAY RESPOND TO THEIR MOTHERS' VOICES AS WELL. IT SEEMS IMPOSSIBLE IN THE ENDLESS CACOPHONY OF THE MIGRATION, BUT IS DOES WORK – MOST OF THE TIME. SOME EXPERTS OPINE THAT MORE CALVES DIE FROM GETTING LOST THAN FROM PREDATION. HOWEVER, THOSE THAT GET LOST ARE BOUND TO GET EATEN BEFORE THEY DIE OF STARVATION, SO THE PREDATORS HAVE THE LAST WORD.

(1) For those readers interested in a definitive scientific treatment of the migration, the Serengeti ecosystem and its plants and animals, apart from the species-specific works cited below, there are no finer resources than: Sinclair, A.R.E. and M. Norton-Griffiths, Eds. (1979). *Serengeti: Dynamics of an Ecosystem*. Chicago, University of Chicago Press, and Sinclair, A.R.E. and P. Arcese, Eds. (1995). *Serengeti II: Dynamics, Management and Conservation of an Ecosystem*. Chicago, University of Chicago Press.

(2) Hanby, J.P. and J.D. Bygott (1984). *Lions Share: The Story of a Serengeti Pride*. Boston, Houghton Mifflin.

(3) McNaughton, S.J. and F.F. Banyikwa (1995). "Plant Communities and Herbivory". *Serengeti II: Dynamics, Management and Conservation of an Ecosystem*. A.R.E. Sinclair and P. Arcese. Chicago, University of Chicago Press. Pages 49–50.

(4) Schaller, G.B. (1974). *Golden Shadows, Flying Hooves*. London, Collins.

(5) Kruuk, H. (1972). *The Spotted Hyena: A study of predation and social behavior*. Chicago.

(6) FitzGibbon, C.D. and J.H. Fanshawe (1989). "The condition and age of Thomson's gazelles killed by cheetahs and wild dogs." *J. Zool* (Lond) 218: 99–107.

(7) Scott, J. (1991). *Painted Wolves: Wild dogs of the Serengeti-Mara*. London, Hamish Hamilton.

(8) Root, A & J. Root. *The Year of the Wildebeeste*. Anglia Films

(9) Norton-Griffiths, M. (1996). "Property rights and the marginal wildebeest: an economic analysis of wildlife conservation options in Kenya." *Biodiversity and Conservation* 5: Pages 1557–77.

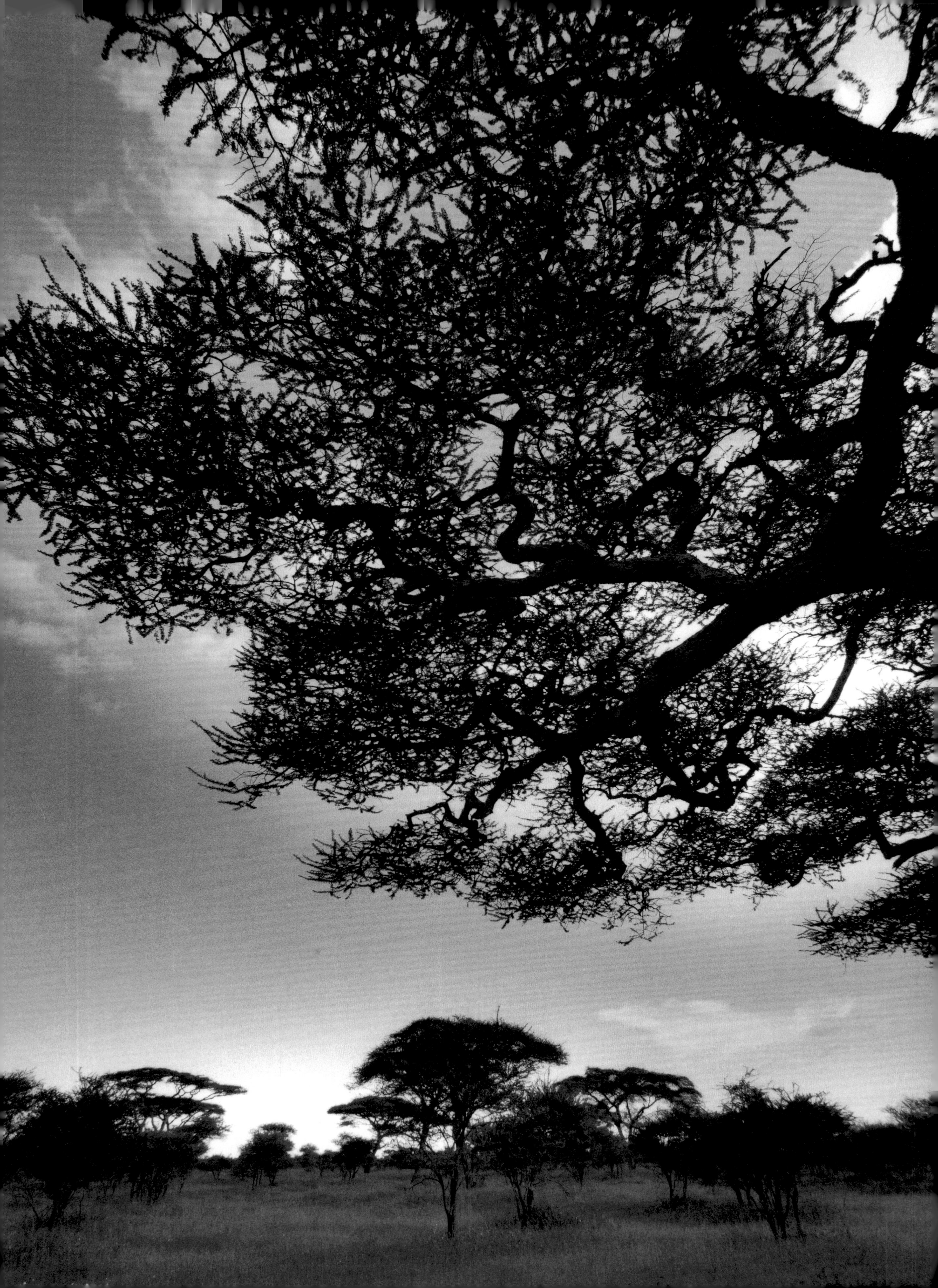

I

WESTERN WOODED-GRASSLANDS

Long Dry Season, June–September

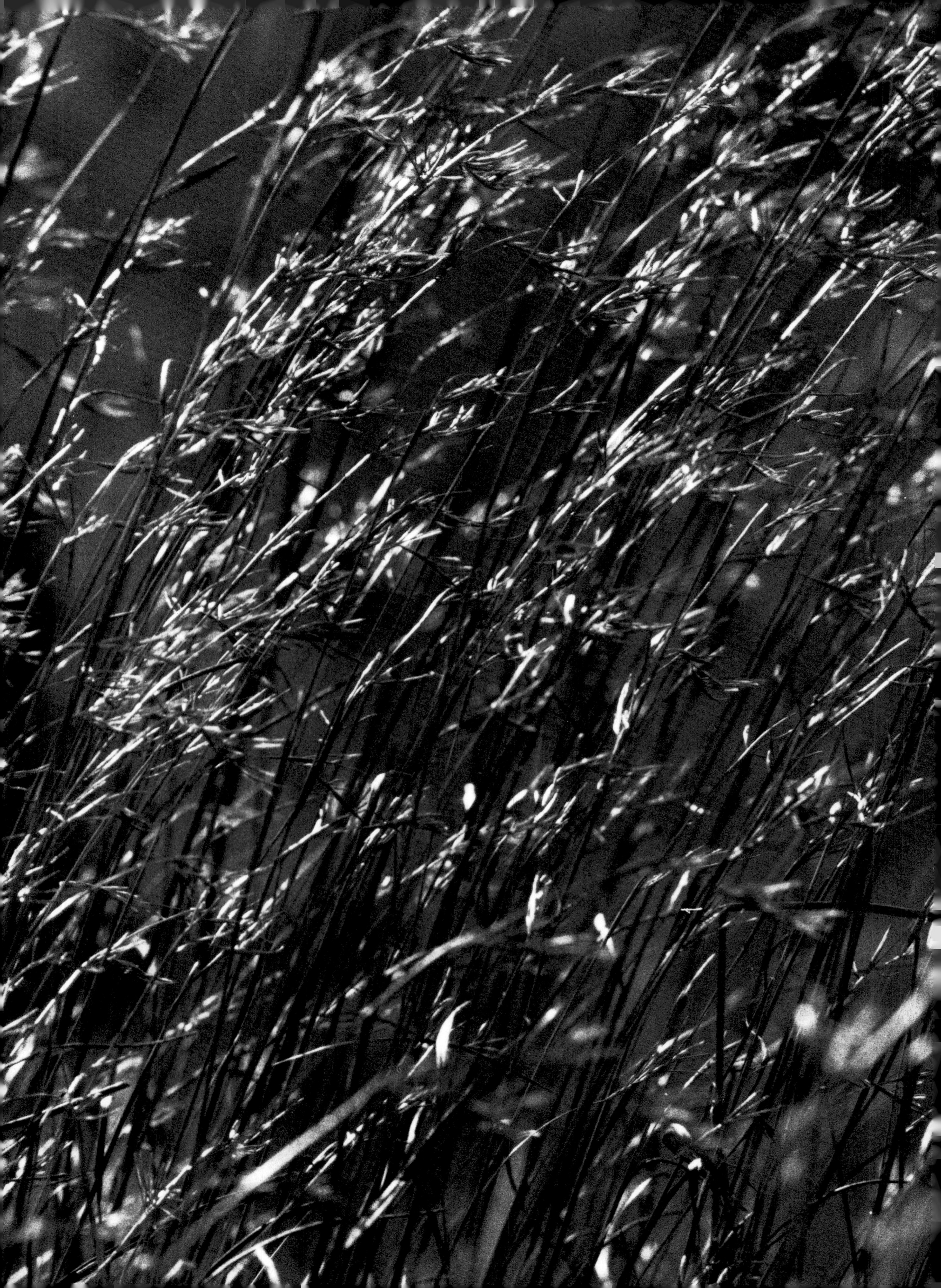

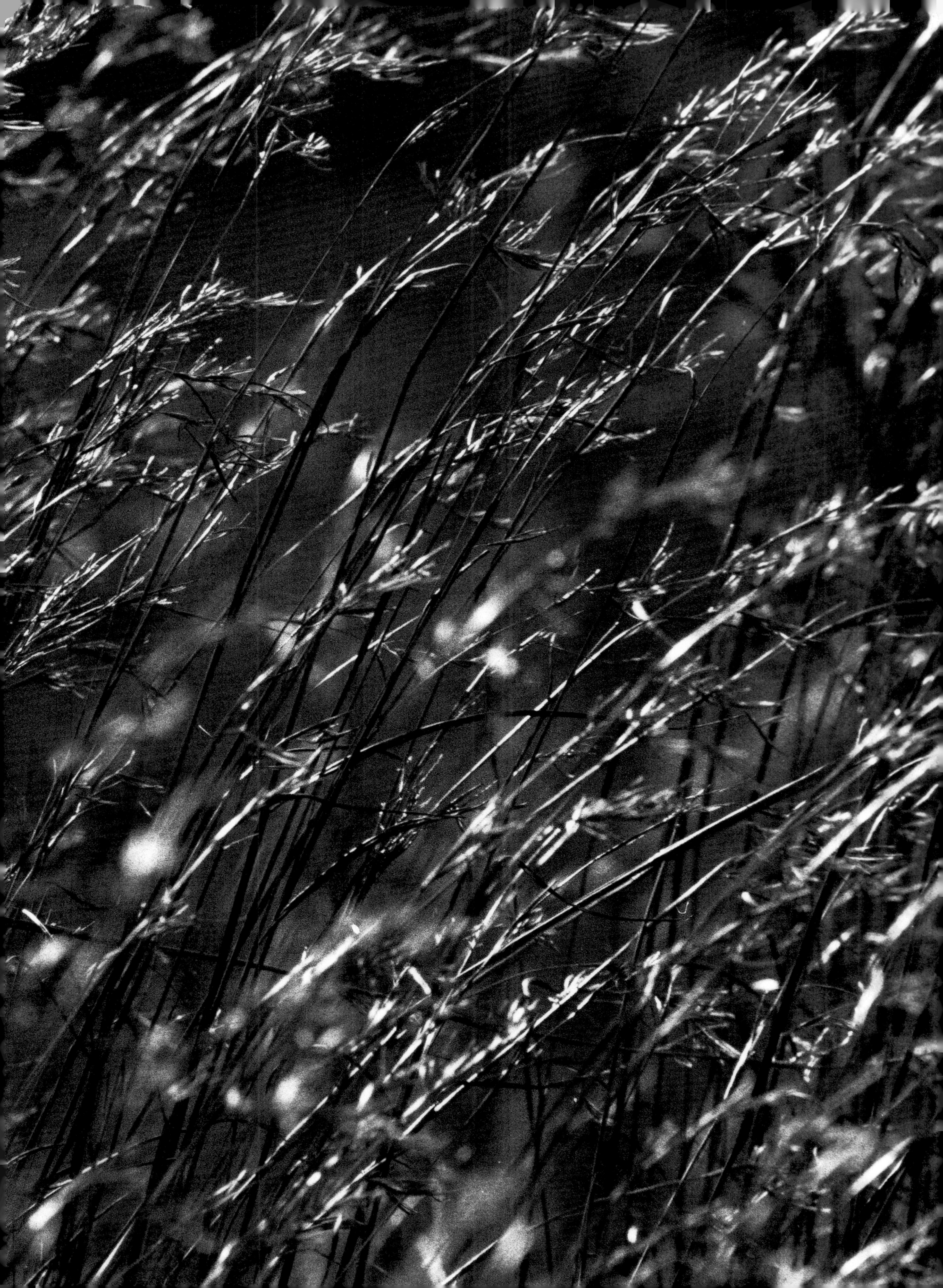

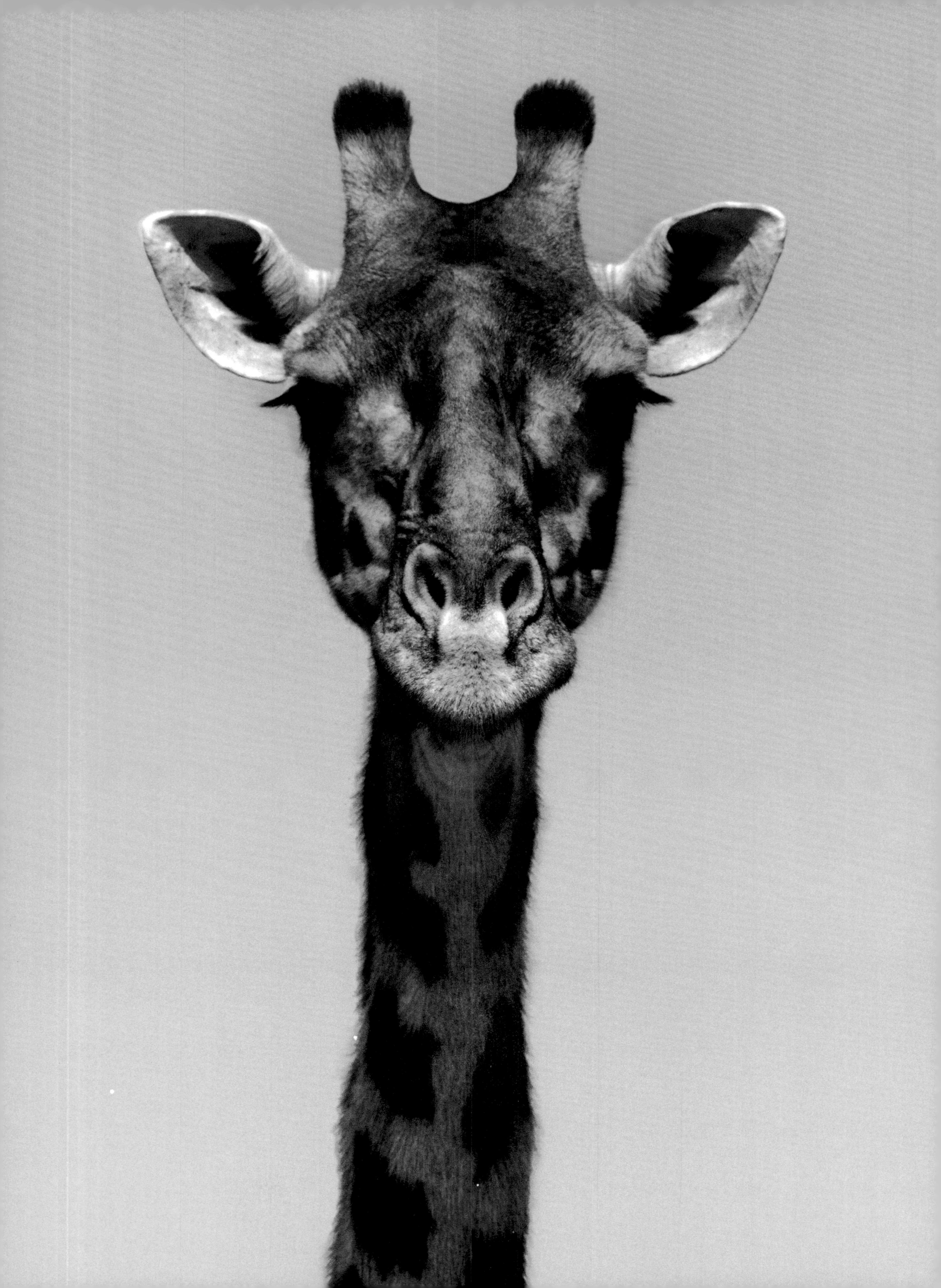

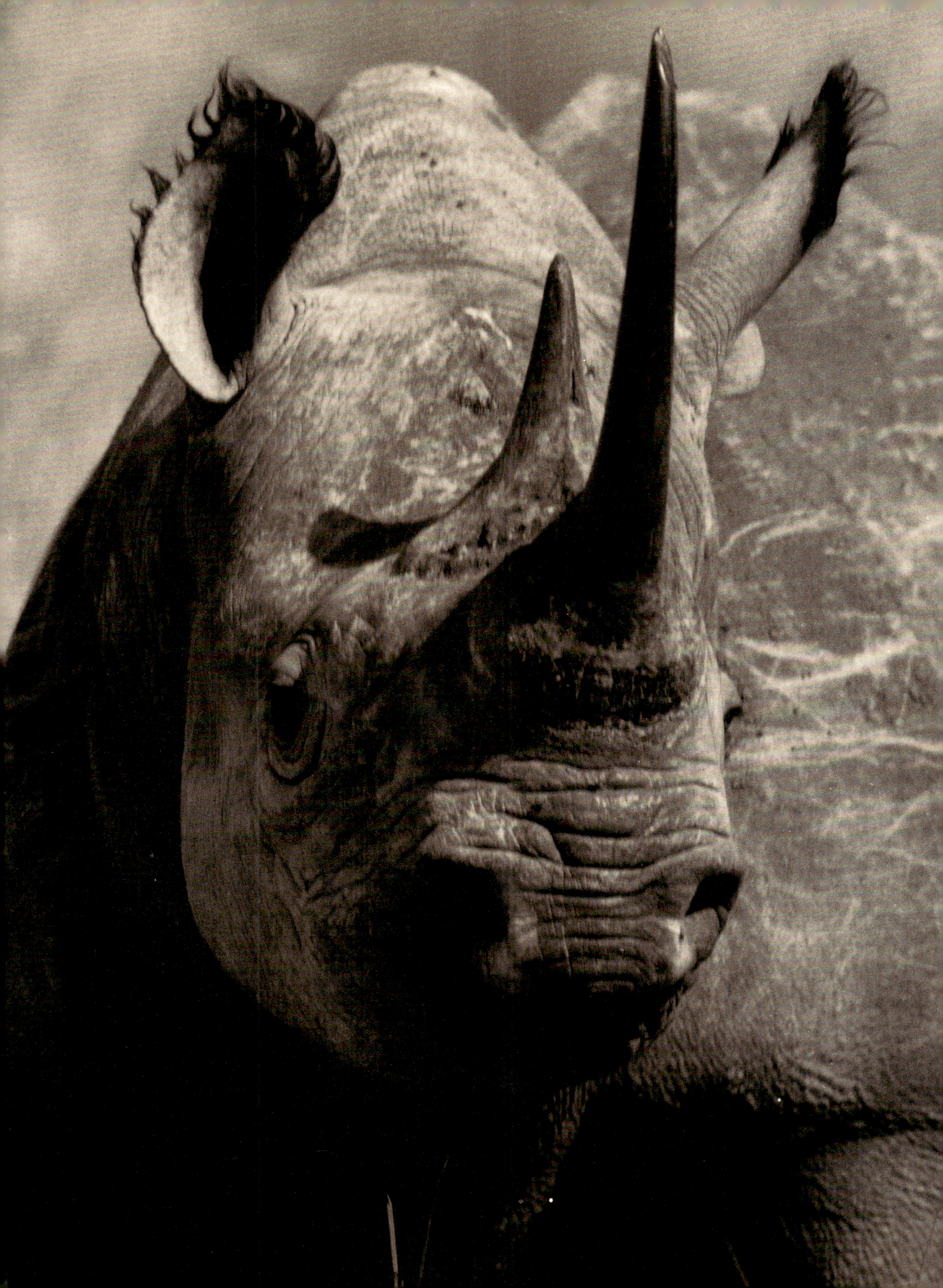

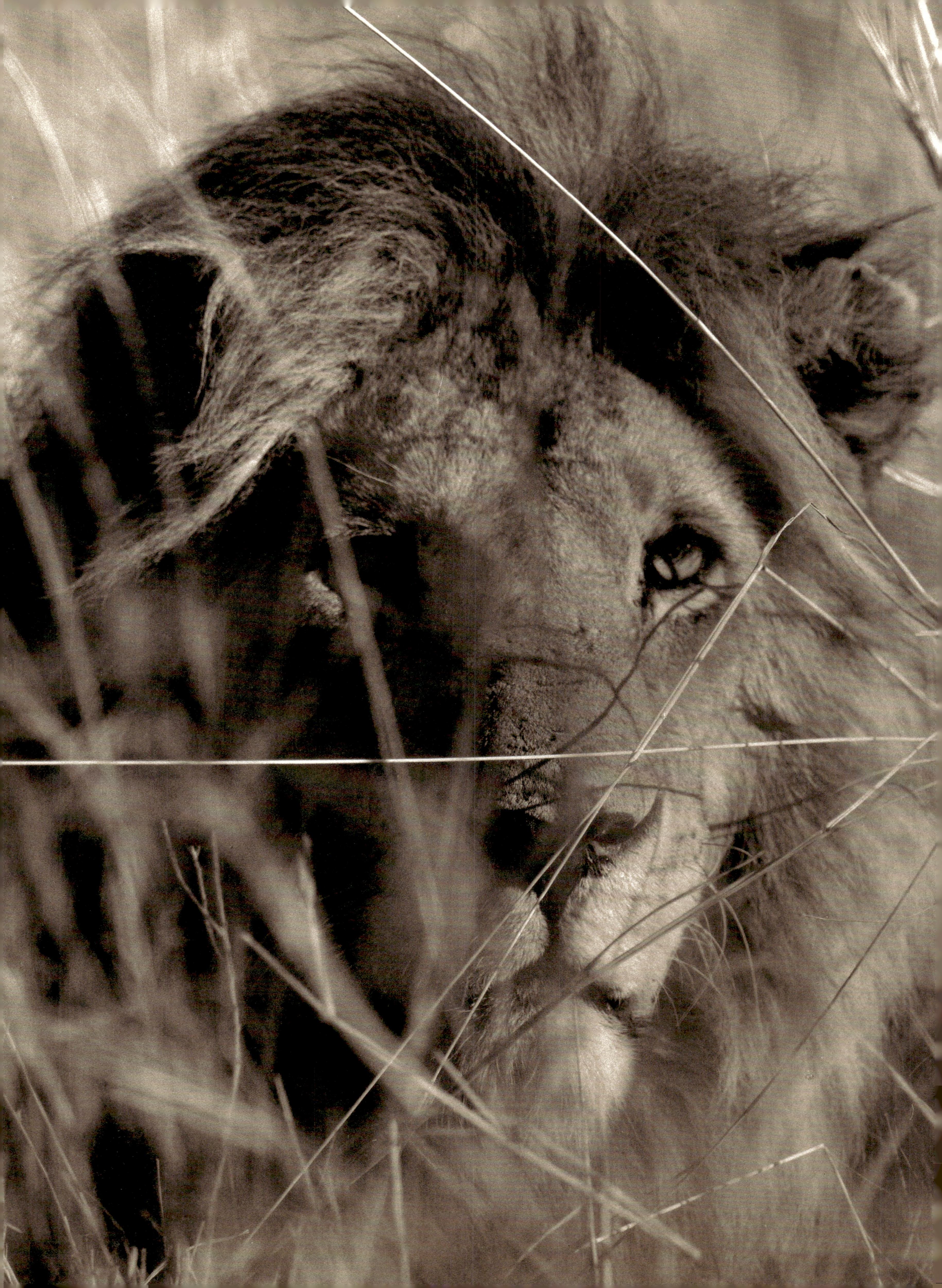

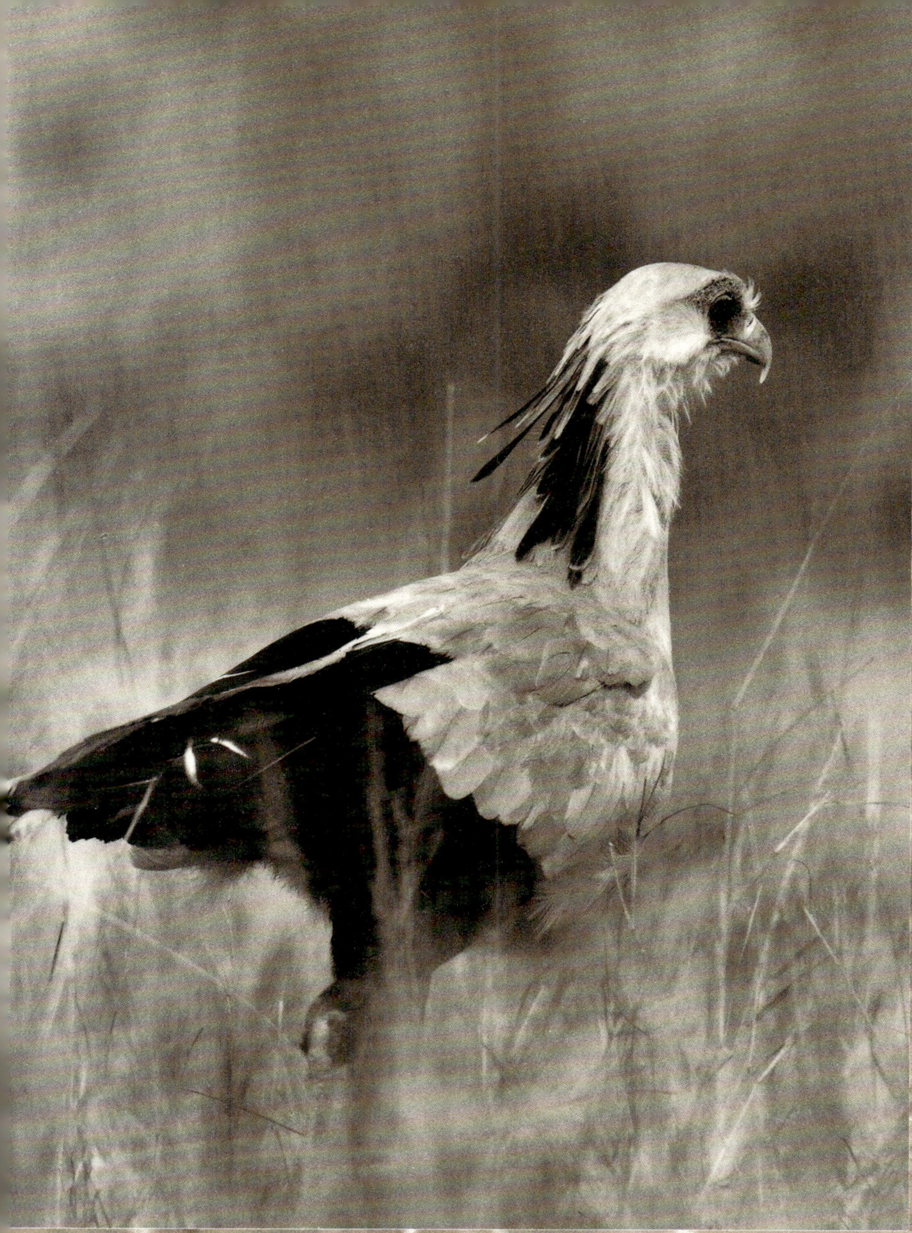
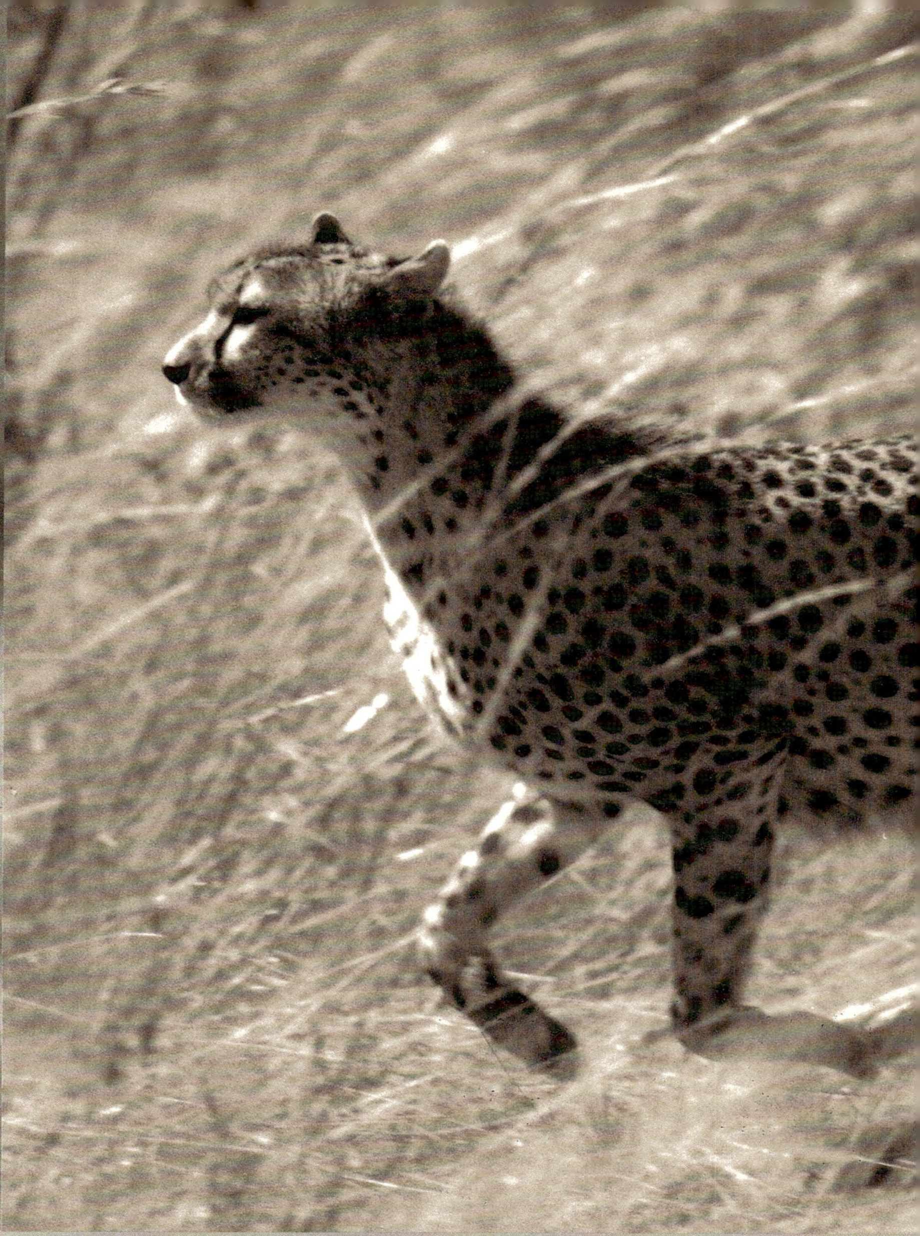
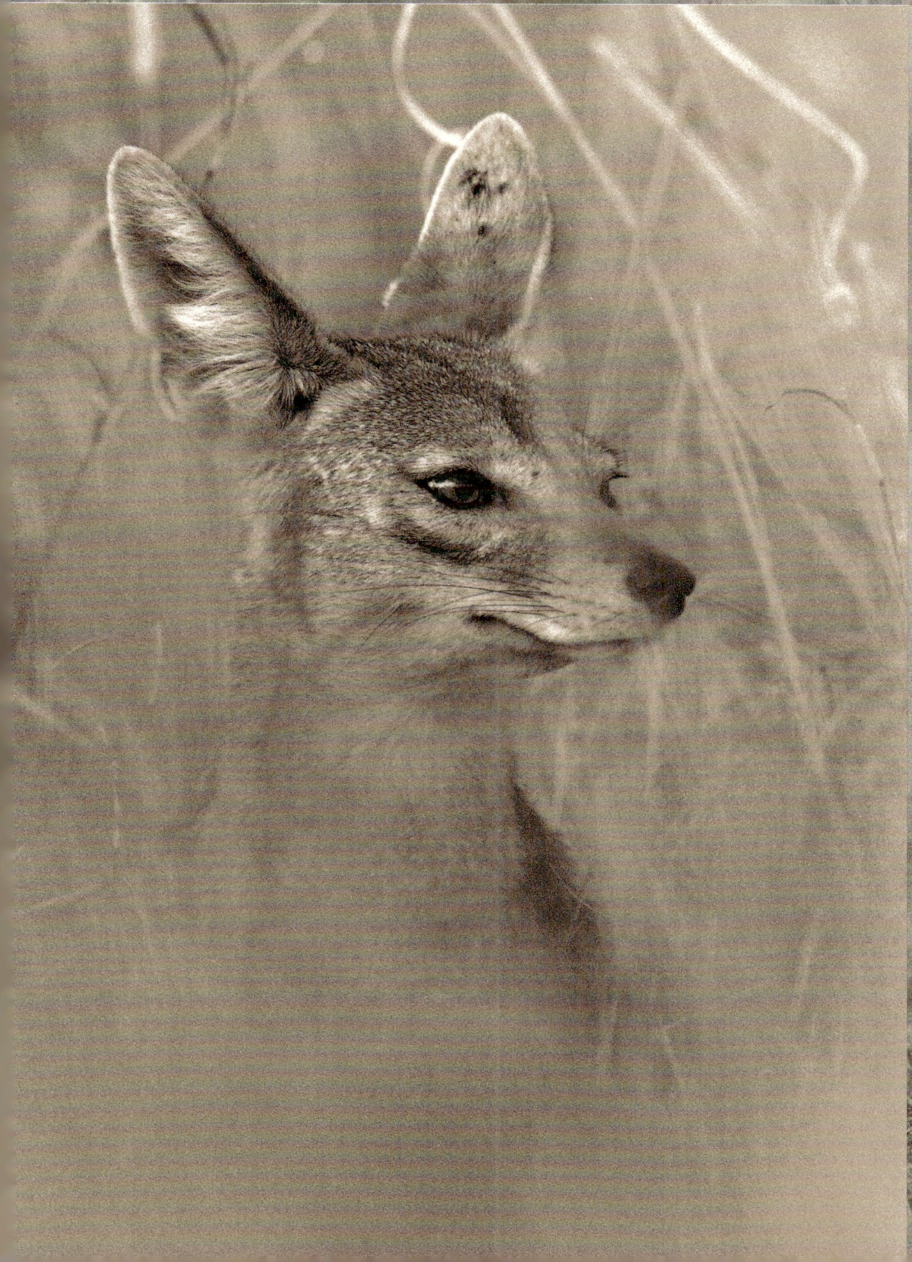
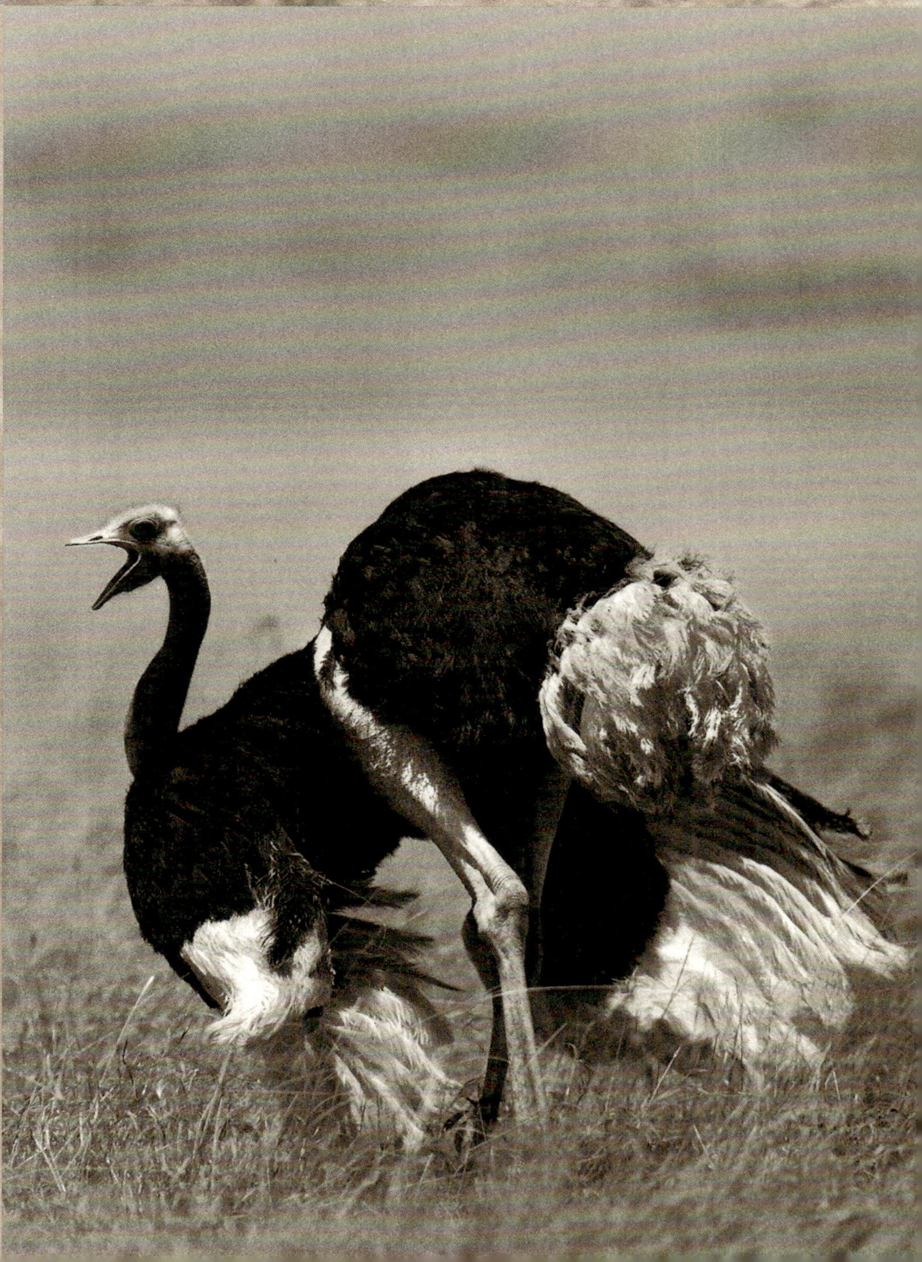

THE GNUS WILL BLUNDER THROUGH THE HABITATS OF RESIDENT HERBIVORES, WHO, IF THEY ARE PRIMARILY BROWSING BEASTS LIKE IMPALA (OPPOSITE), WILL REMAIN INDIFFERENT AND UNAFFECTED BY THE VAST APPETITE OF THEIR COLLECTIVE SEASONAL VISITOR. THEIR SUPPLY OF FOOD REMAINS INTACT; THEY MUST ONLY SUFFER THE NOISE. THE GRAZERS, ON THE OTHER HAND, SUCH AS TOPI (OVERLEAF), MUST NOT ONLY PUT UP WITH THE CLAMOUR BUT ALSO WATCH IN DISMAY AS HERDS OF WILDEBEESTE MOW DOWN THEIR LARDER.

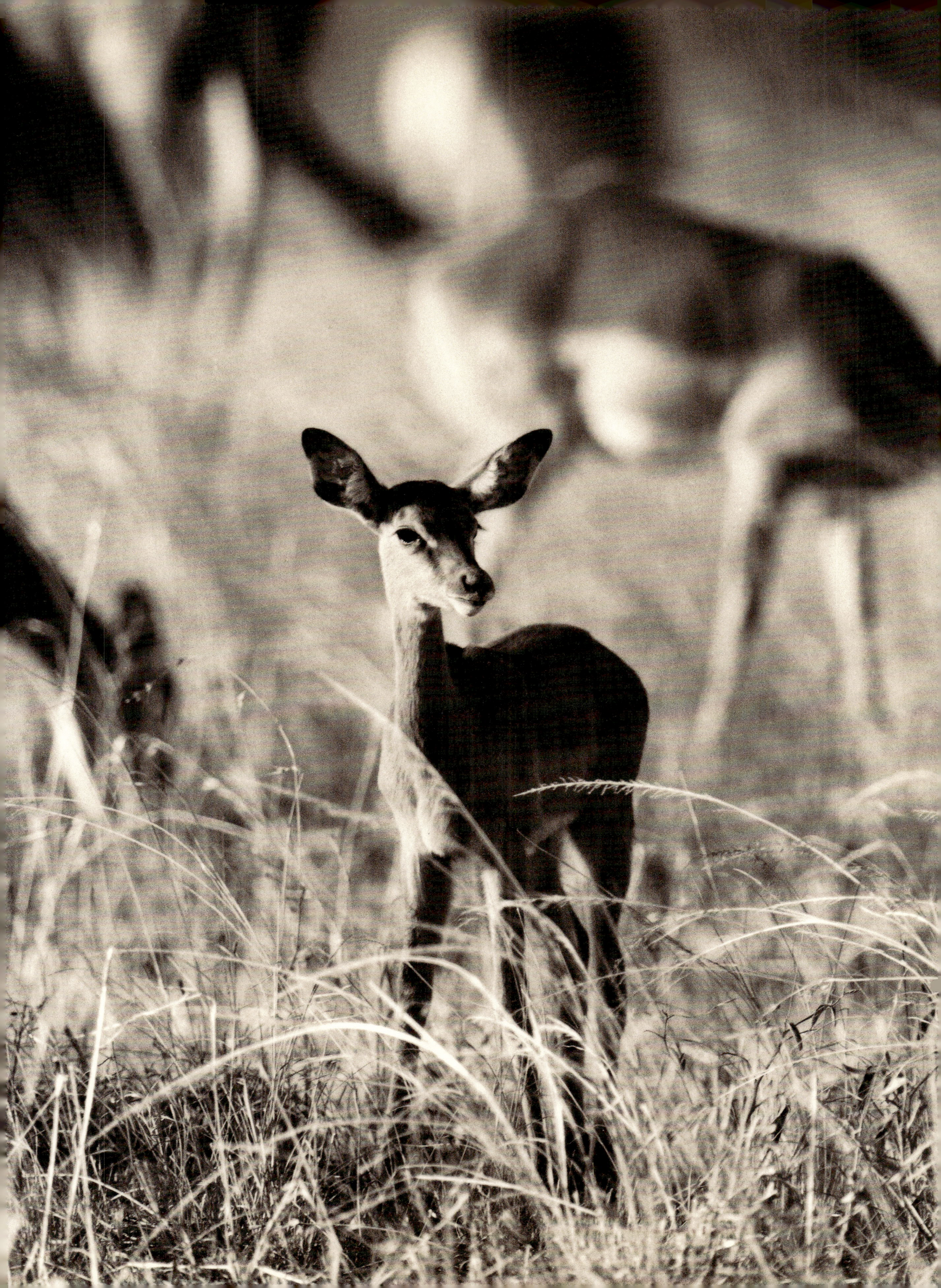

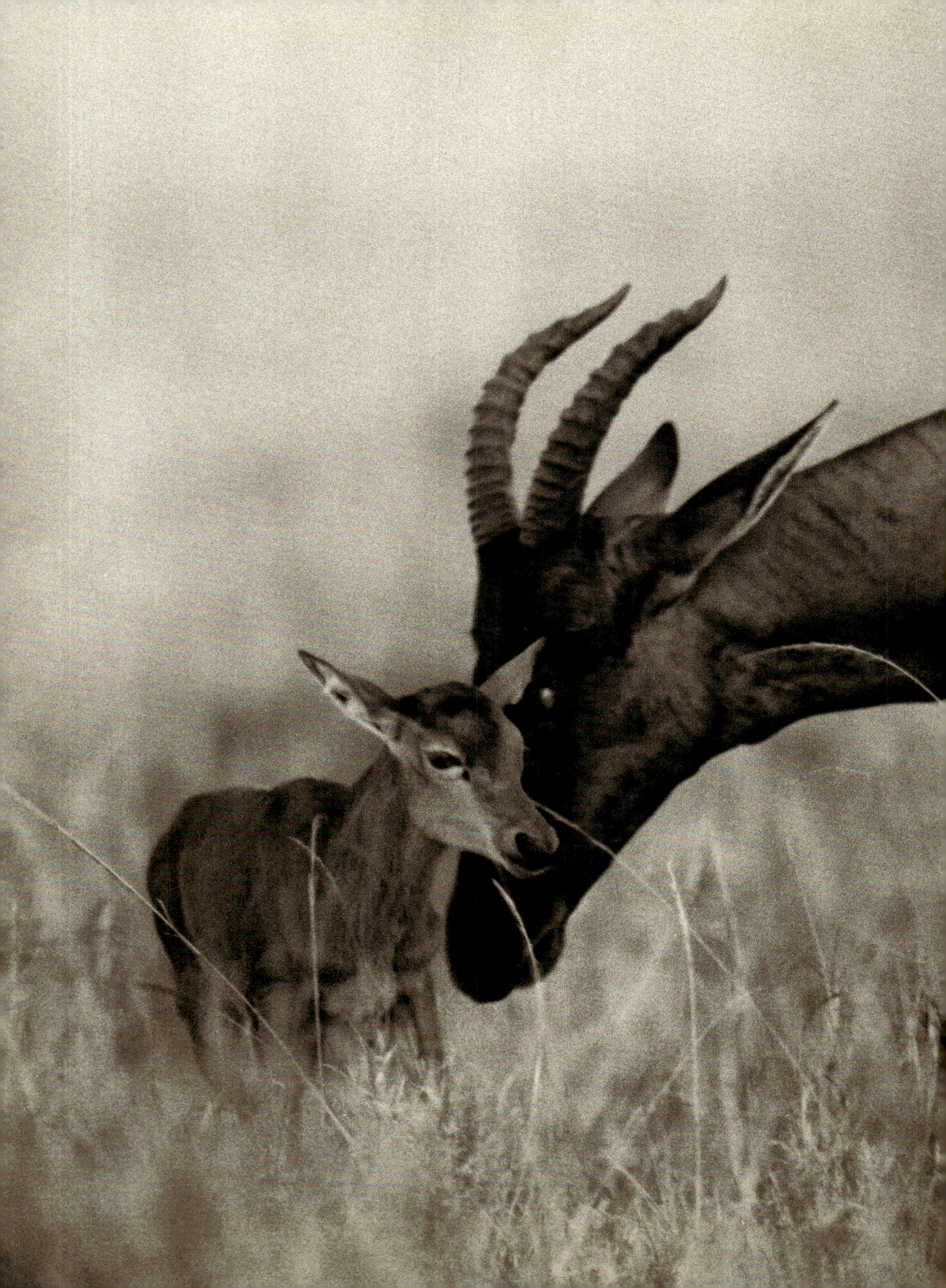

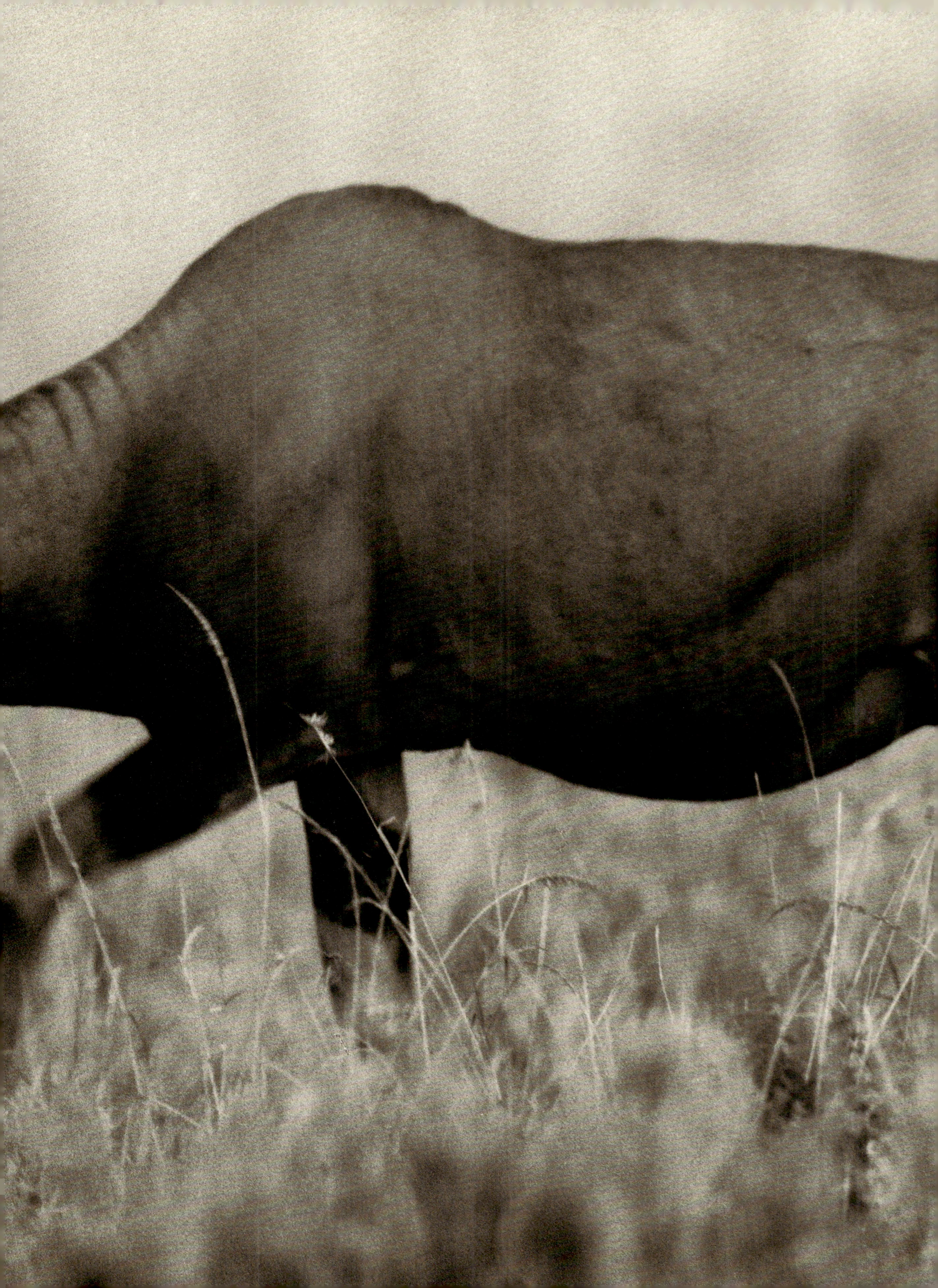

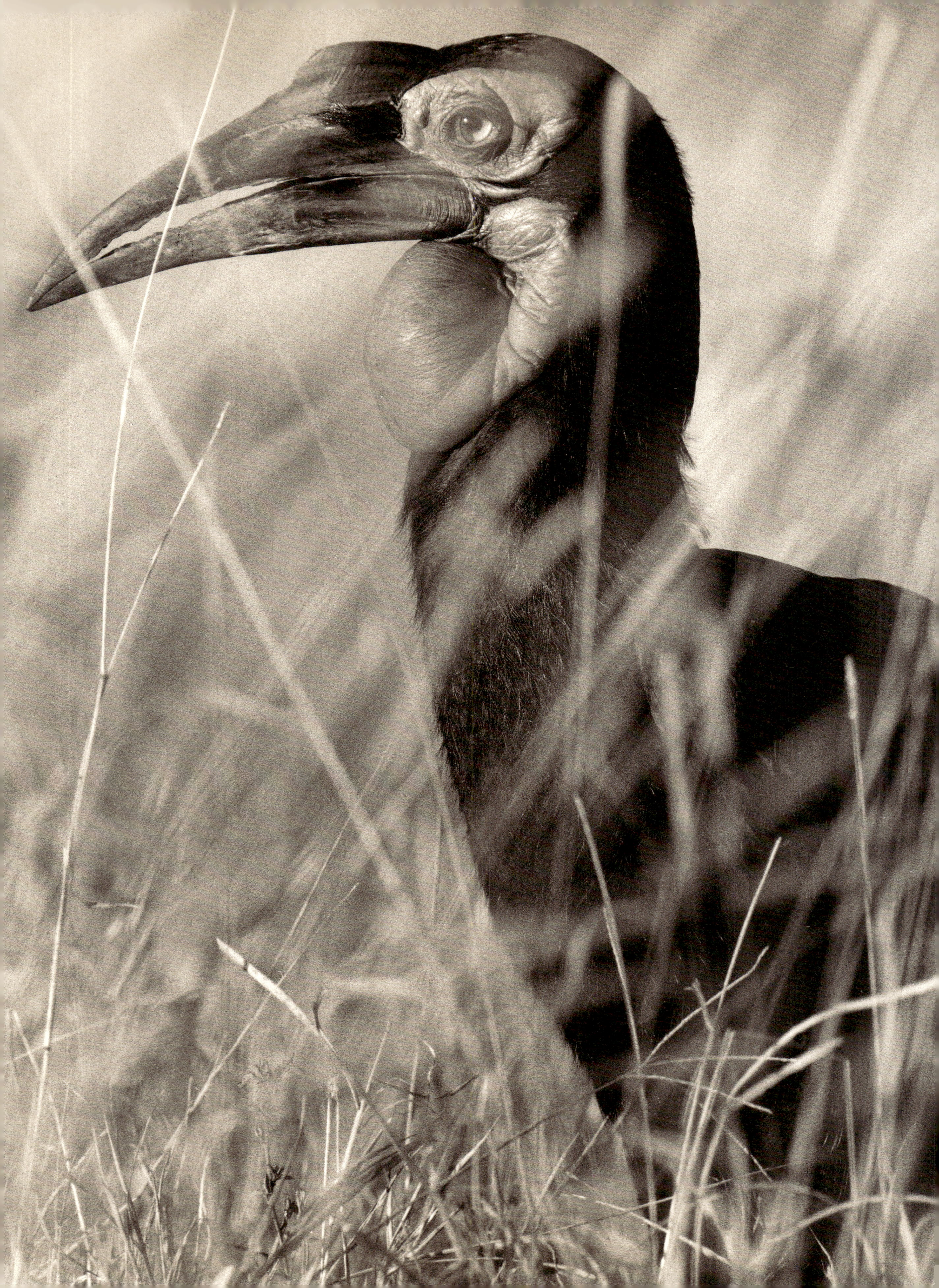

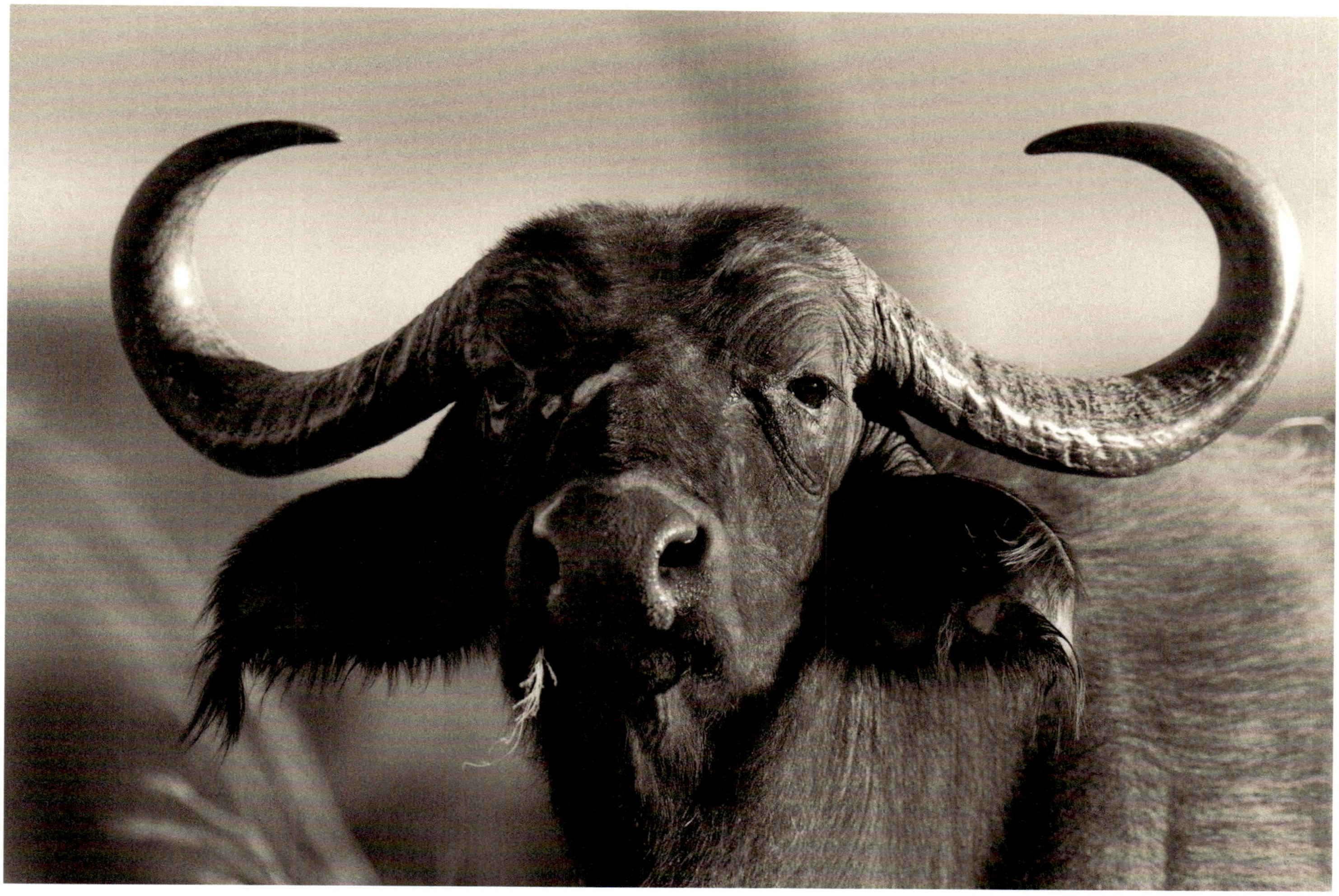

THE SERENGETI ECOSYSTEM. "ECO-" FROM THE GREEK *OIKOS*, HOUSE; "-SYSTEM", DERIVED FROM A ROOT MEANING *TO SET UP*. TWENTY-EIGHT SPECIES OF HERBIVORE, FROM ELEPHANTS TO DIK-DIKS SET UP HOUSE AND MAKE A LIVING IN THE 15,000-SQUARE-FOOT ECOSYSTEM OF THE SERENGETI THAT STRADDLES THE NORTHERN TANZANIA–SOUTHERN KENYA BORDER, INDIFFERENT TO POLITICS. THEIR BIOMASS IS IN THE REGION OF 550 MILLION POUNDS, ROUGHLY THE SAME AS THE WEIGHT OF THE INHABITANTS OF ROME. THIS EXCLUDES THE FIVE SPECIES OF PRIMATES, 26 SPECIES OF CARNIVORE, MORE THAN 500 SPECIES OF BIRDS AND THE UNCOUNTED THOUSANDS OF INVERTEBRATE SPECIES WHICH THEMSELVES PROBABLY WEIGH AS MUCH AS ALL THE VERTEBRATES PUT TOGETHER.

WITHIN THE ECOSYSTEM BOUNDARIES – DEFINED SIMPLY BY WHERE MOST OF THE BEASTS SPEND MOST OF THE TIME – LIVINGS ARE MADE IN MANY WAYS. GRASS AND TREES FEED ON SOIL AND SUN. HERBIVORES LIKE OSTRICHES (P.57, BOTTOM RIGHT) AND BUFFALO (ABOVE) GRAZE; GIRAFFE BROWSE. CARNIVORES ARE MORE DIVERSE: LIONS HUNT WILDEBEESTE, GROUND HORNBILLS (OPPOSITE) HUNT SMALL MAMMALS, SECRETARY BIRDS (P.57, TOP LEFT) HUNT SNAKES, JACKALS (P.57, BOTTOM LEFT) HUNT SMALL MAMMALS TOO BUT SCAVENGE WHENEVER THEY CAN. EACH HAS A PARTICULAR ROLE TO PLAY IN KEEPING MATERIALS MOVING FROM SOIL TO PLANT TO HERBIVORE TO CARNIVORE TO SCAVENGER AND INEVITABLY THROUGH DEATH OR EXCRETION BACK TO THE SOIL. ANOTHER GREAT CIRCULAR ROUTE.

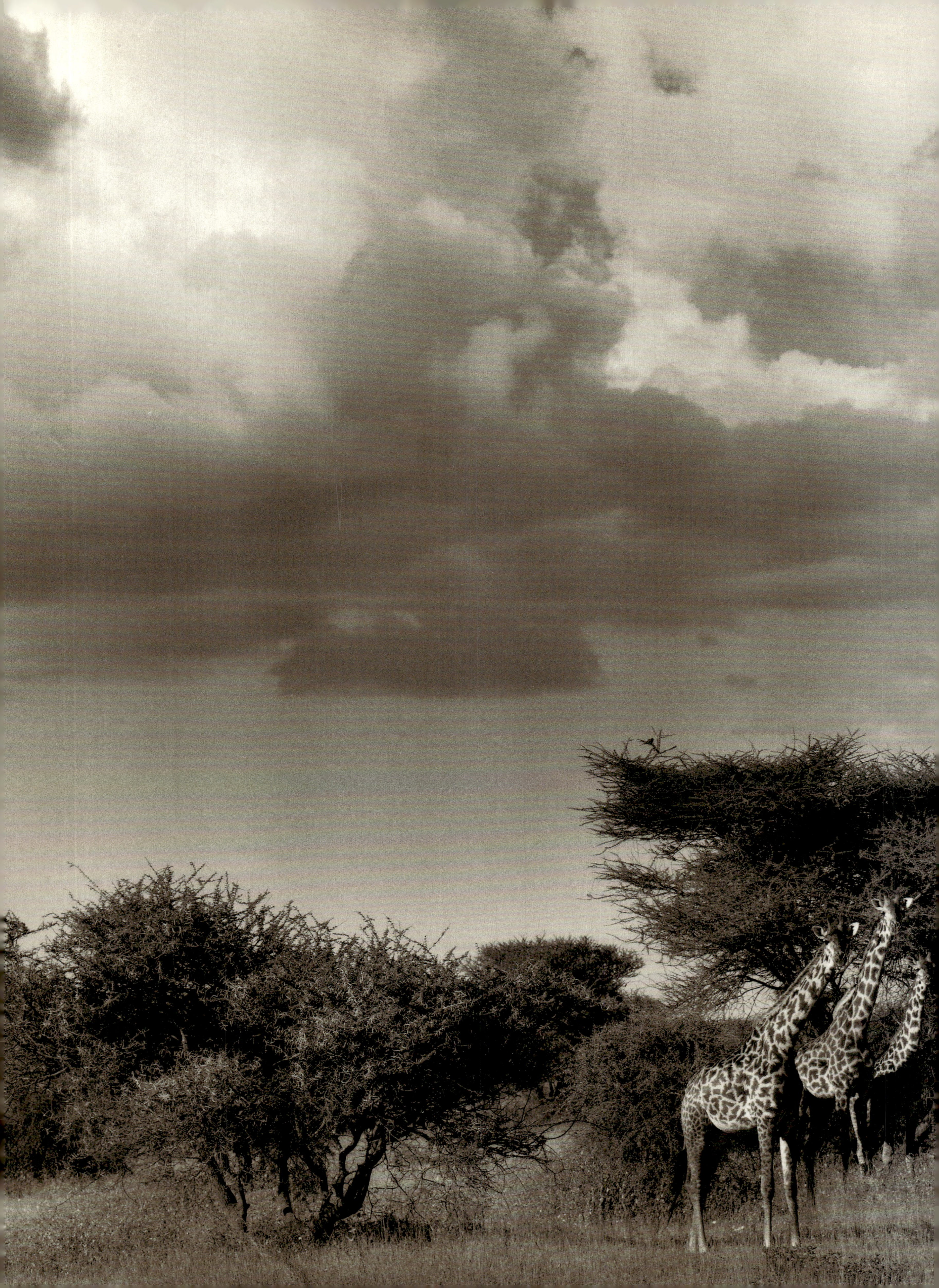

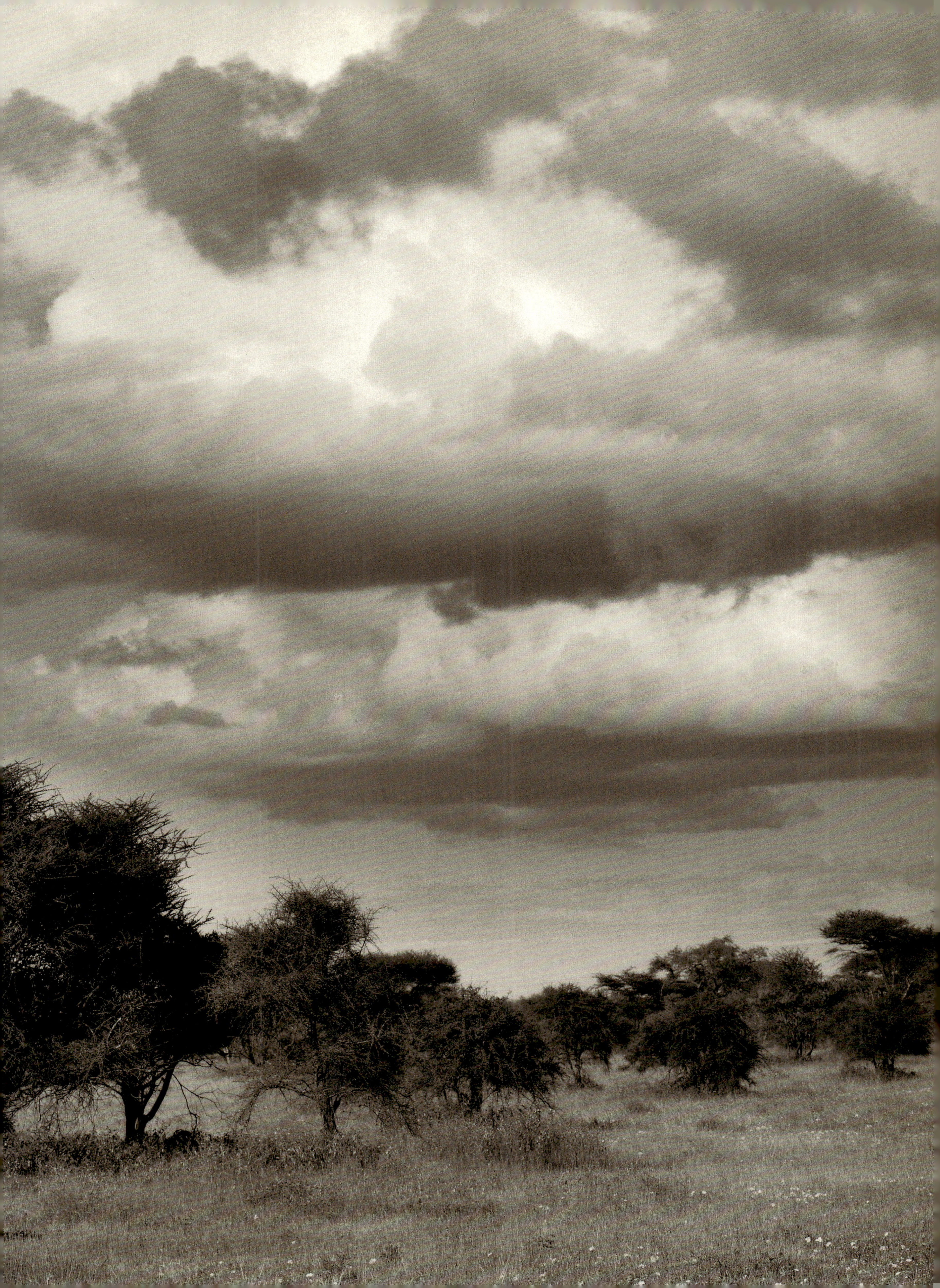

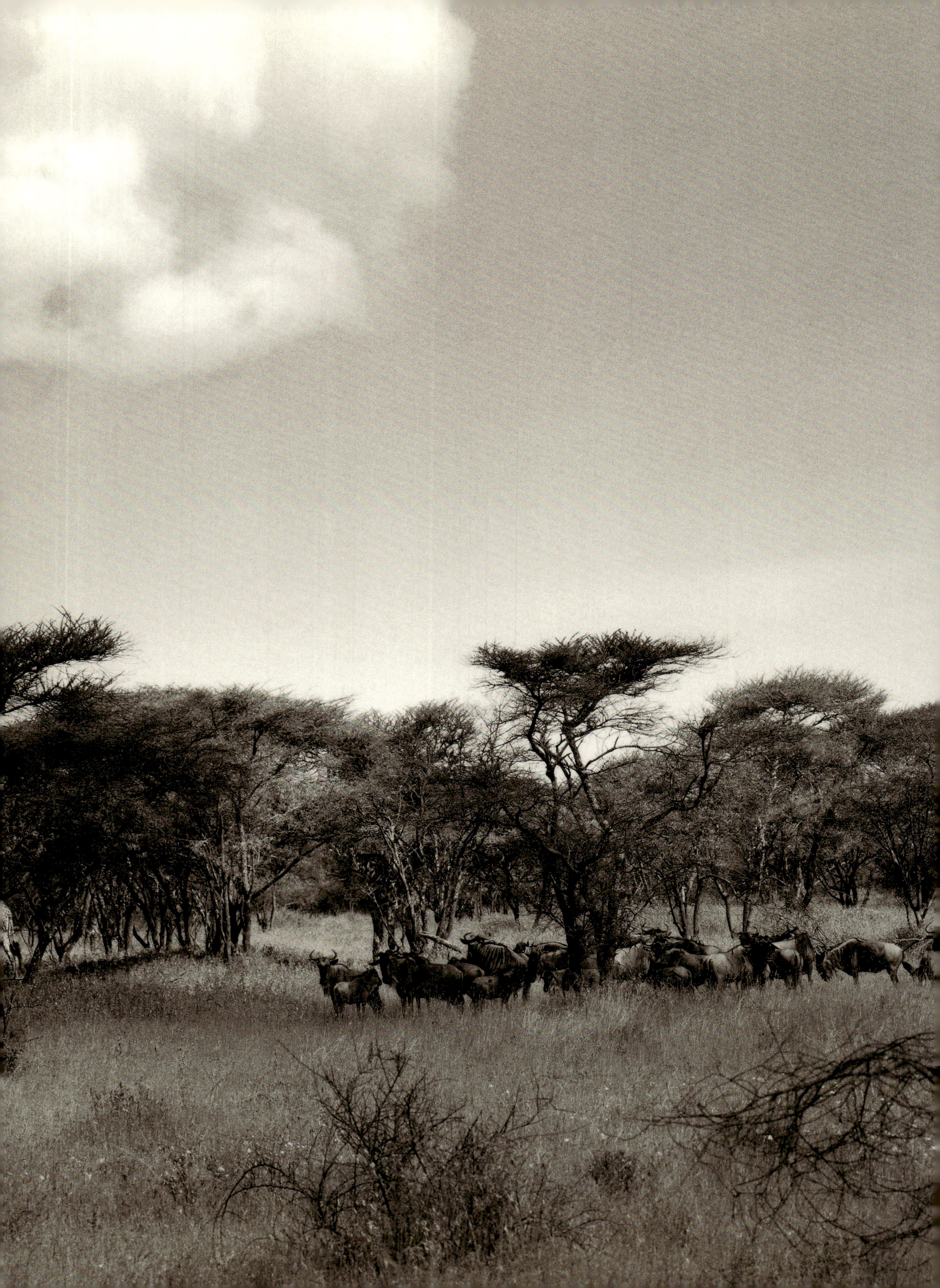

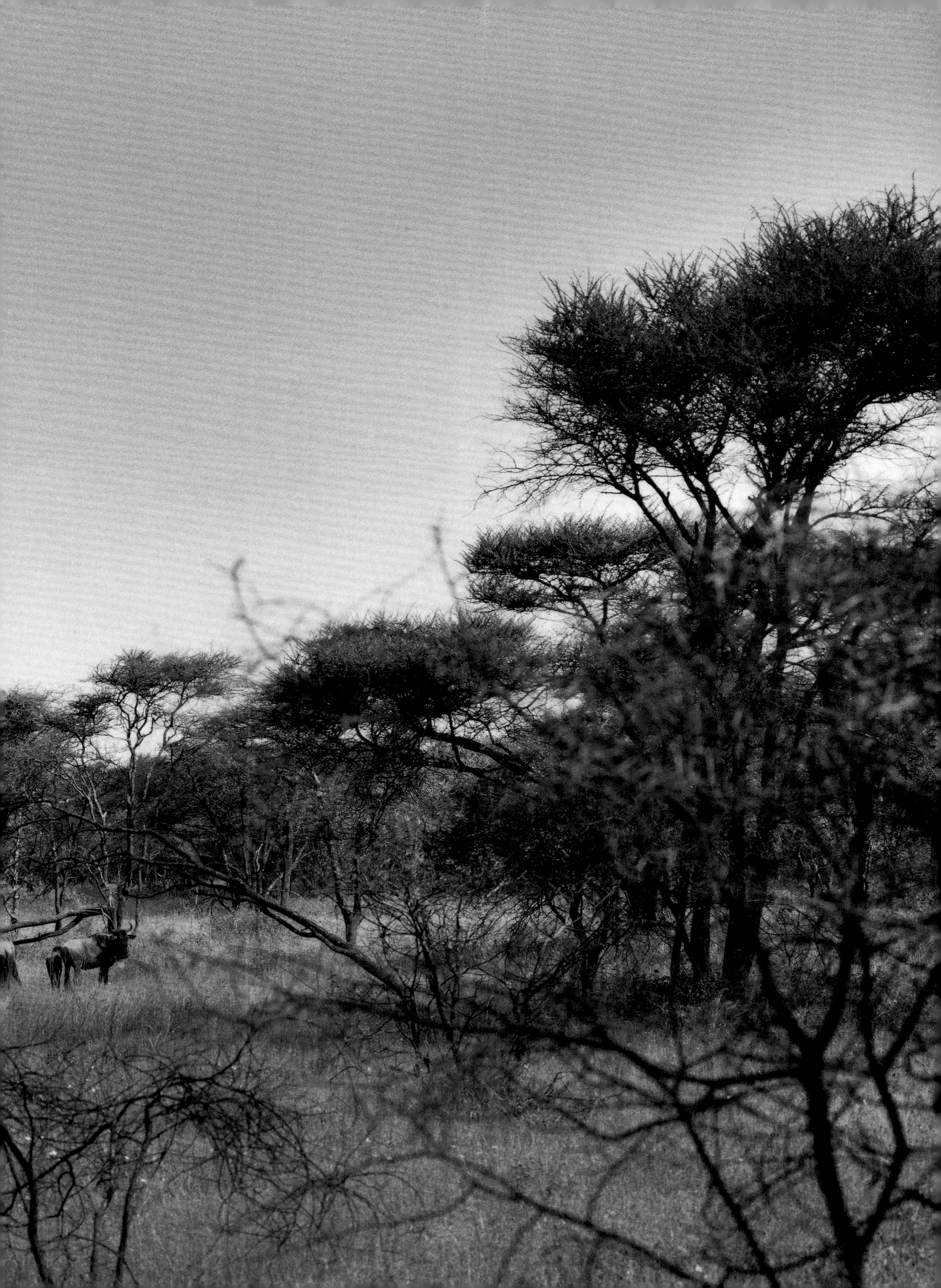

THE WHOLE POINT OF GOING ANYWHERE IS TO SEE SOMETHING NEW, TO GET SOMETHING DONE, OR TO PROTECT YOUR INVESTMENT IN LIFE. THE GNU'S INVESTMENT, LIKE OURS, IS IN PRODUCING COPIES OF ITSELF. TO PARAPHRASE OTHERS, A WILDEBEEST IS GNU DNA'S WAY OF MAKING MORE GNU DNA. IN THE LITTLE MIGRATION THAT DNA MAKES EVERY SEASON FROM GONADS TO EGG OR SPERM TO IMMMATURE ADULT, THE MOST VULNERABLE COURIERS ARE THE YOUNG. THE BEGINNING OF THE LONG DRY SEASON: TIME TO MOVE CALVES LIKE THESE, WITH UMBILICALS STILL ATTACHED, OFF INTO THE WOODLANDS WHERE GRASS IS STILL GROWING AND GREEN.

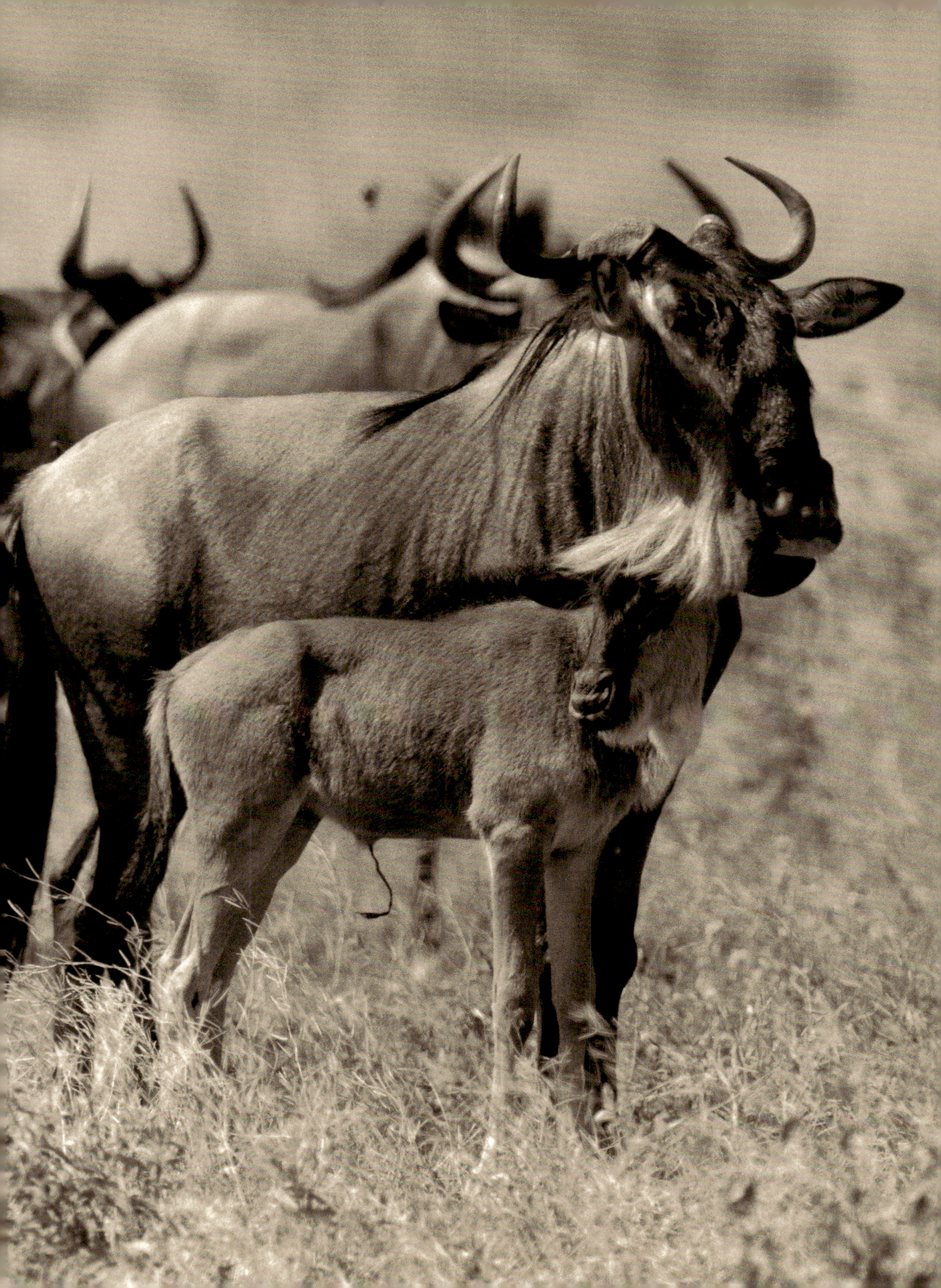

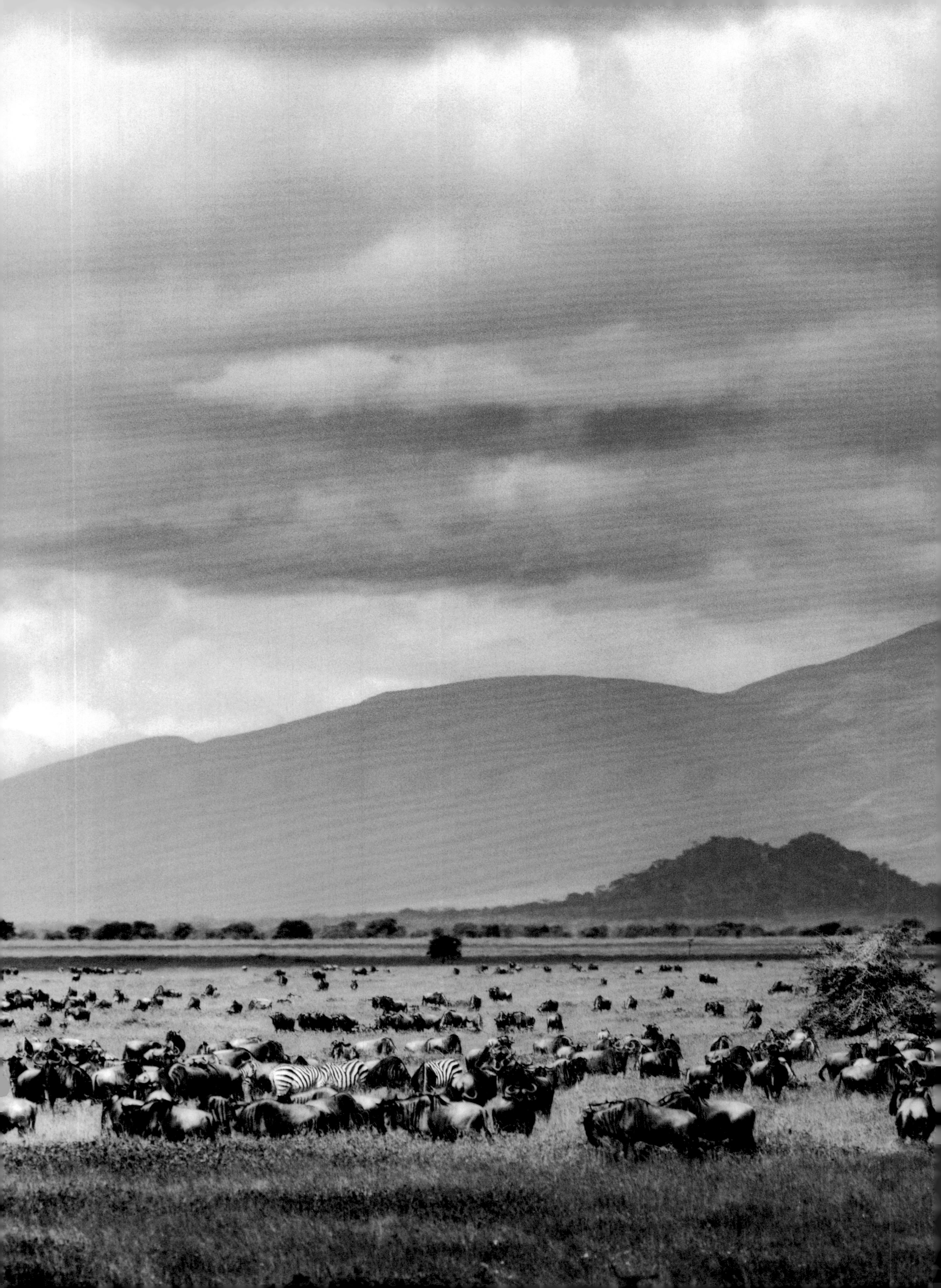

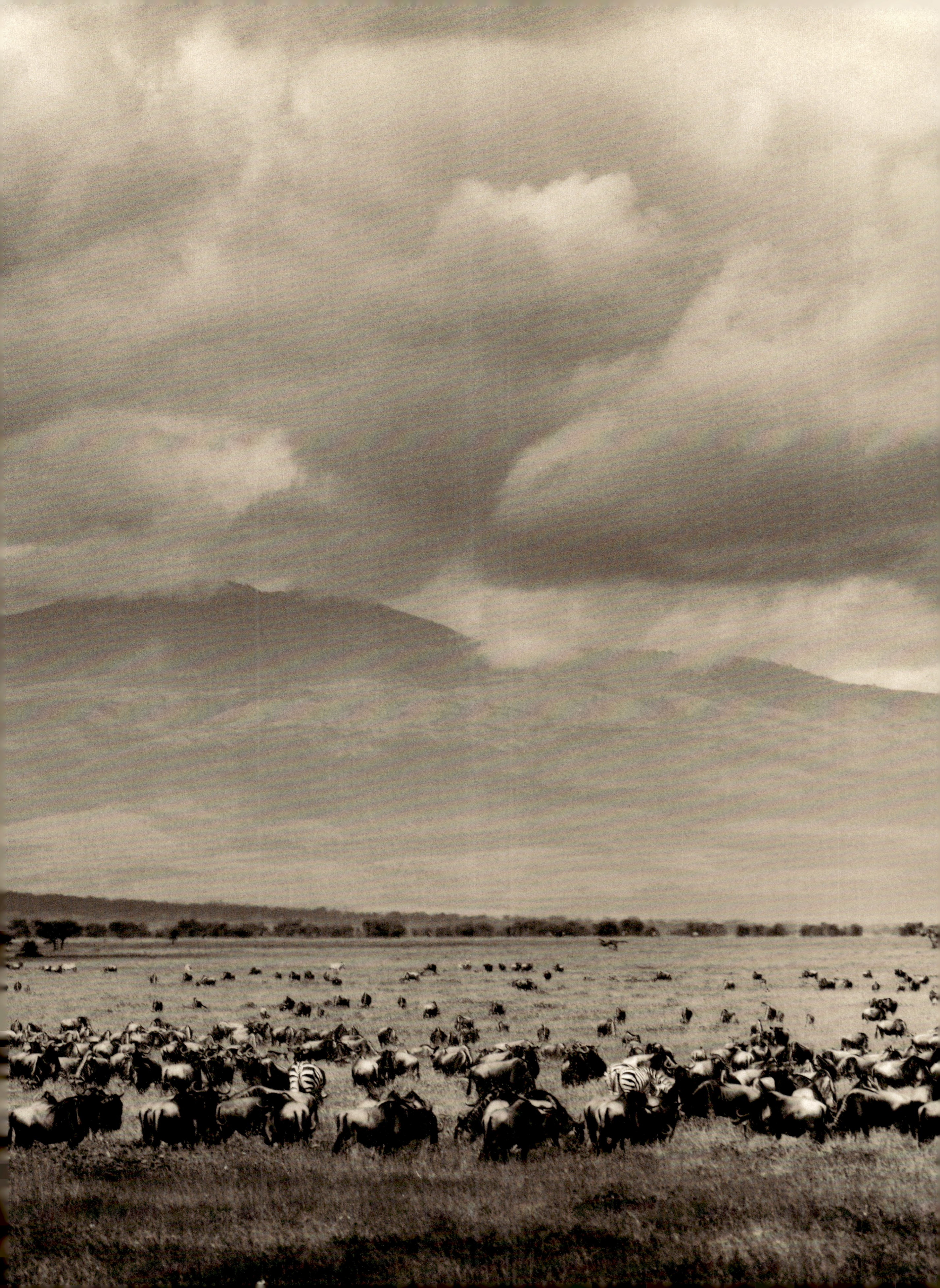

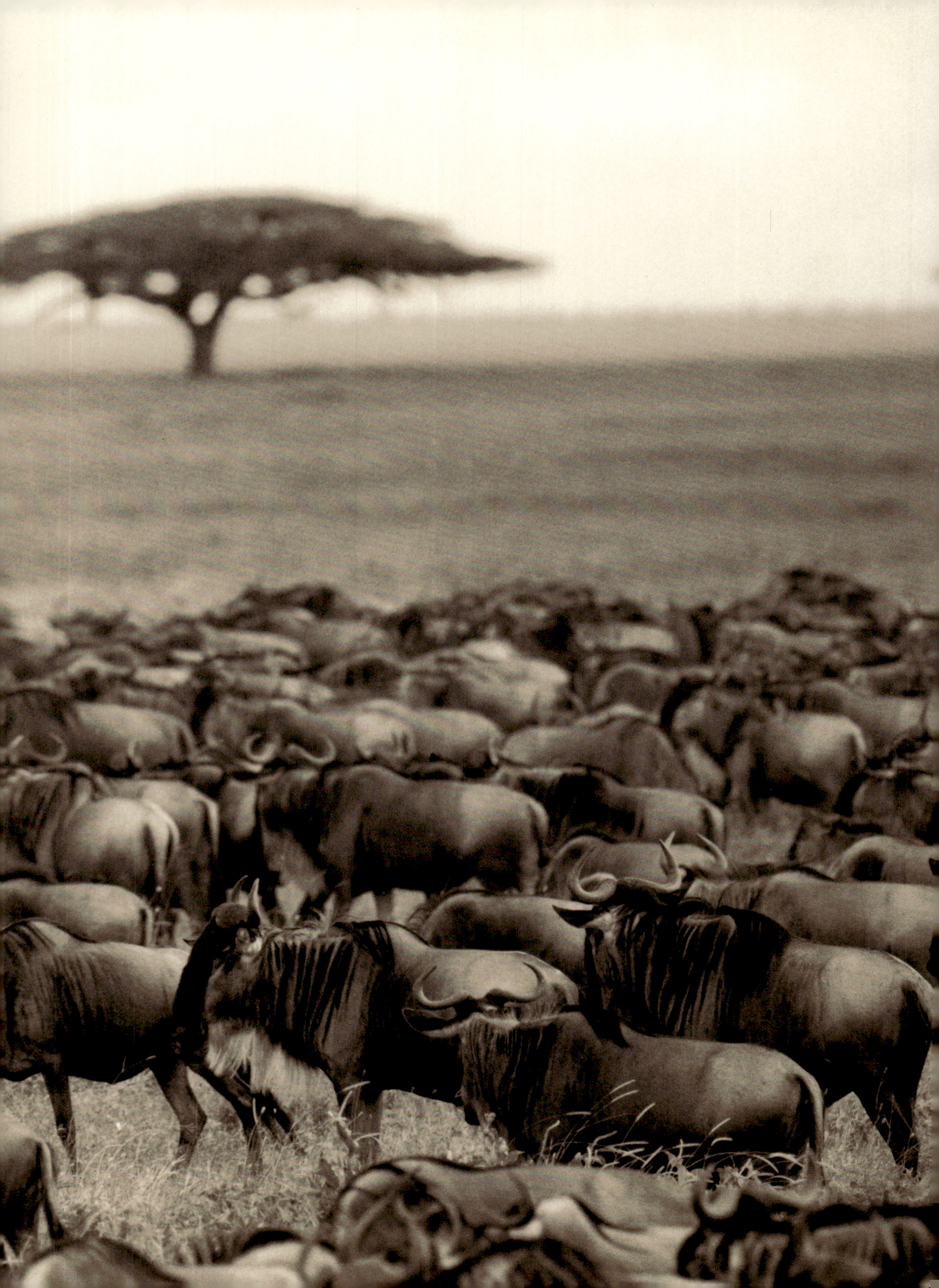

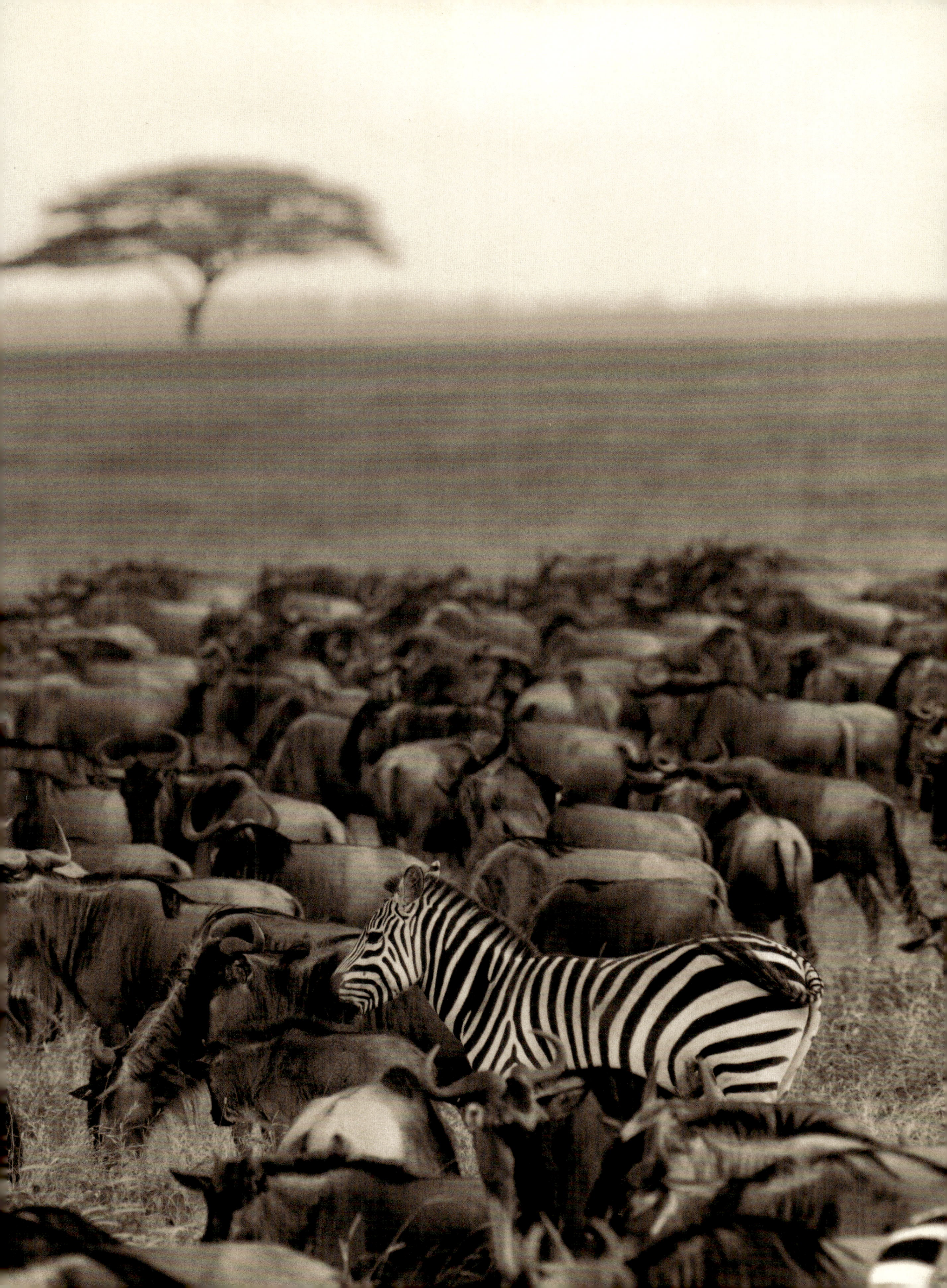

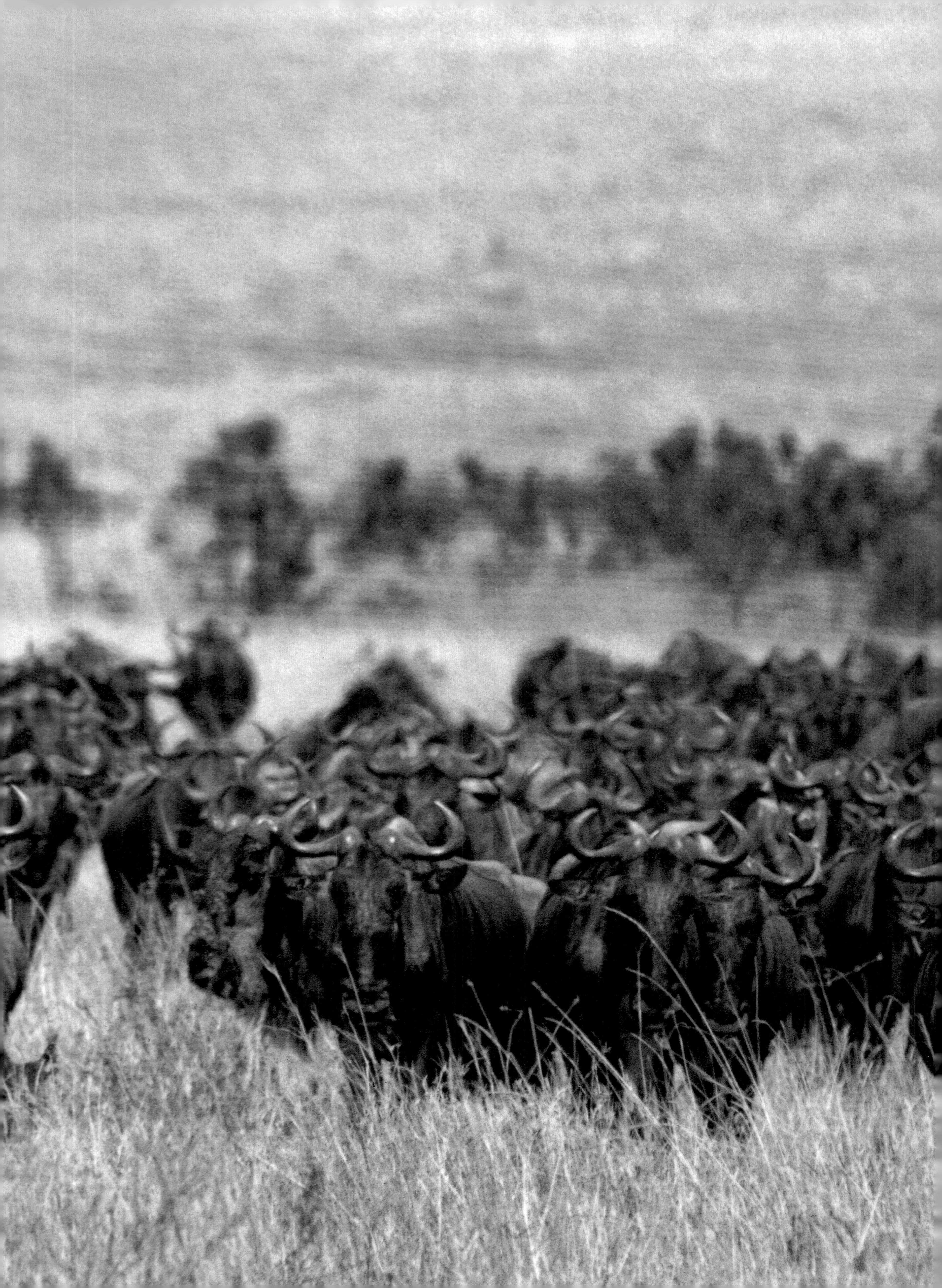

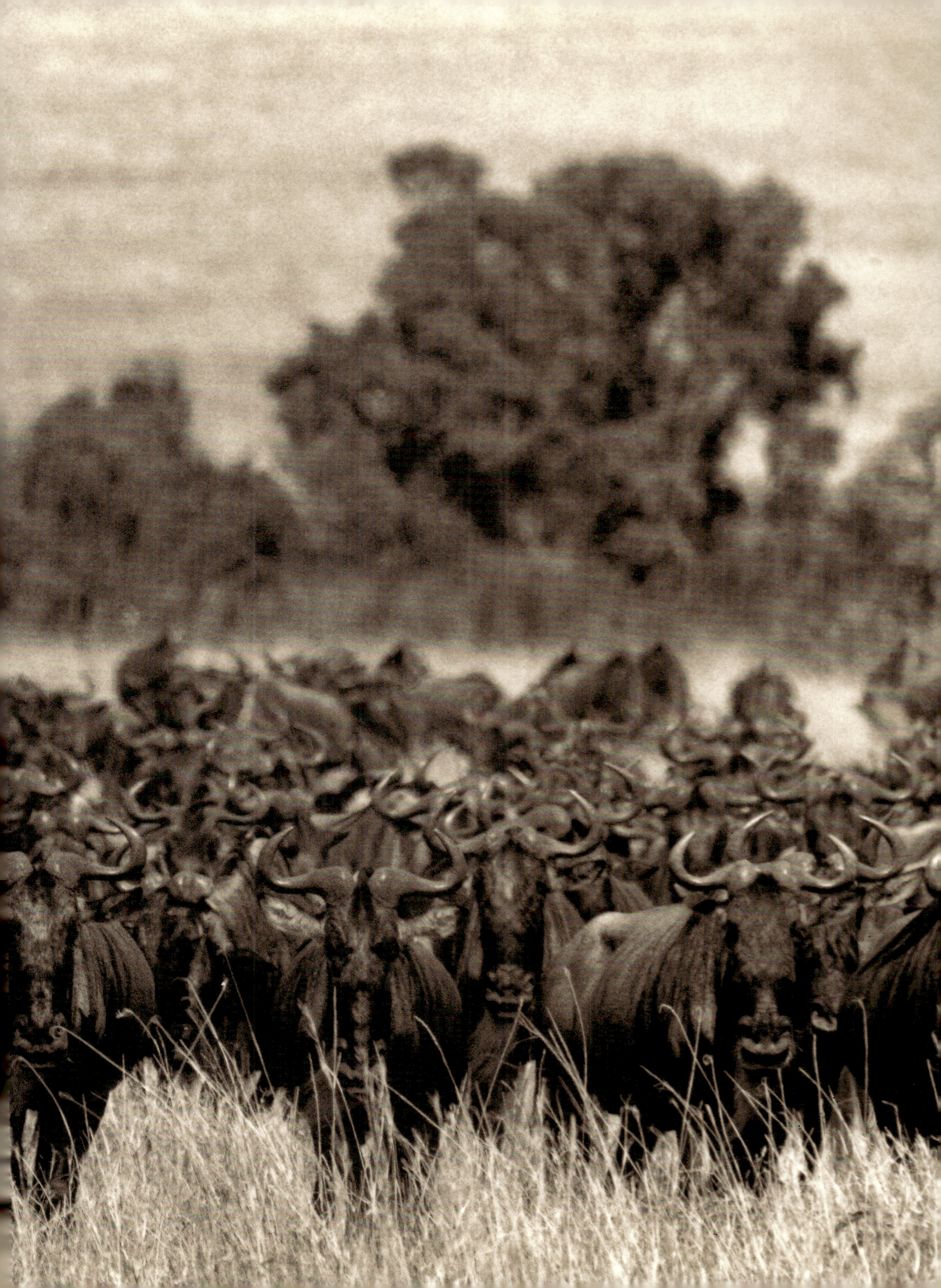

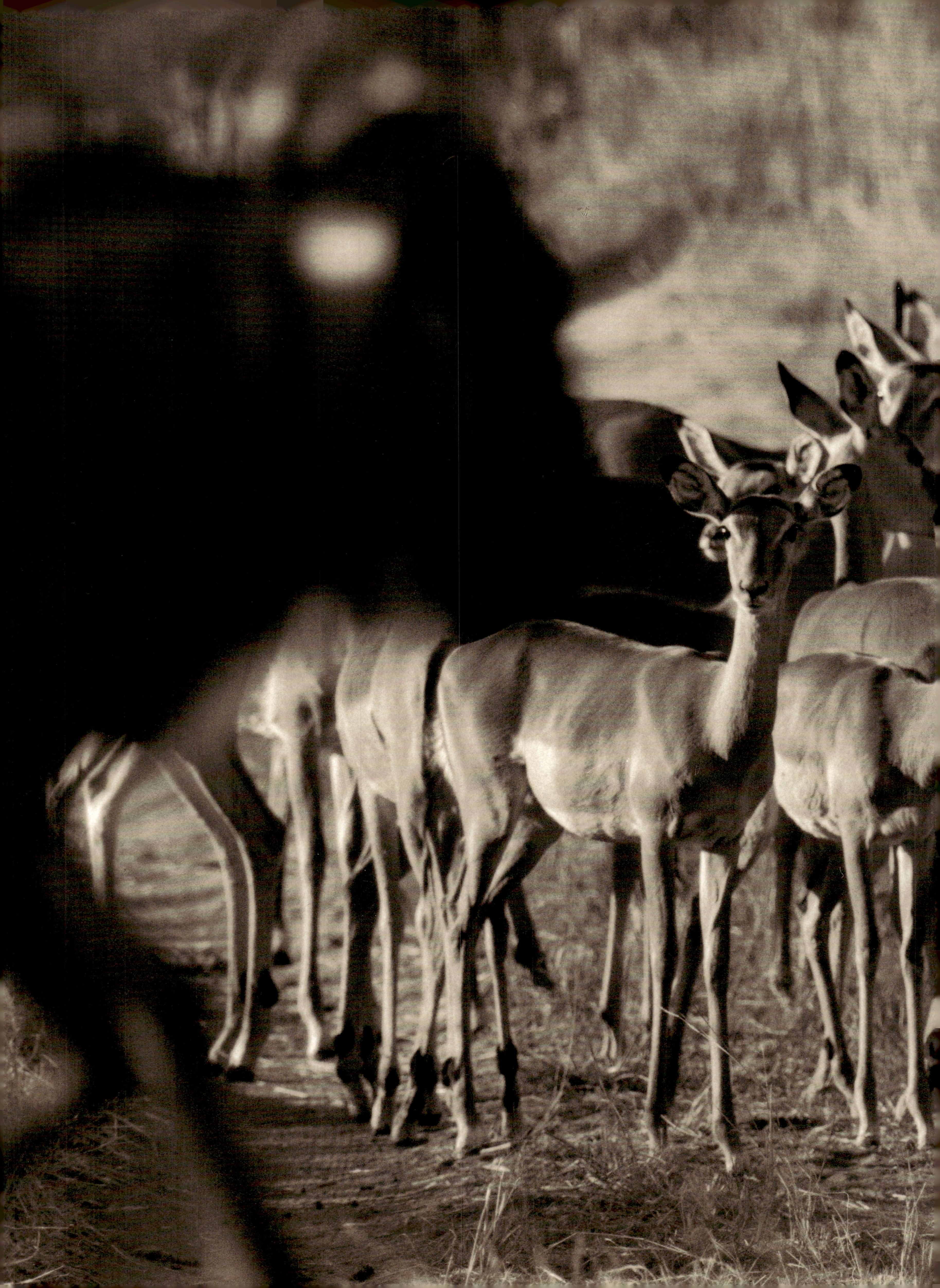

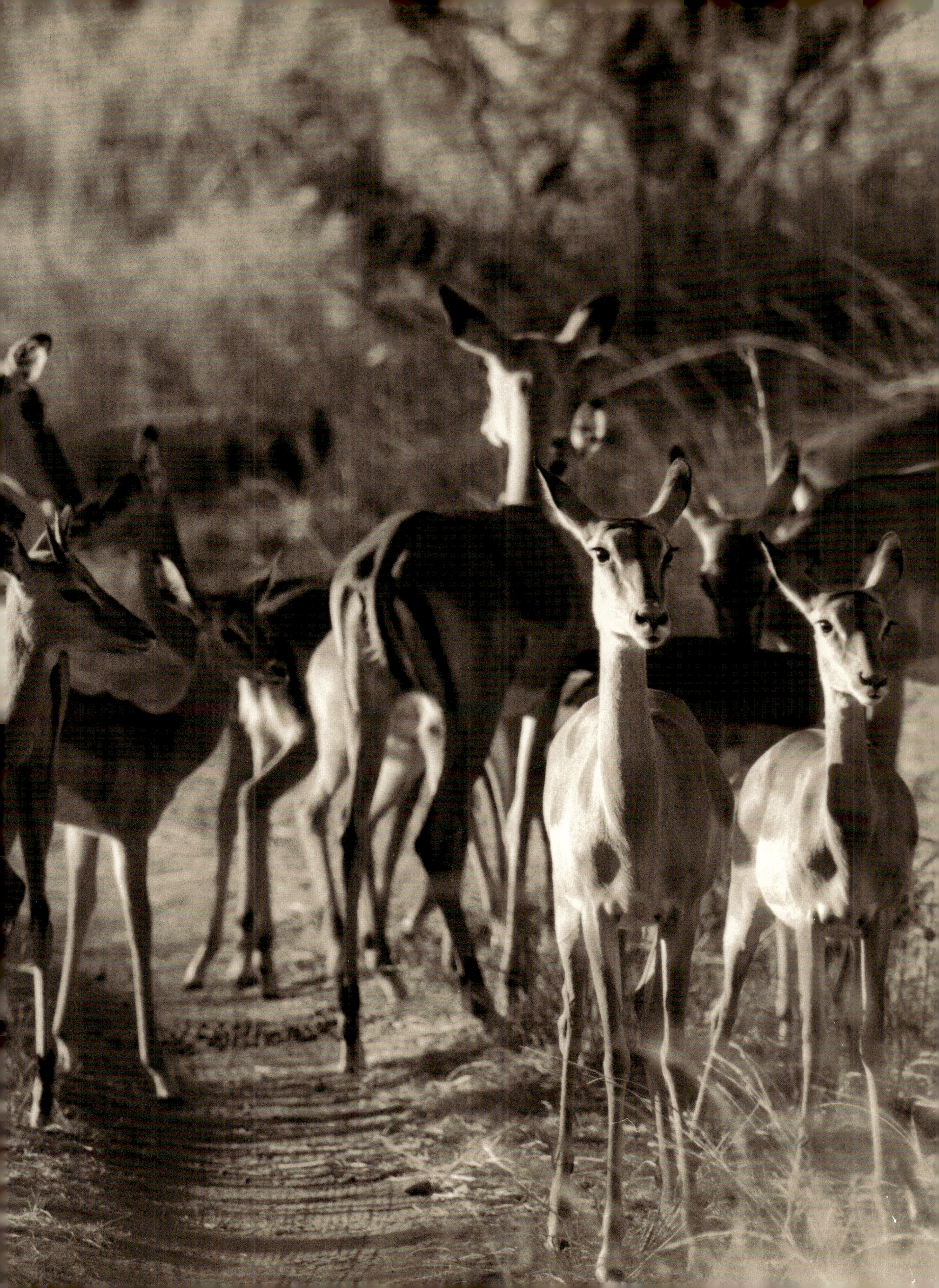

ANIMALS COME INTO SEASON, BECAUSE THE SEASON GETS INTO THEM. ABUNDANT FOOD AND THE STIMULUS OF MANY OTHER ANIMALS SOUNDING AND SMELLING AND SHOWING THAT THEY ARE HEALTHY AND INTERESTED IN PROCREATION GETS EVERYONE GOING. THE SVELTE IMPALA MALE UTTERS AN INCONGRUOUSLY GUTTURAL ROAR TO WARN RIVALS AWAY FROM HIS HAREM. THE WILDEBEEST SNORTS AND SAYS "GNU" OVER AND OVER AGAIN, AND PUMPS HIMSELF UP FOR THE "RUT". DISPLAYING AND FIGHTING, CORRALLING FEMALES, MATING AND REPELLING OTHER MALES FROM TEMPORARY *EN ROUTE* TERRITORIES, THE MALE WILDEBEEST IS THE "MAD COW" OF THE PLAINS. THE EARLY MIGRATION IS NOT A RELAXING RAMBLE: IT IS A DEMANDING SEX TOUR.

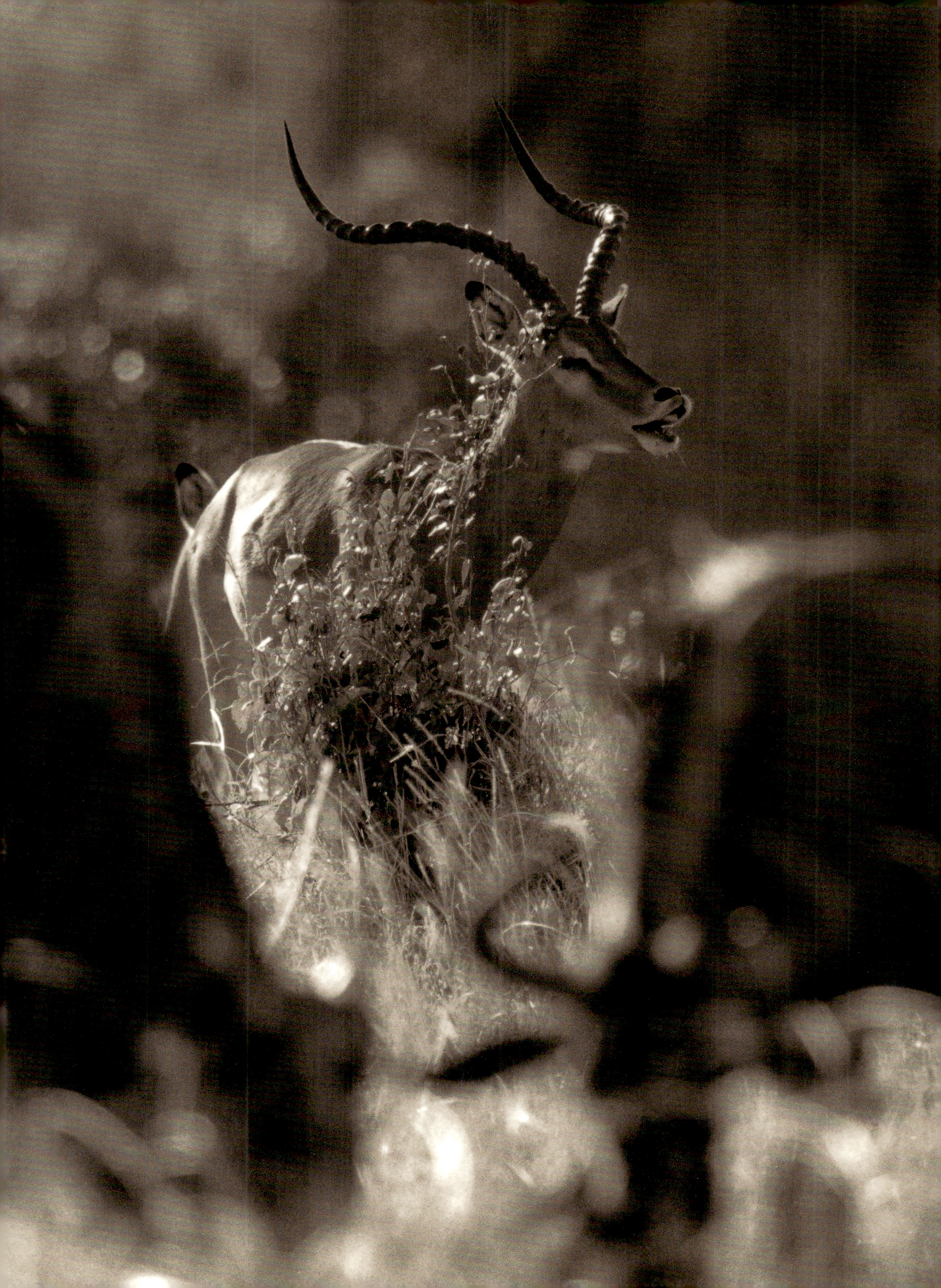

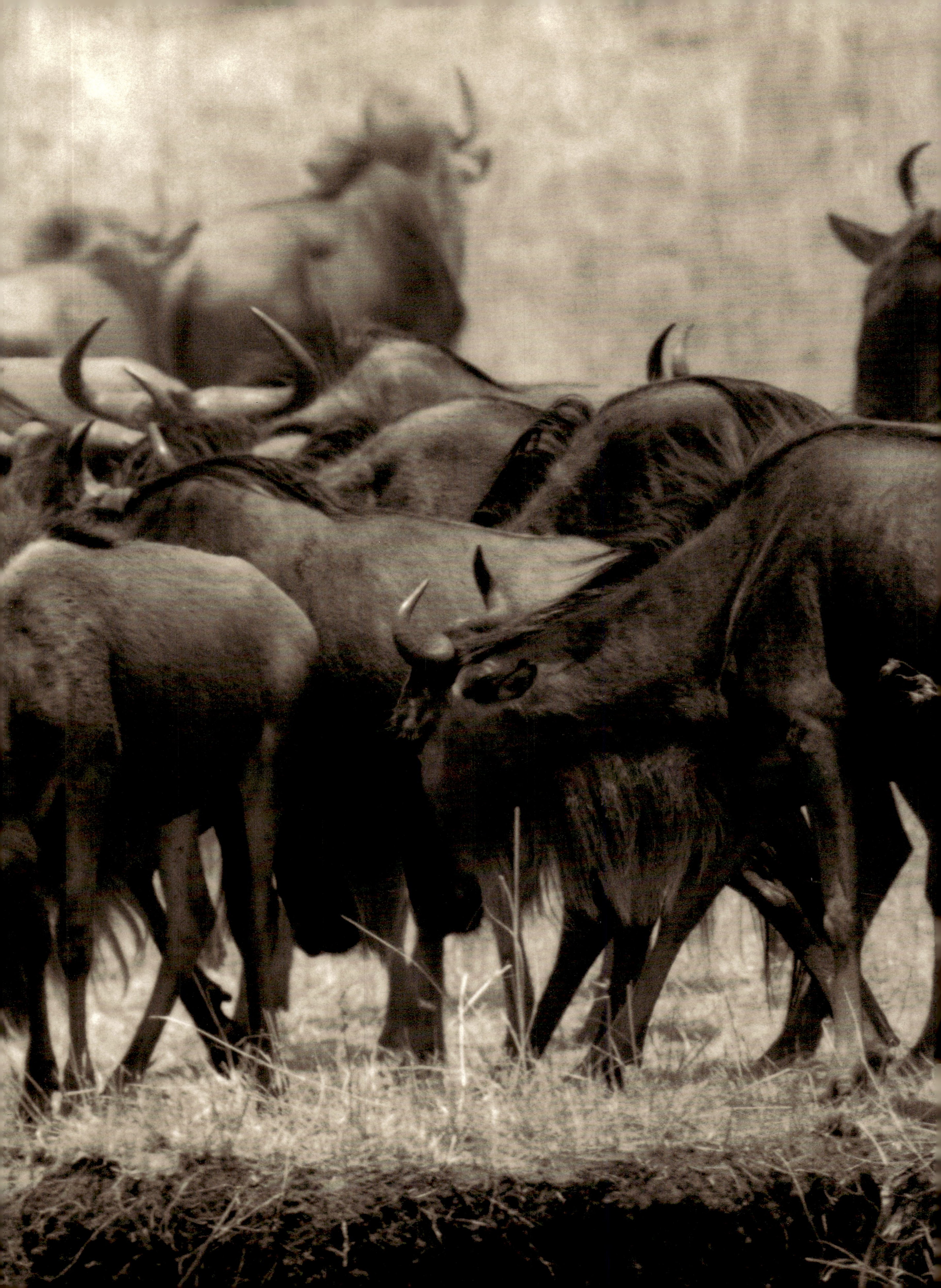

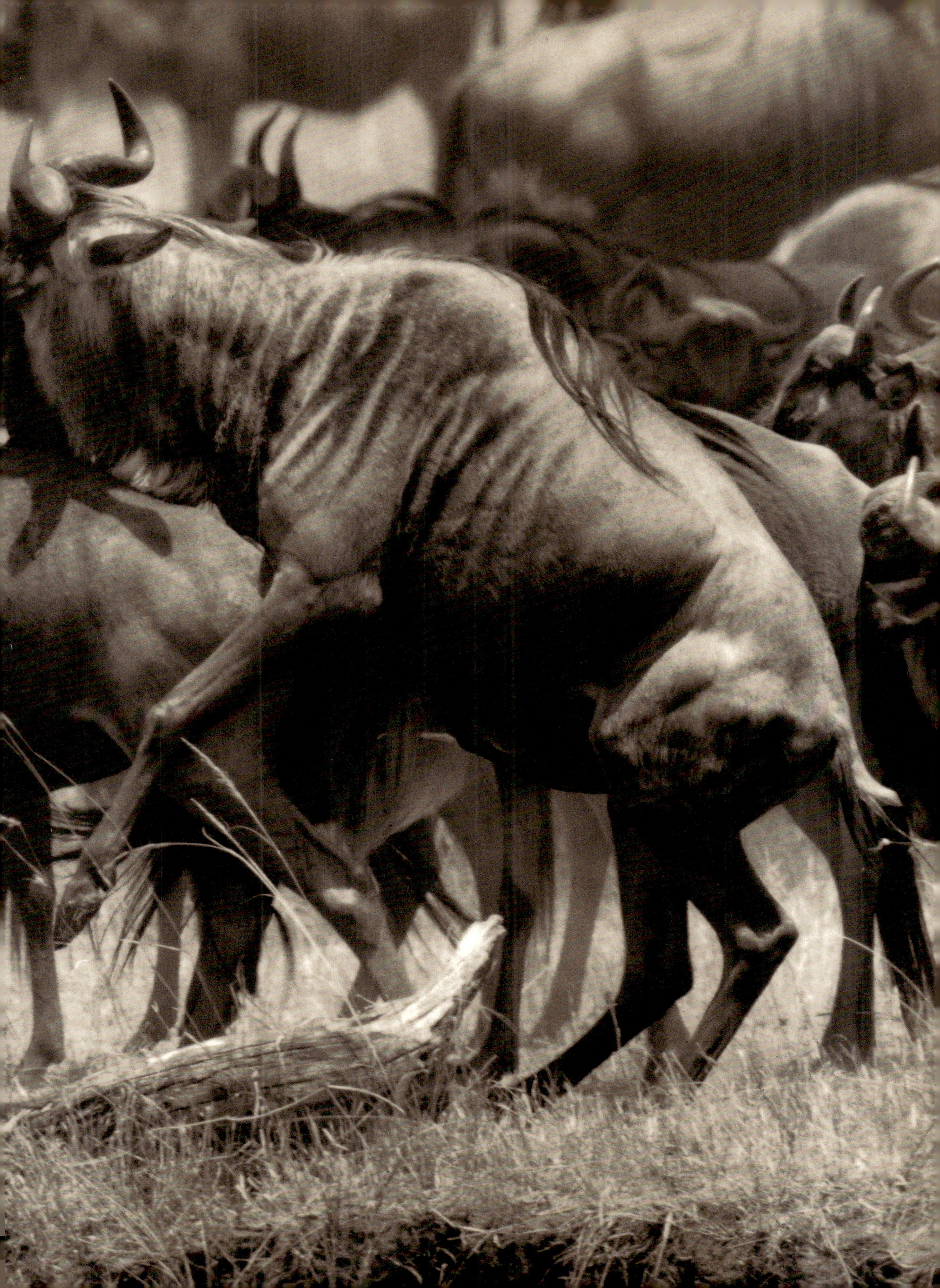

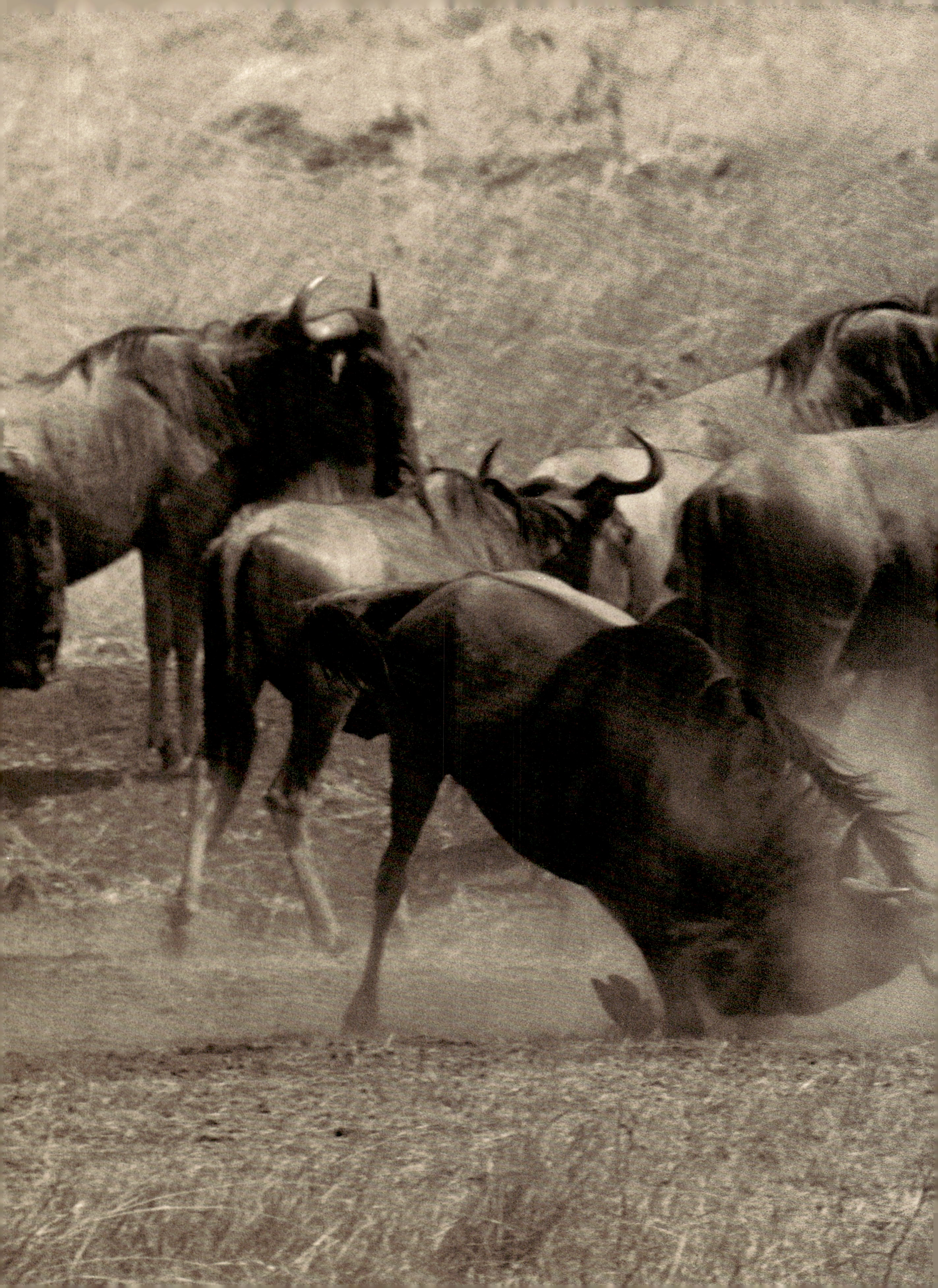

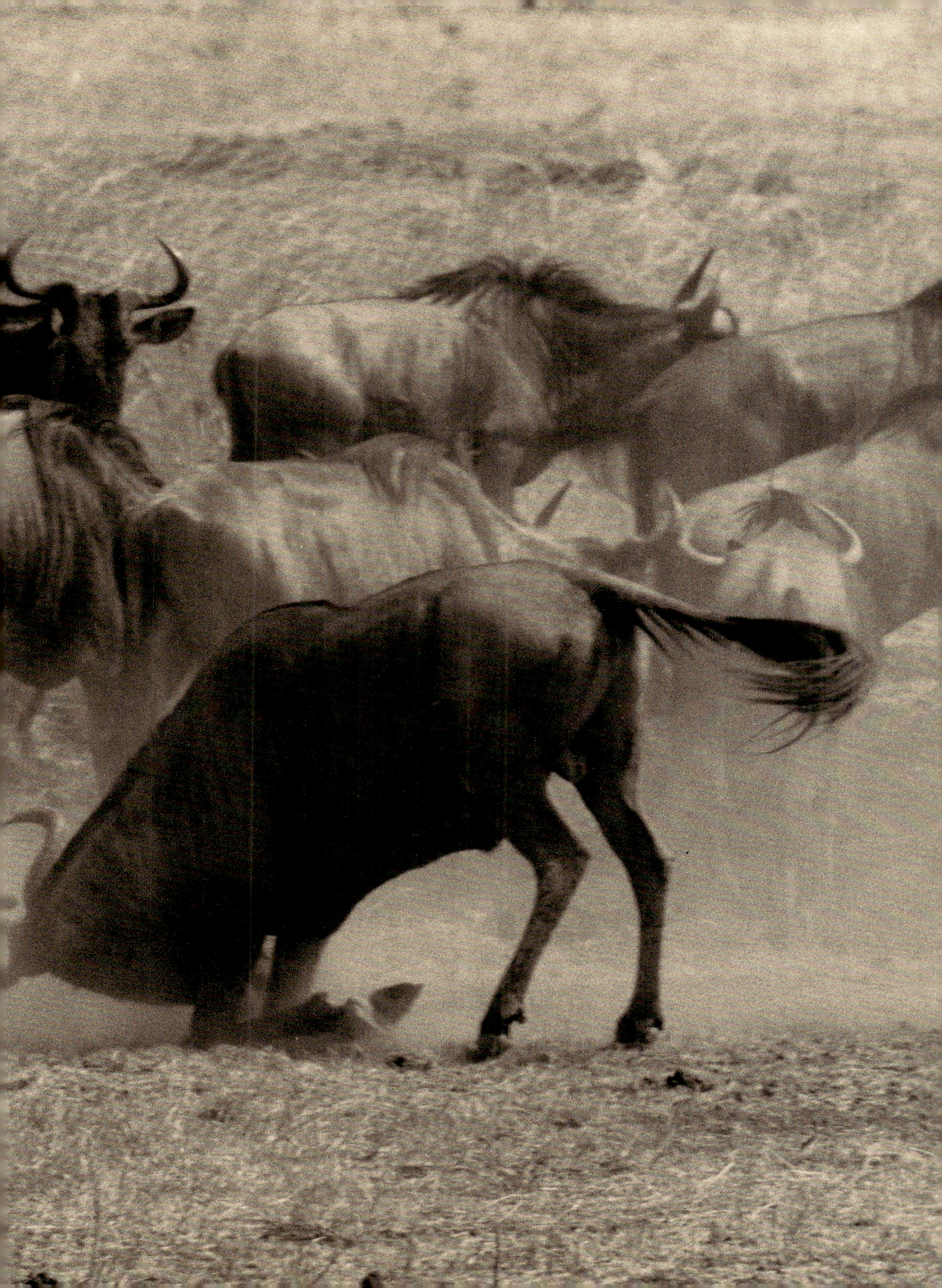

SO ATTRACTIVE AND PROFITABLE IN TERMS OF SURVIVAL IS THE MIGRATION, THAT TENS OF THOUSANDS OF OTHER HERBIVORES – MAINLY ZEBRA AND THOMSON'S GAZELLE – JOIN THE MILLION AND A HALF GNUS IN THEIR ANNUAL 310-MILE MARCH. THEY EAT DIFFERENT PARTS OF THE GRASS SWARD AND SO DO NOT COMPETE. THEY ALSO PROVIDE ADDITIONAL TARGETS FOR LOCAL PREDATORS: THERE IS SOME SAFETY IN NUMBERS FOR ALL CONCERNED. WITH A LITTLE MORE DECORUM THAN THE GNUS BUT WITH EQUAL FEROCITY, MALE ZEBRA (OPPOSITE) FIGHT TO DEFEND THEIR HAREMS OR TO POACH FEMALES FROM OTHERS. THE FIGHTING SKILL, PARTICULARLY THEIR KICKS, CAN KEEP PREDATORS AT BAY WHILST FEMALES AND FOALS ESCAPE.

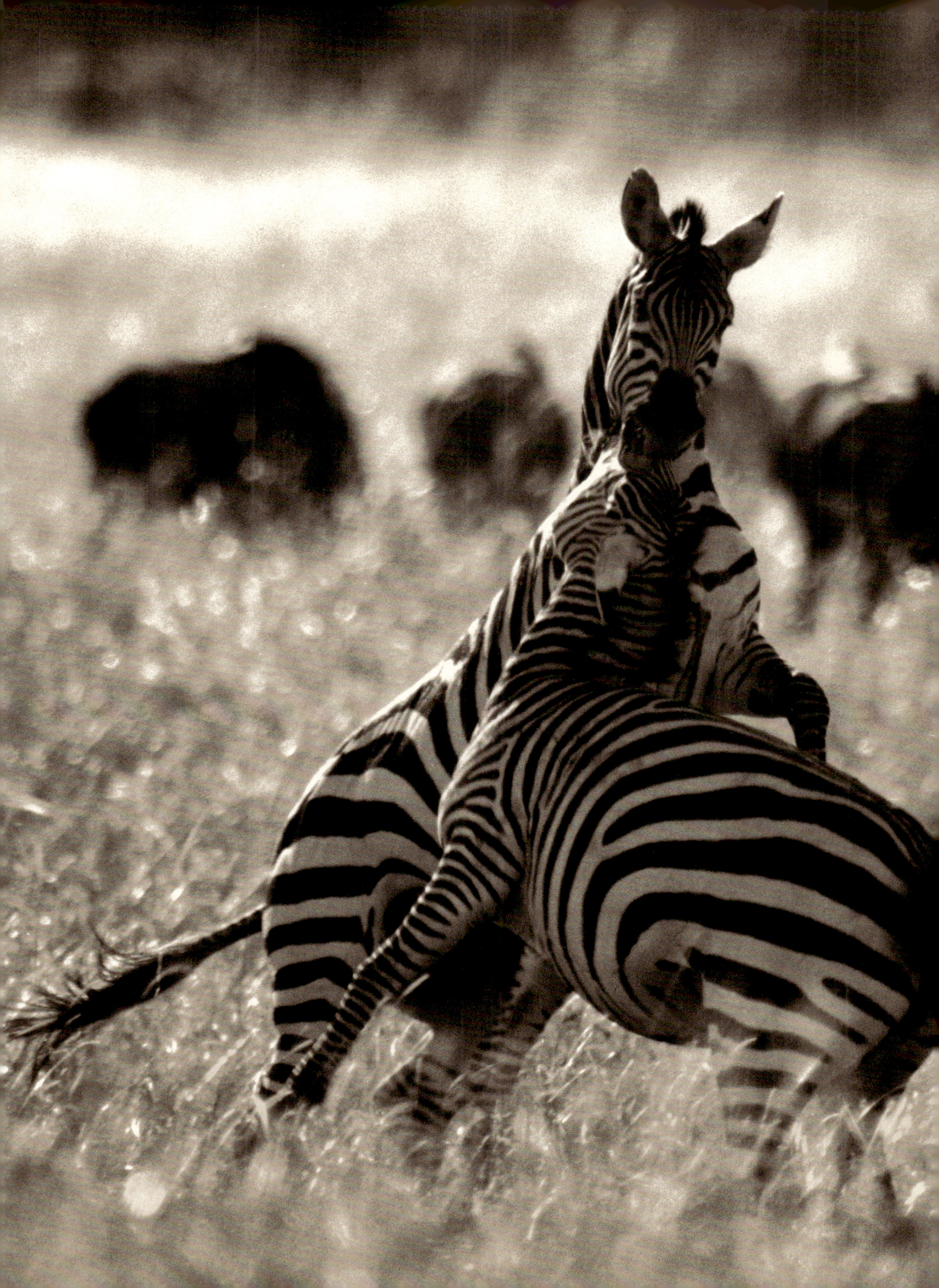

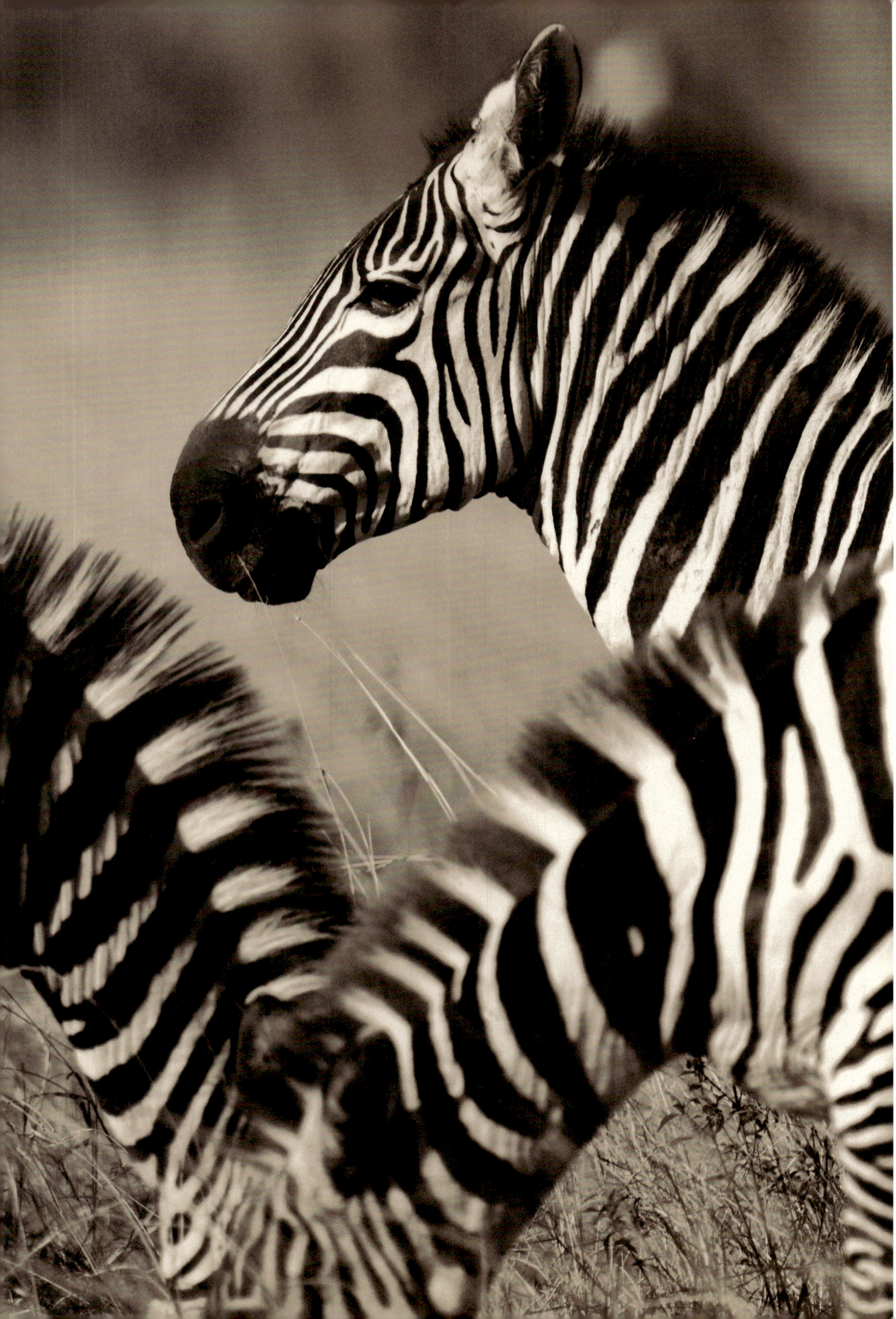

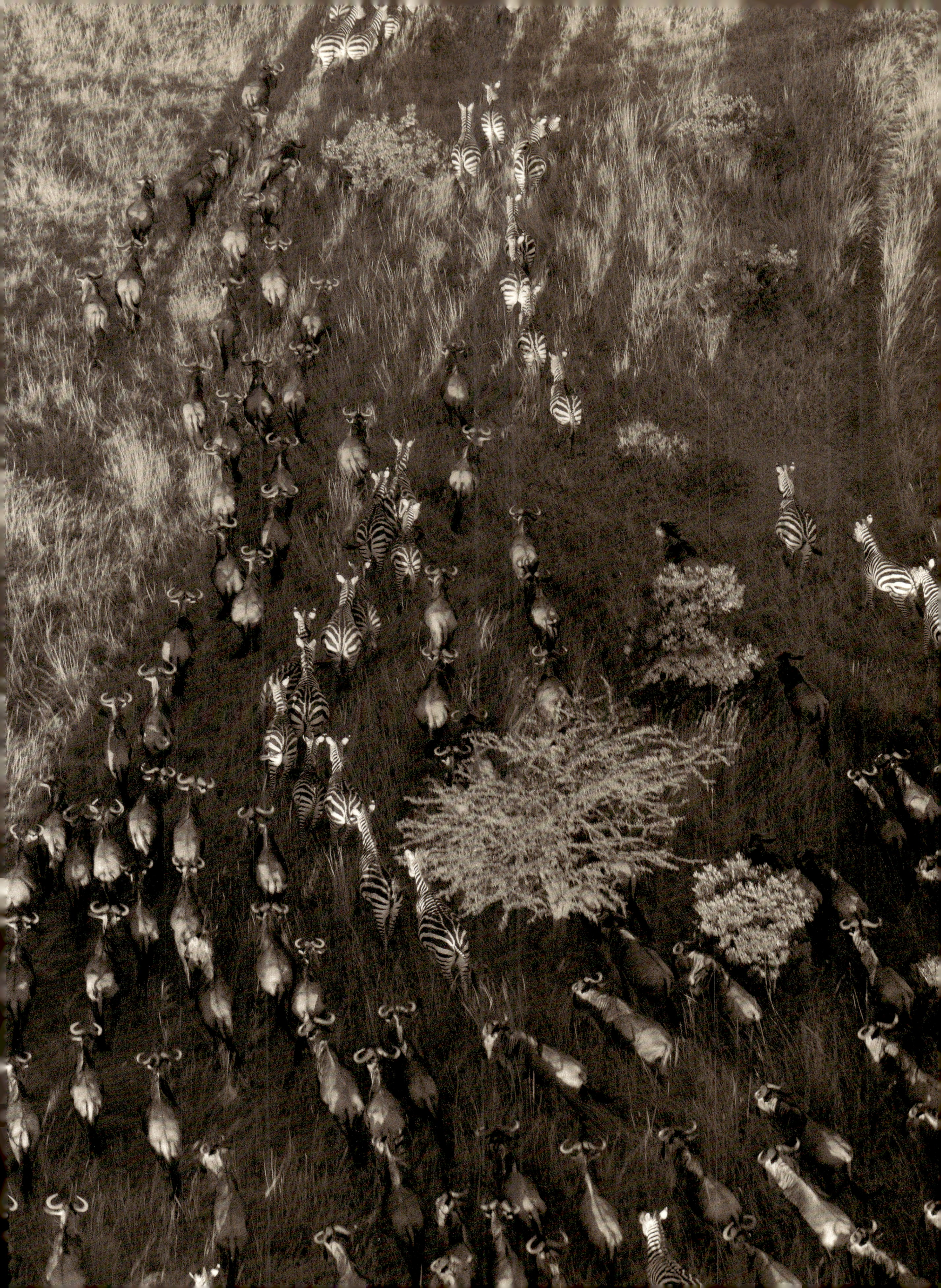

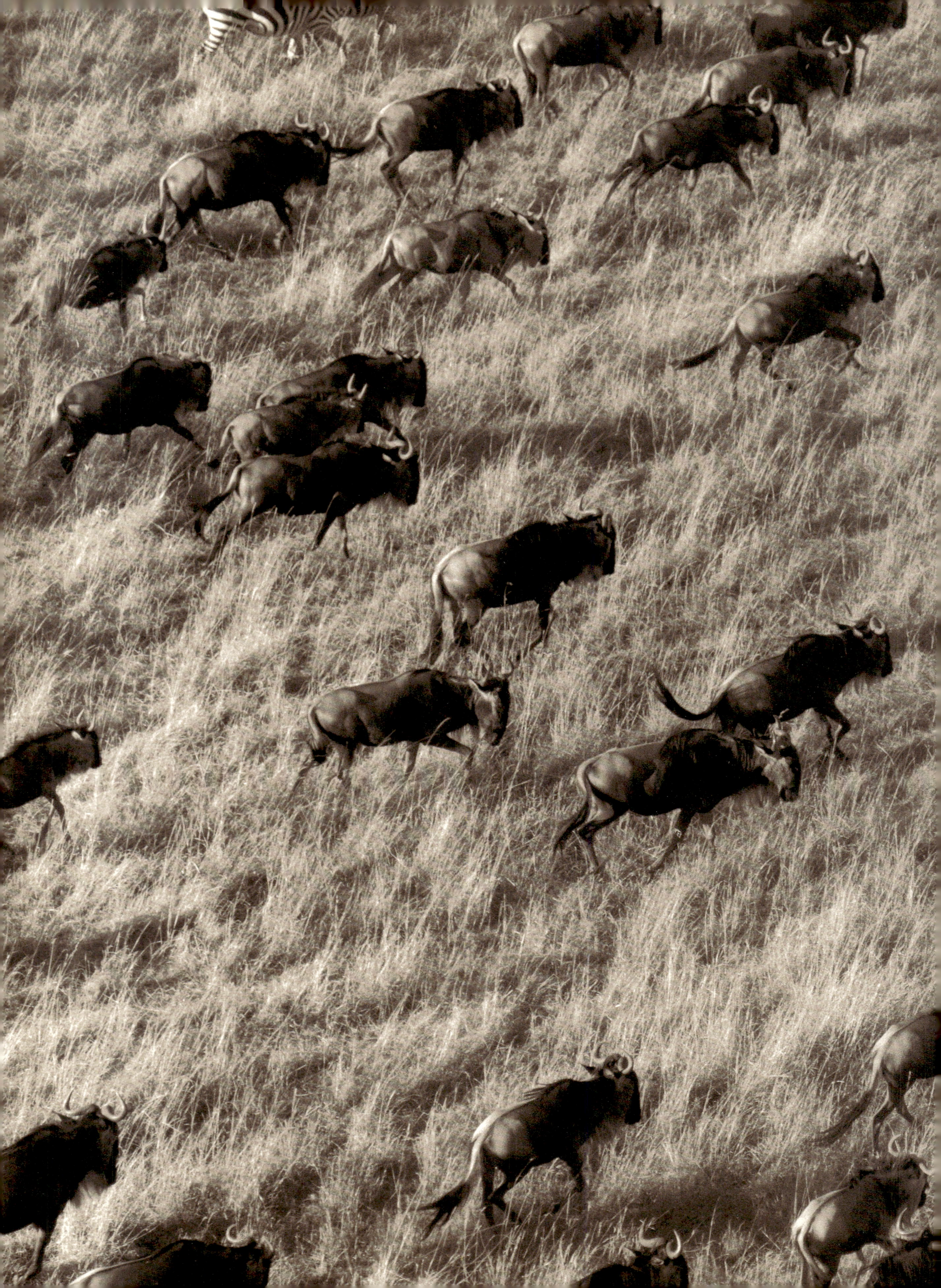

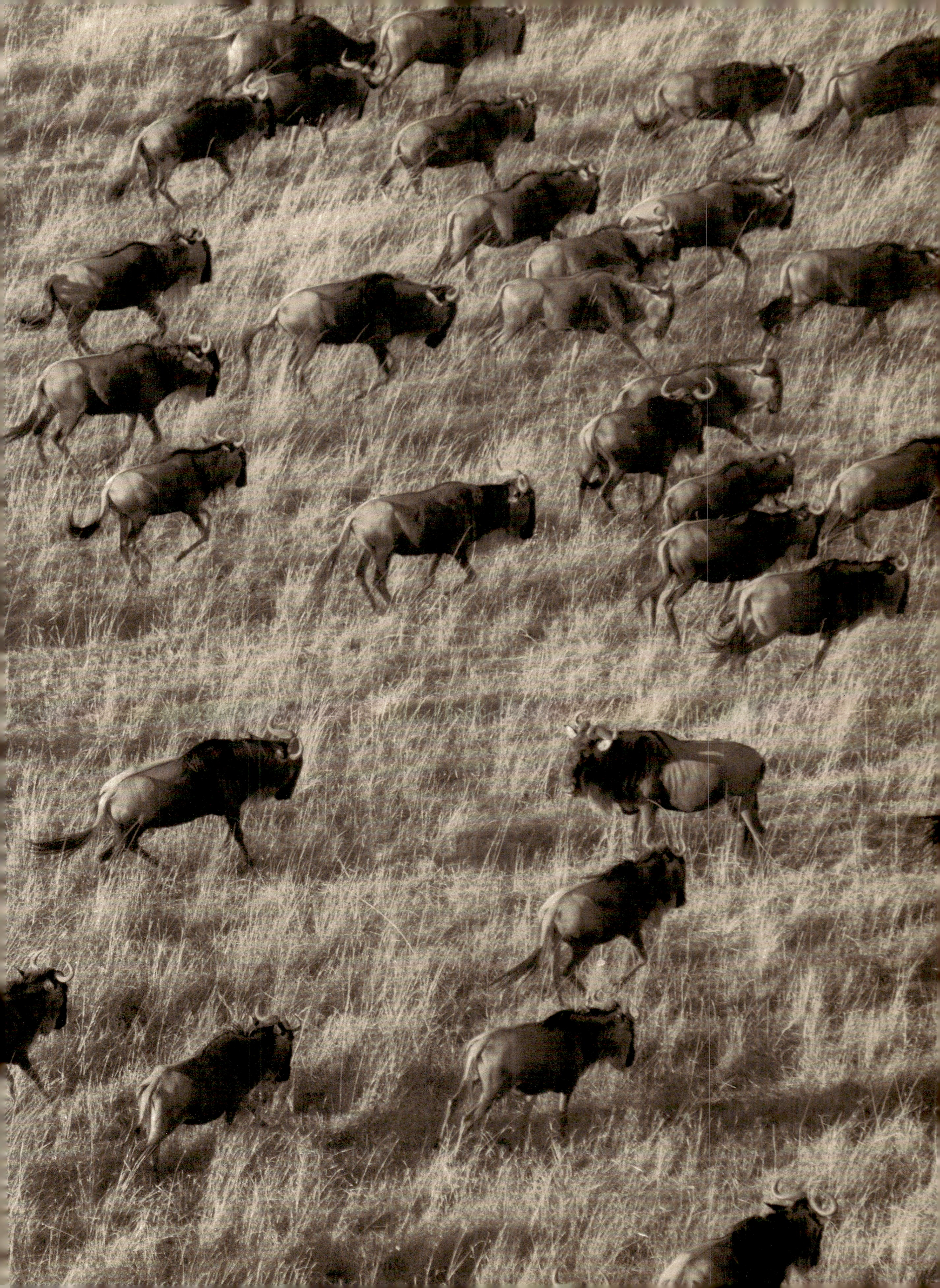

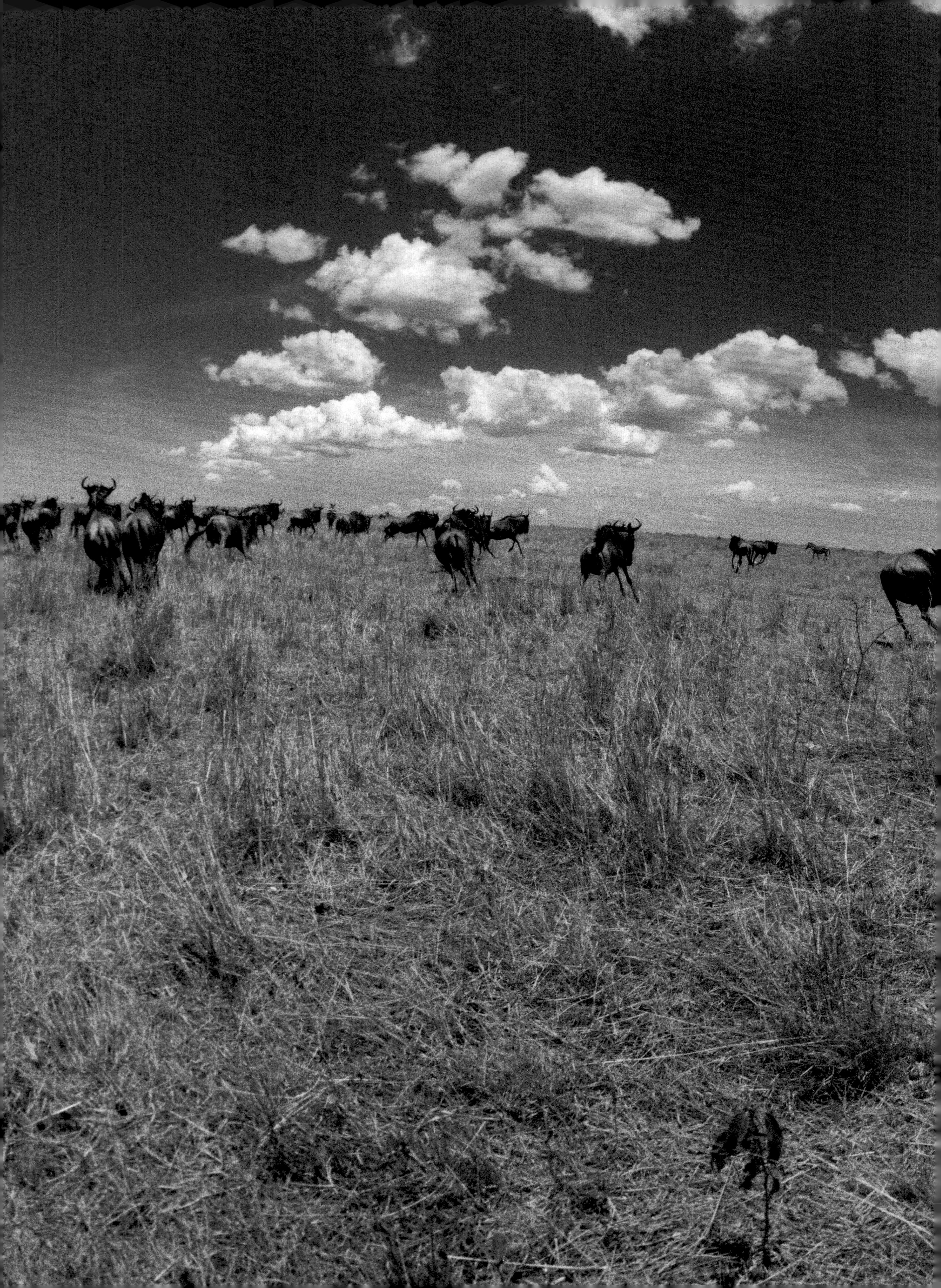

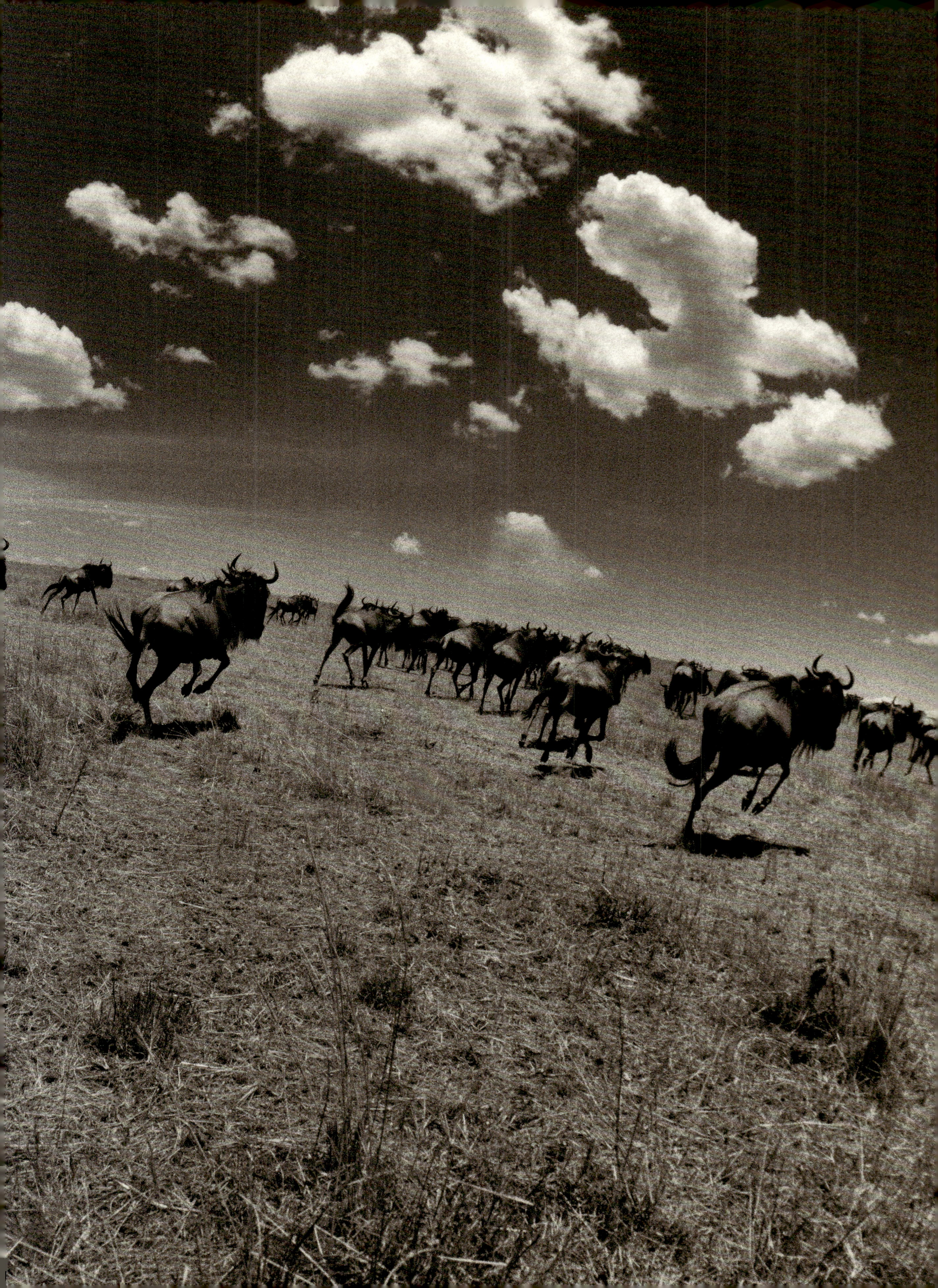

2

NORTHERN MARA BUSHED GRASSLAND

Short Rains, October–November

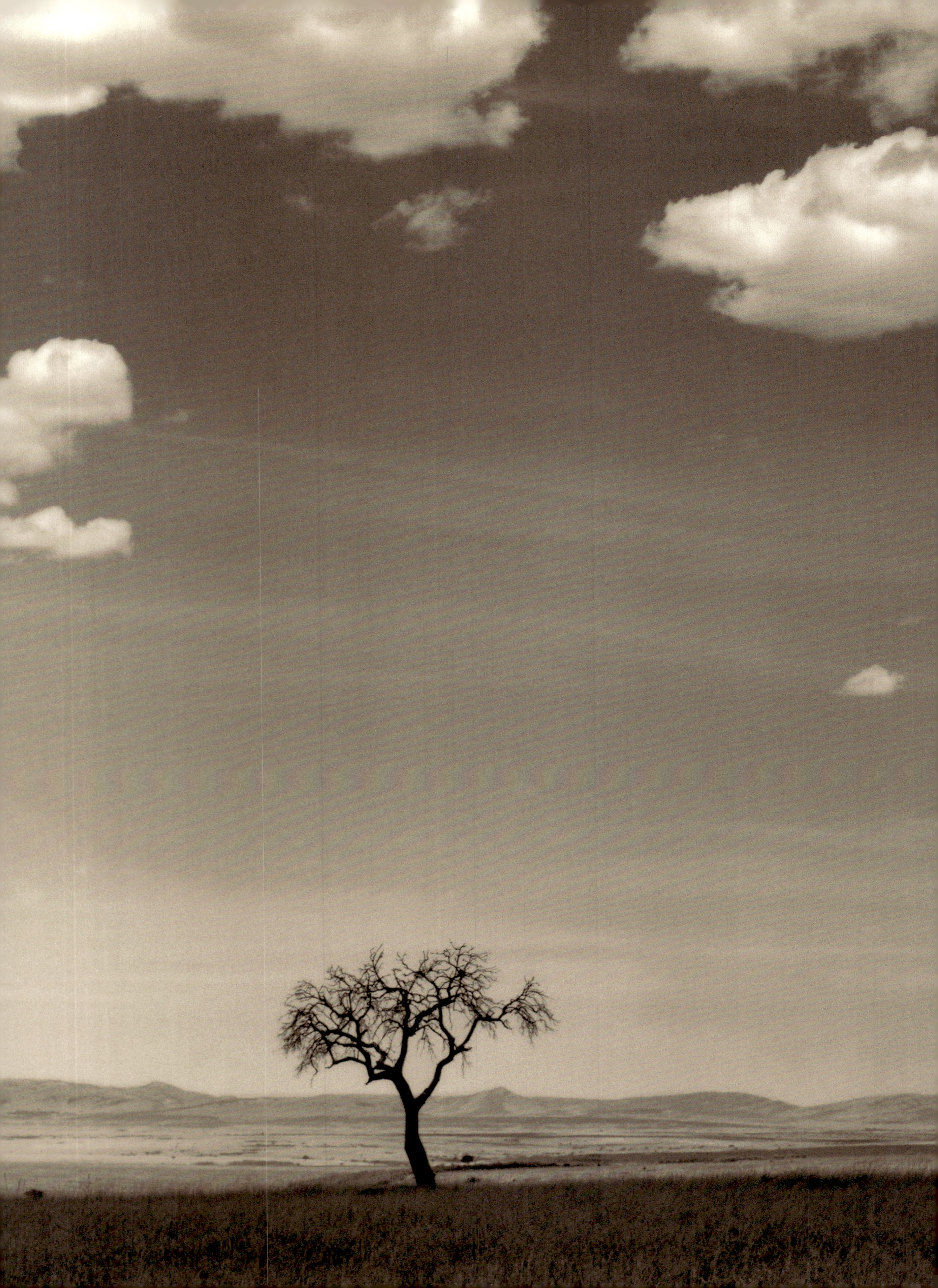

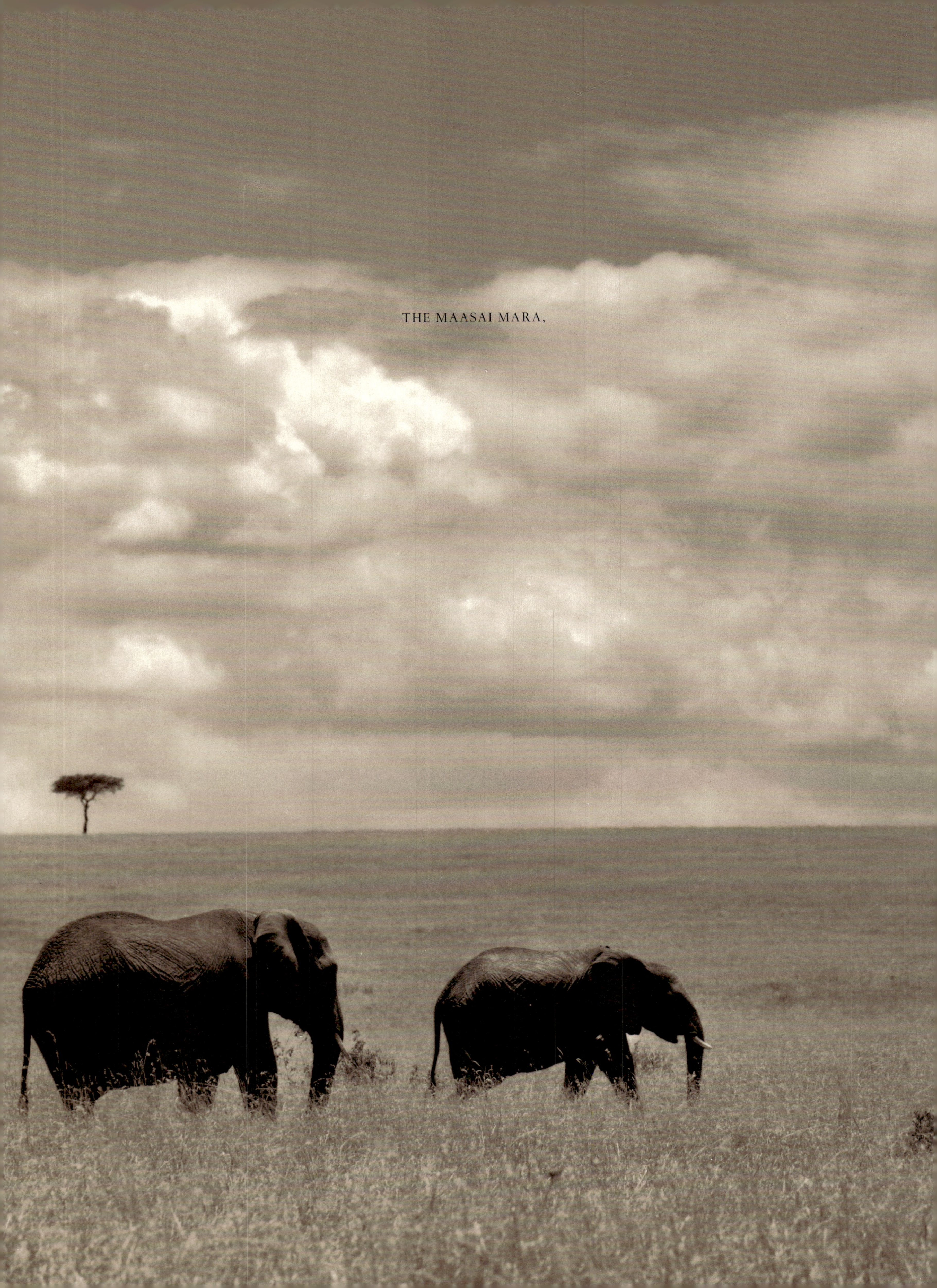
THE MAASAI MARA,

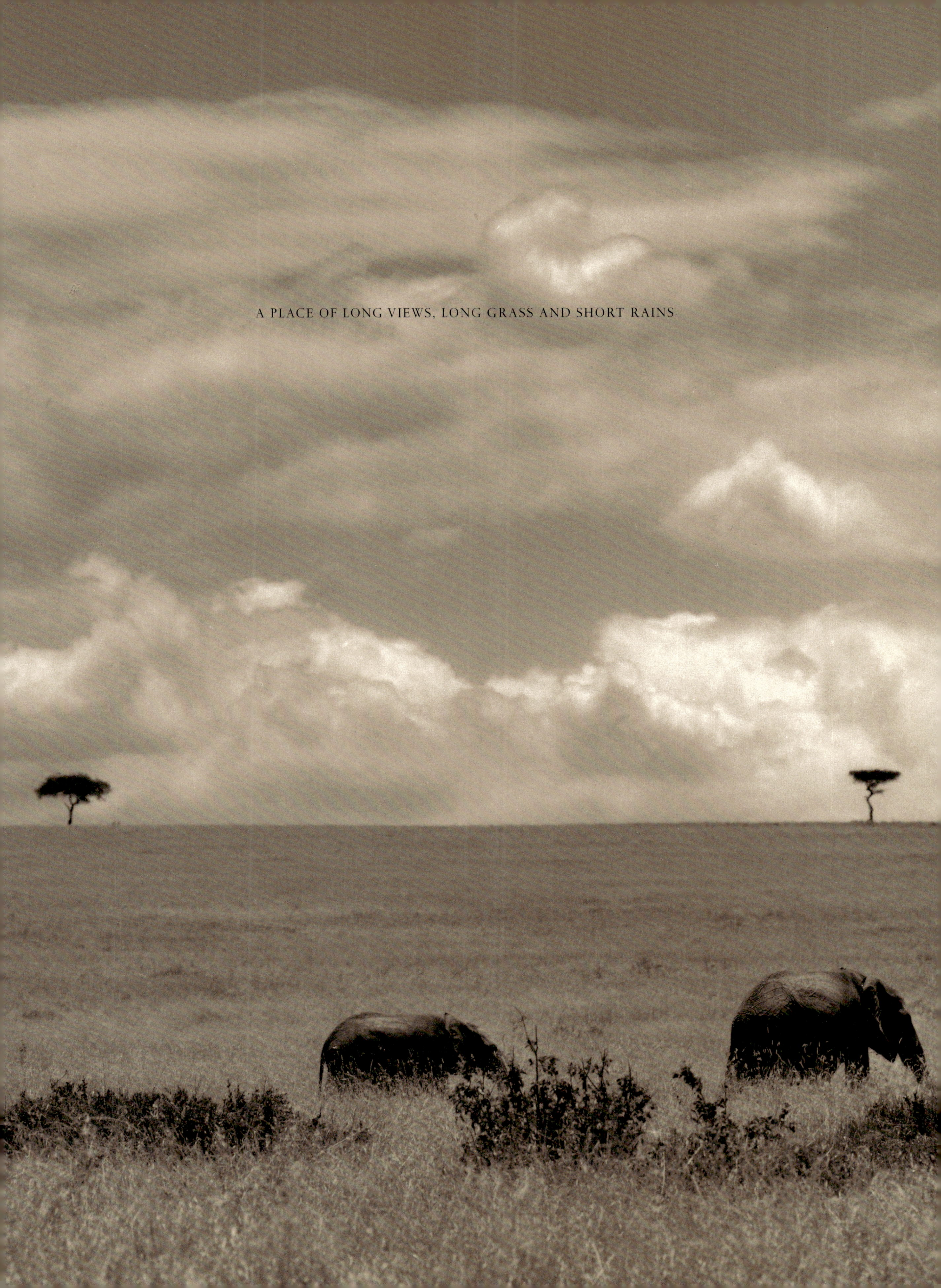

A PLACE OF LONG VIEWS, LONG GRASS AND SHORT RAINS

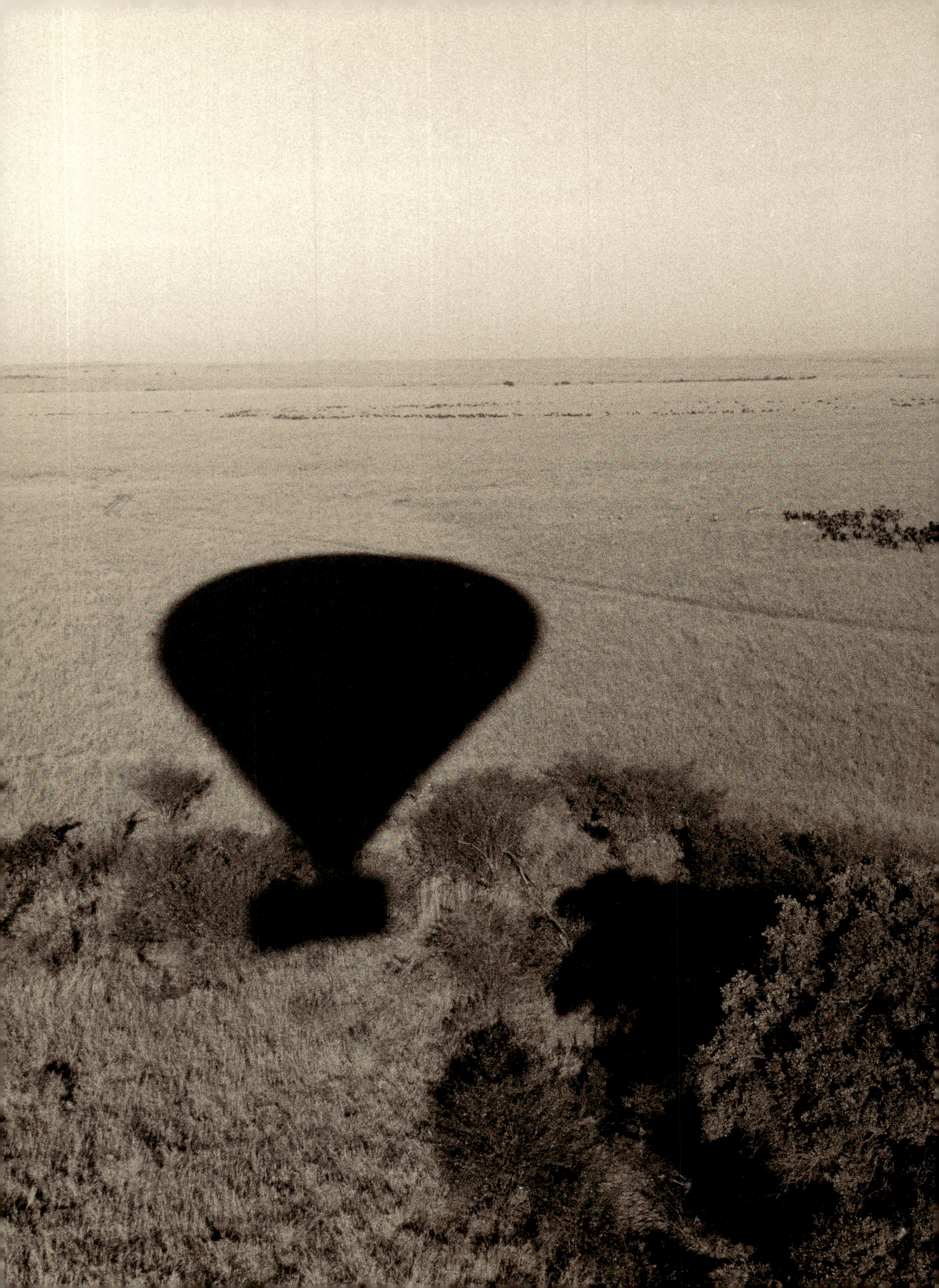

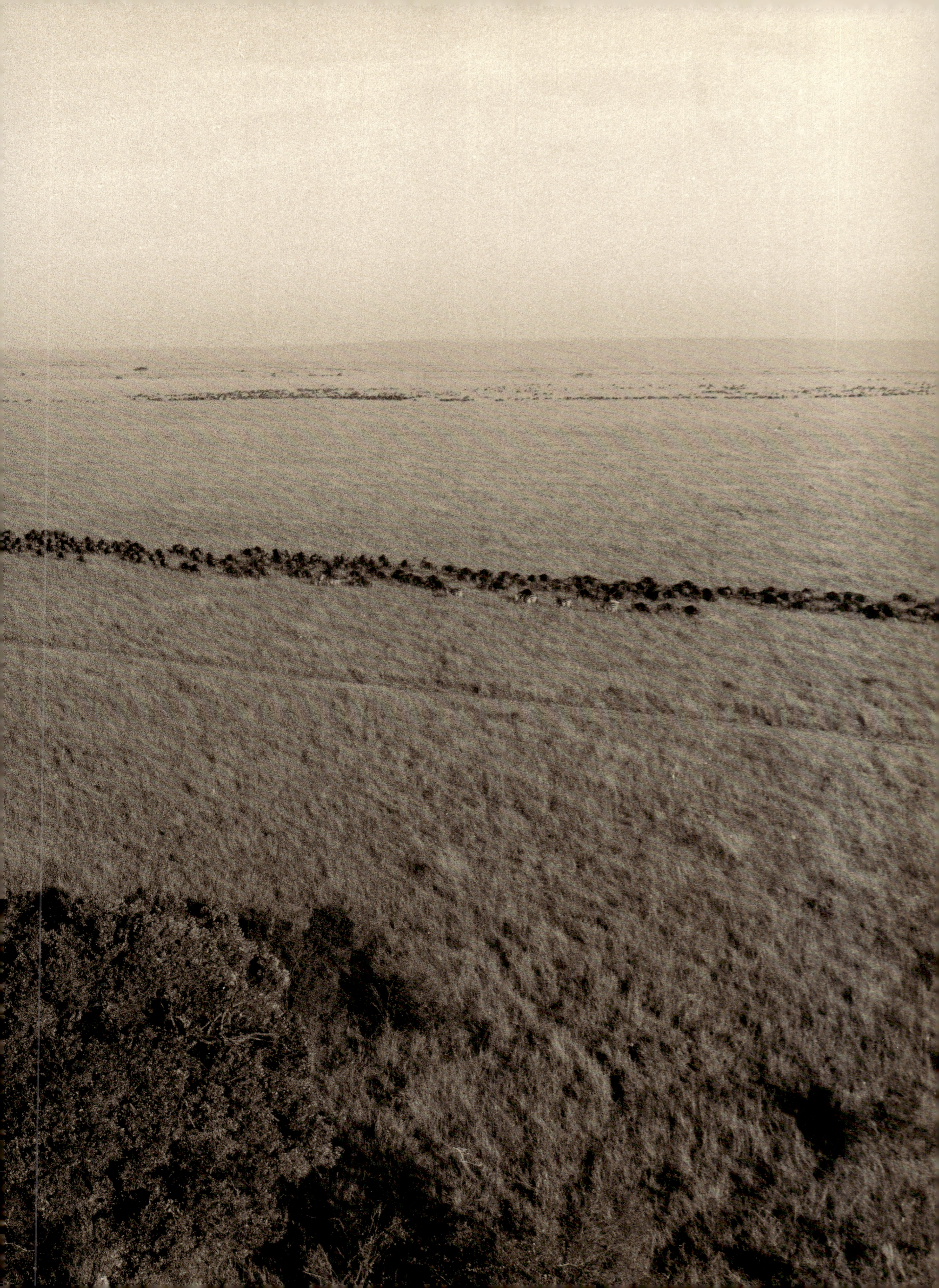

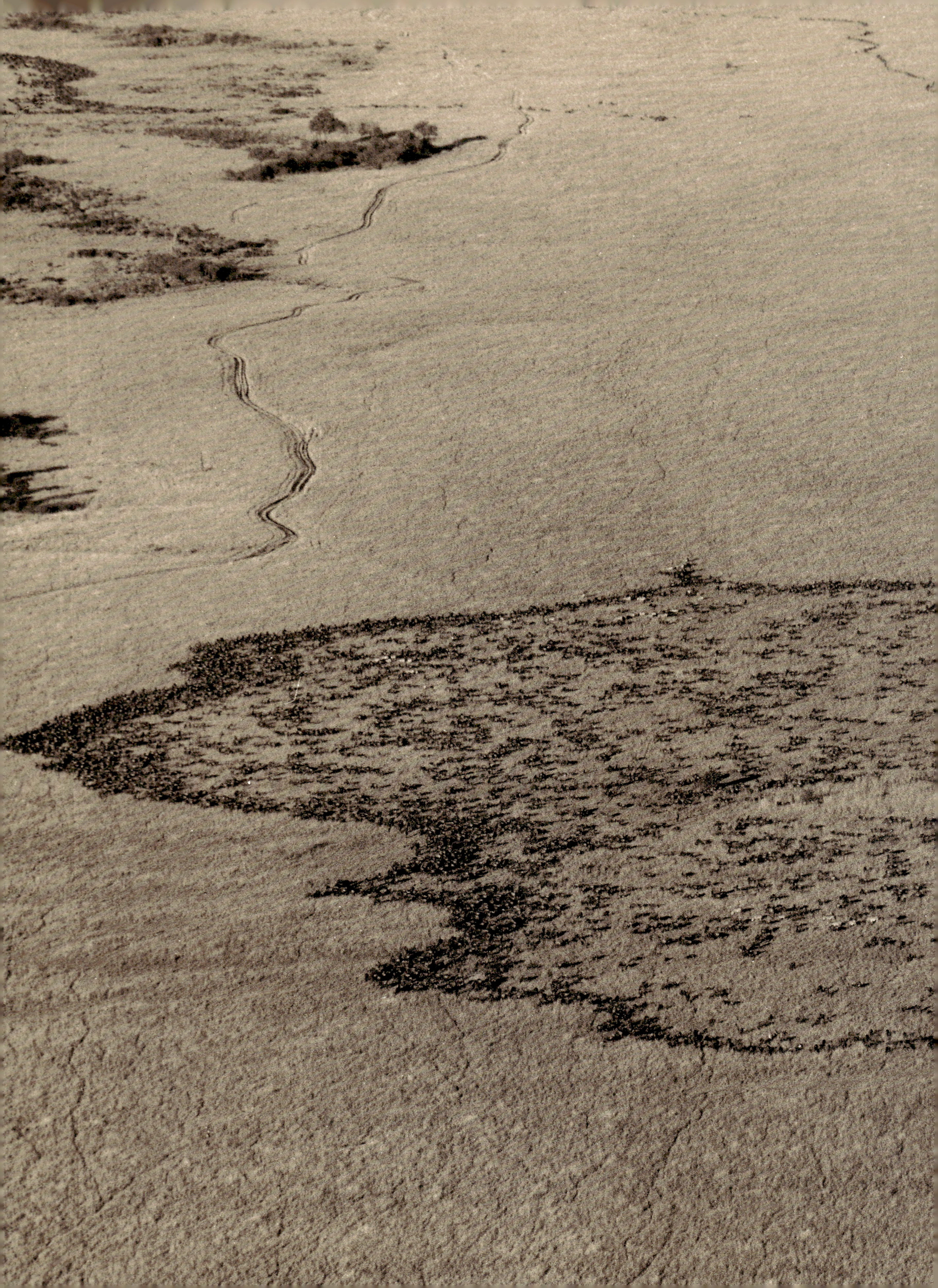

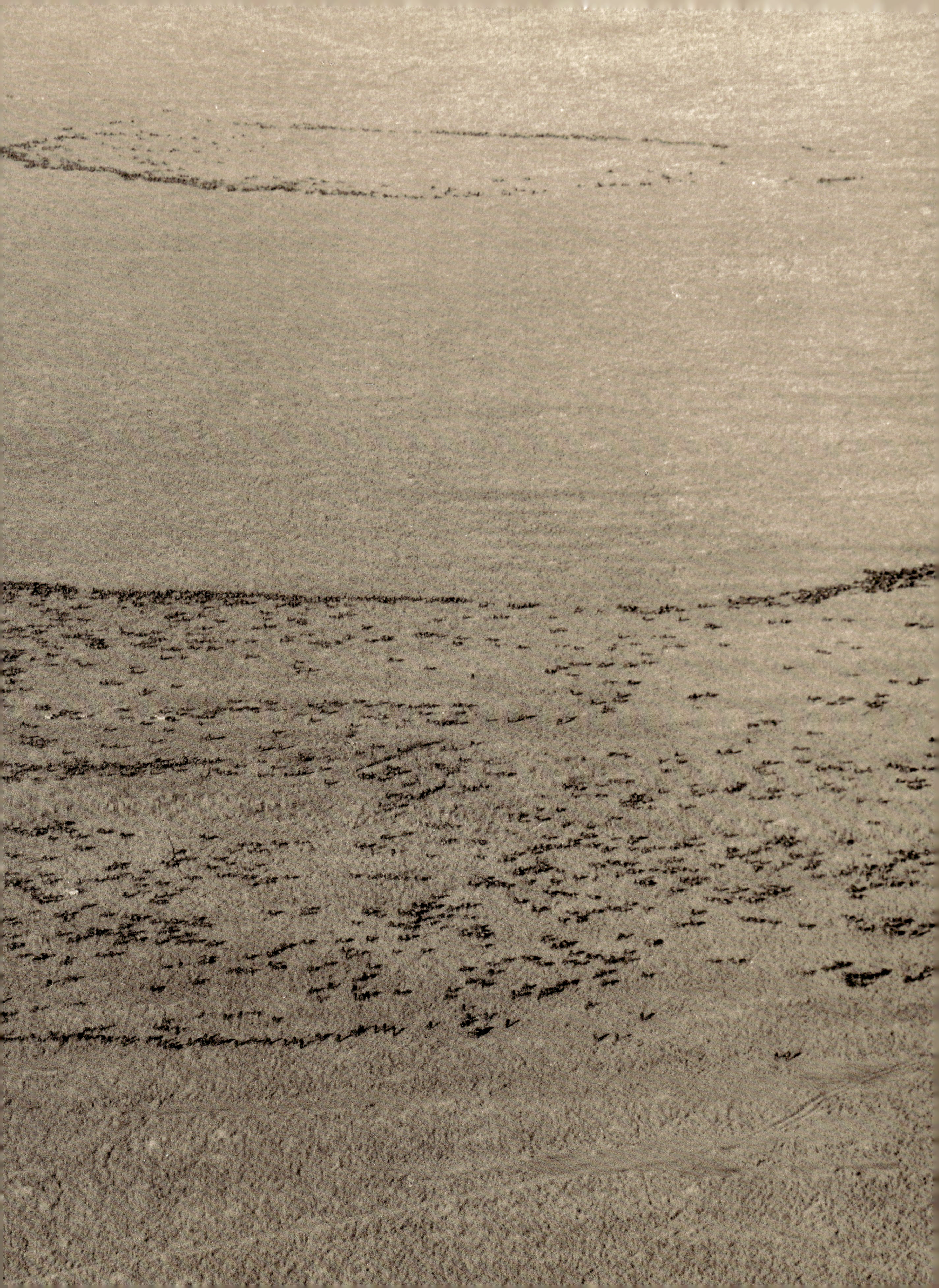

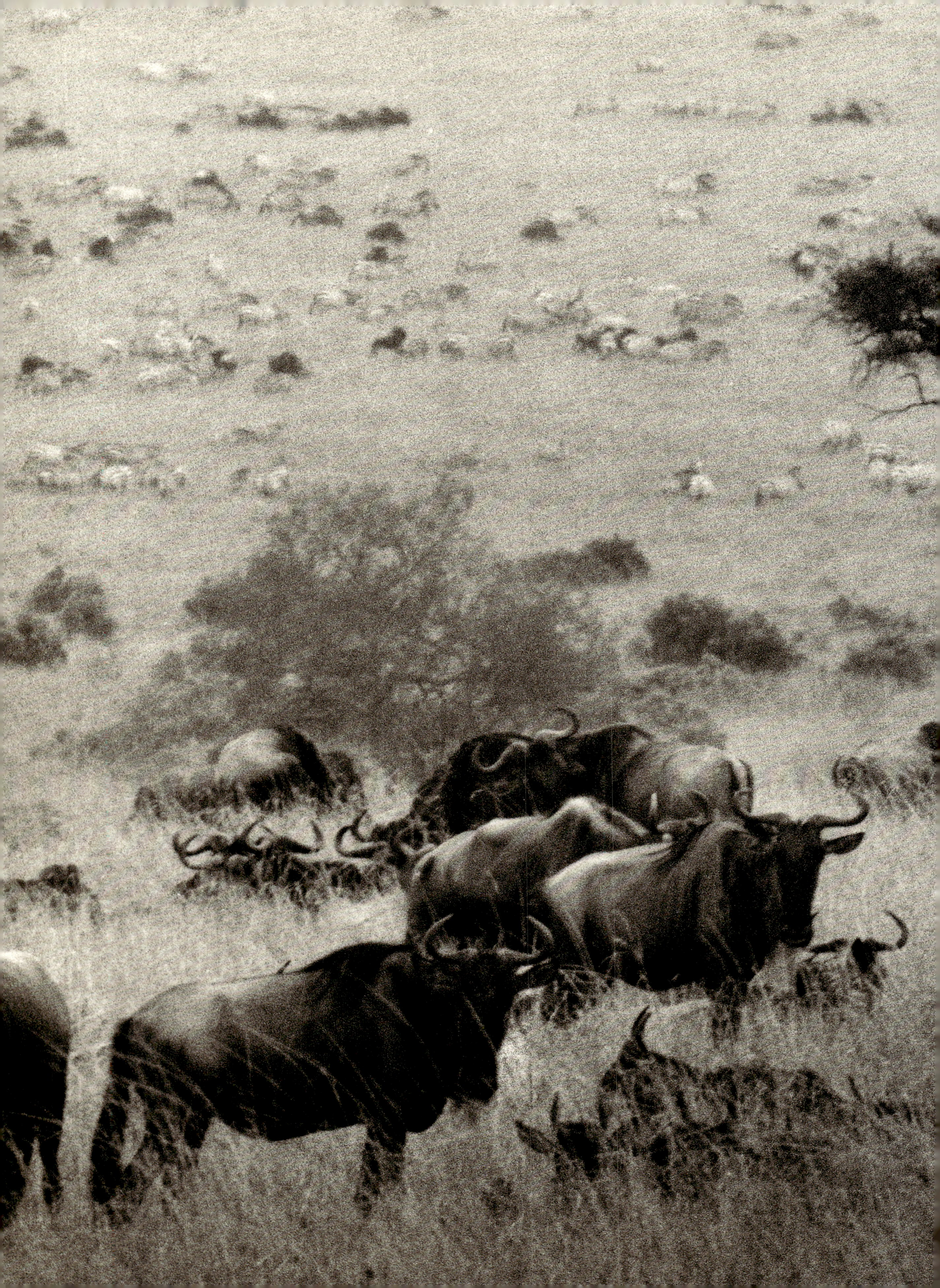

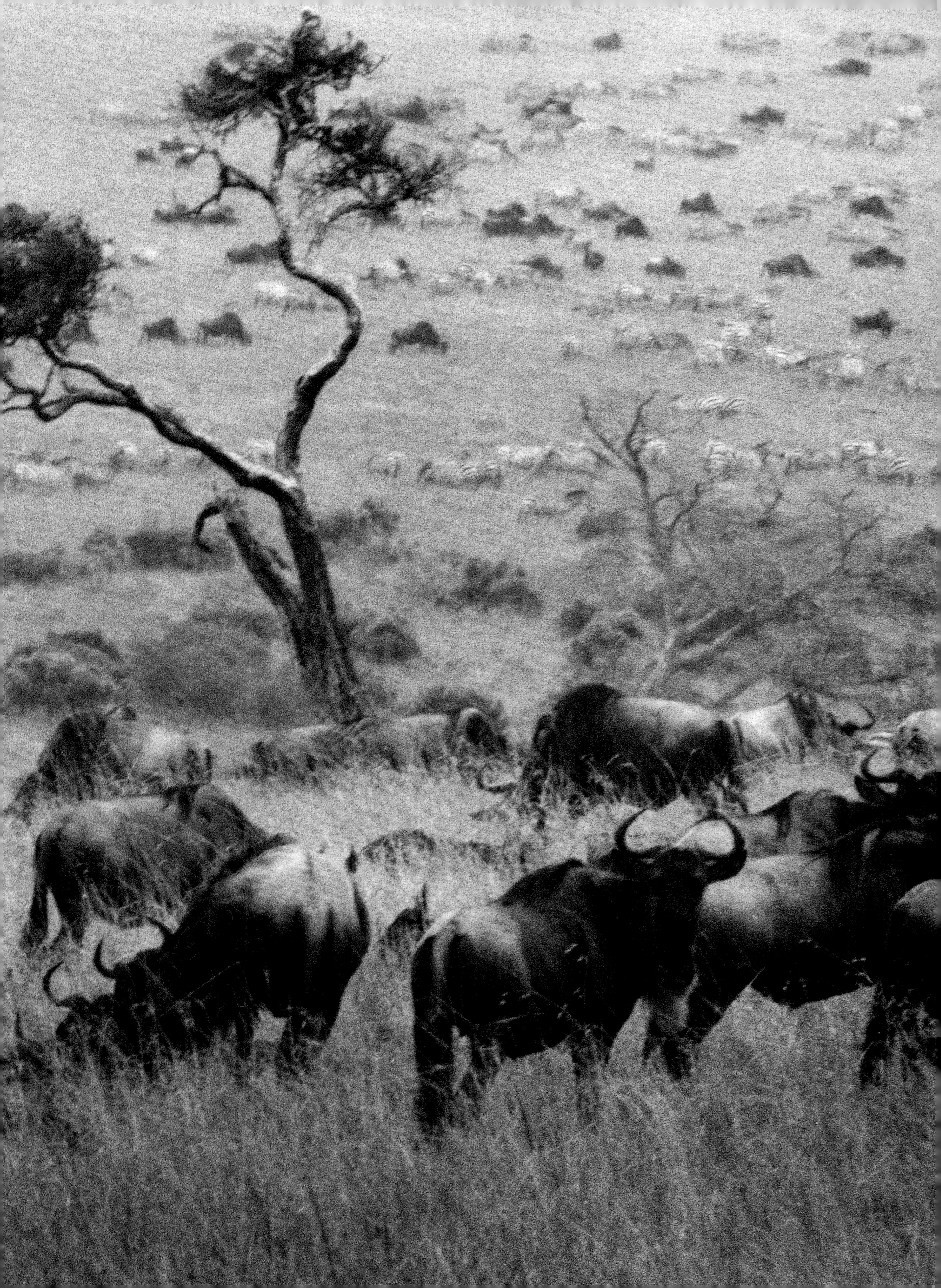

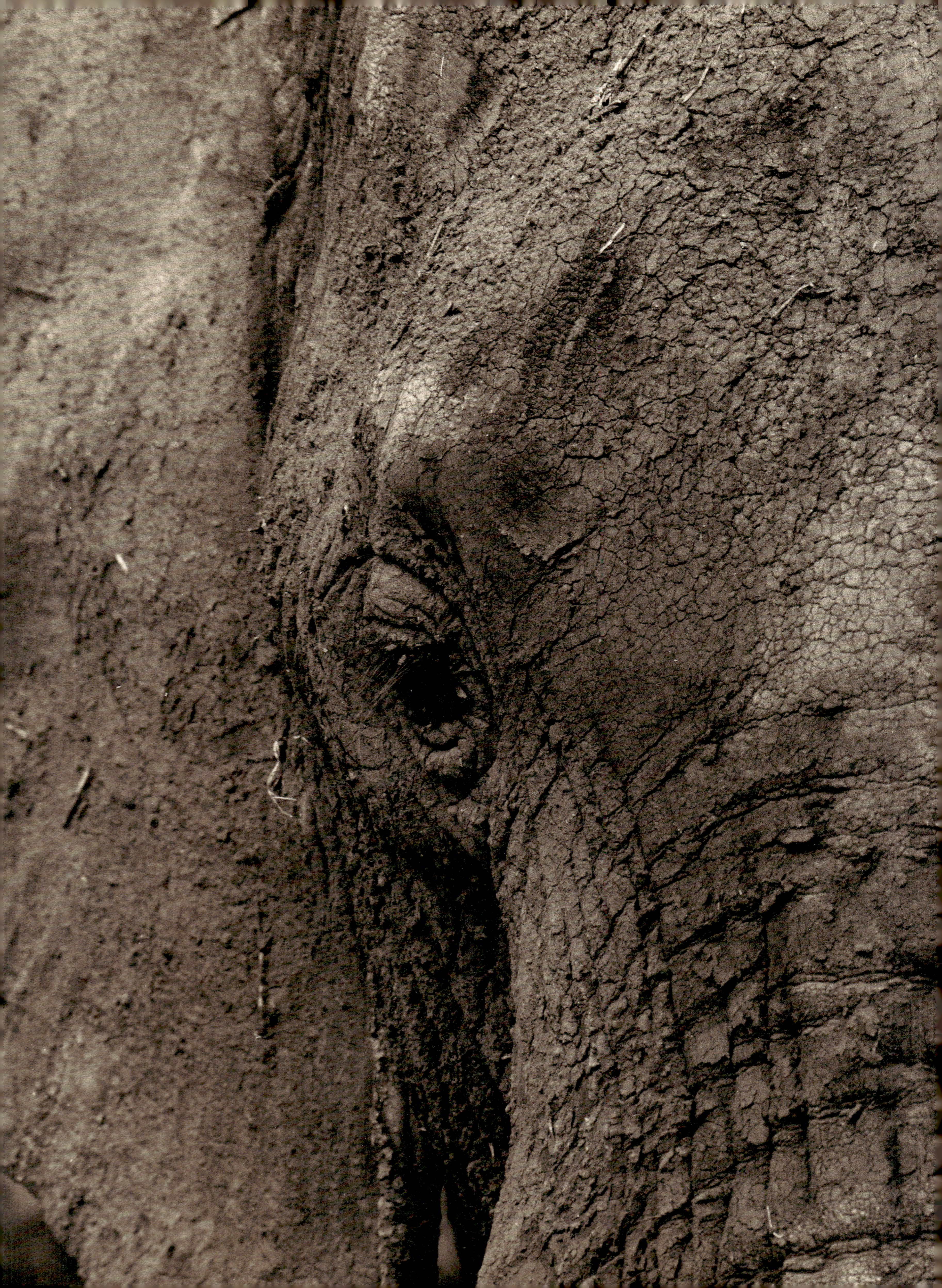

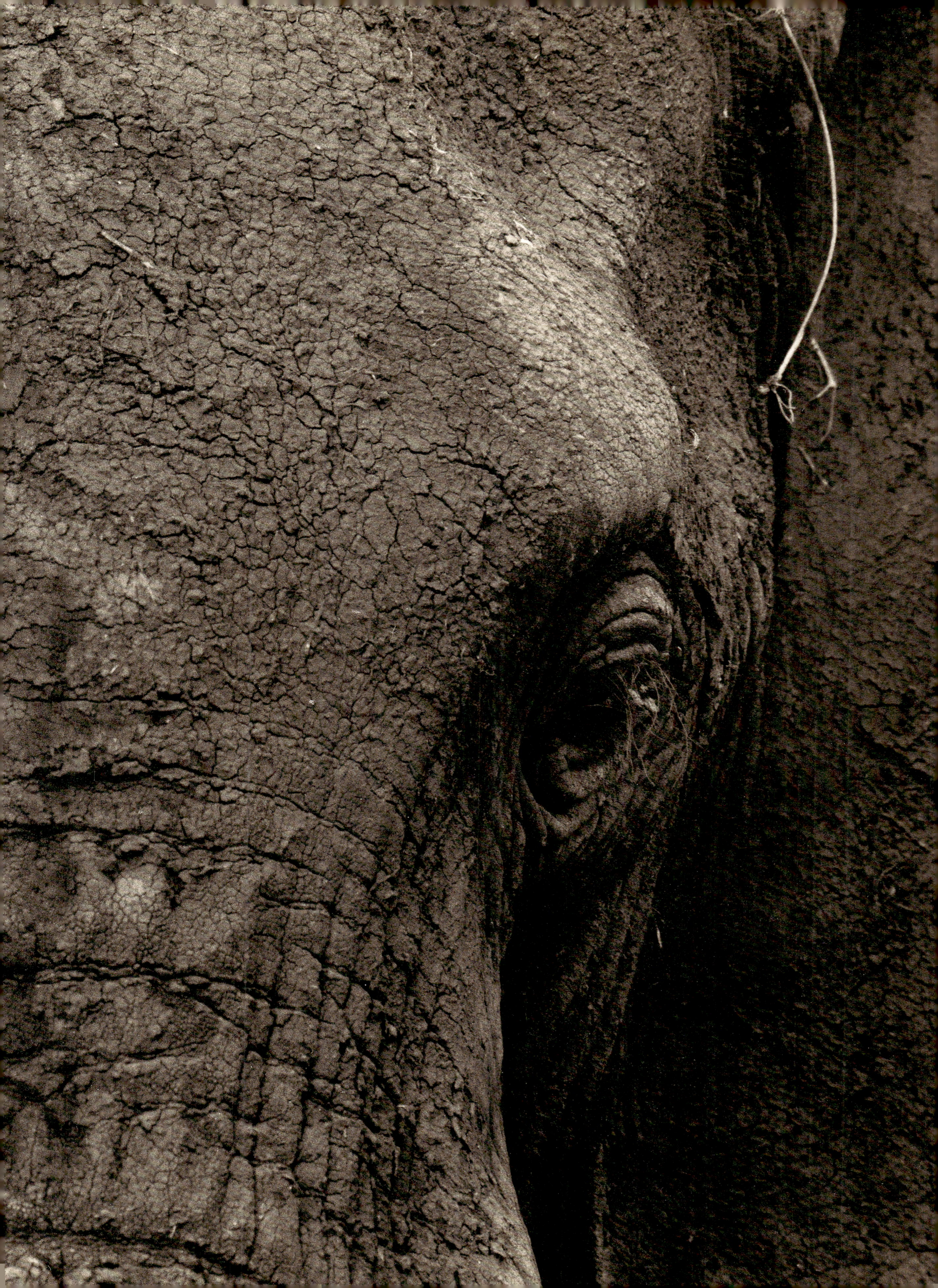

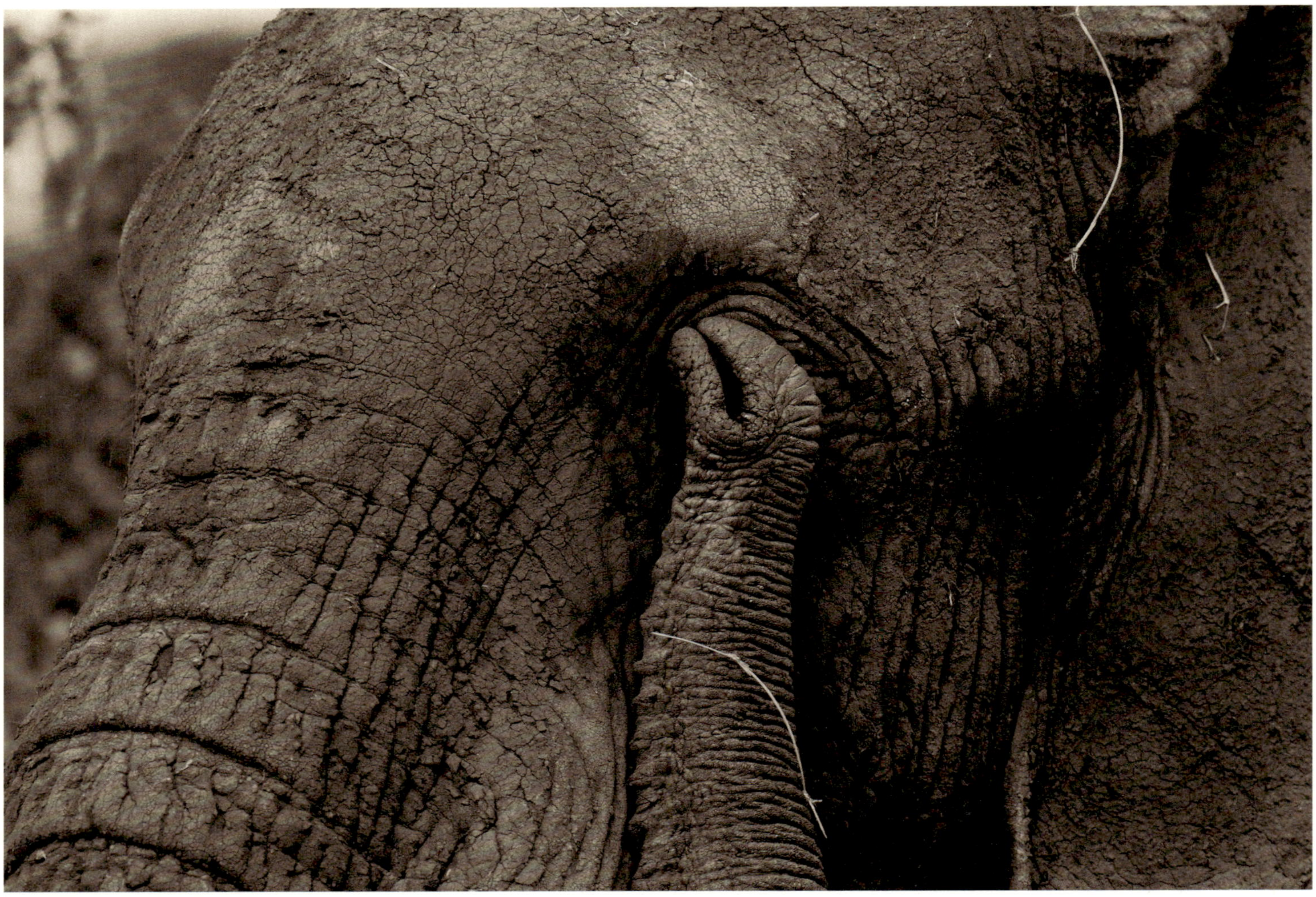

THE GNUS, OF COURSE, HAVE NO IDEA WHAT A DEBT THEY OWE TO ELEPHANTS. THE MIGRATION PASSES EACH YEAR, WELL-FED AND UNHINDERED, INDIFFERENT TO THE WORK OF THE ECOSYSTEM'S GREAT HABITAT MODIFIERS, WHO, TOGETHER WITH BUSH-RETARDING GRASS FIRES, KEEP THE GRASSLANDS OPEN, A PLACE FOR GNU FAMILIES TO SPEND THE YULETIDE SEASON.

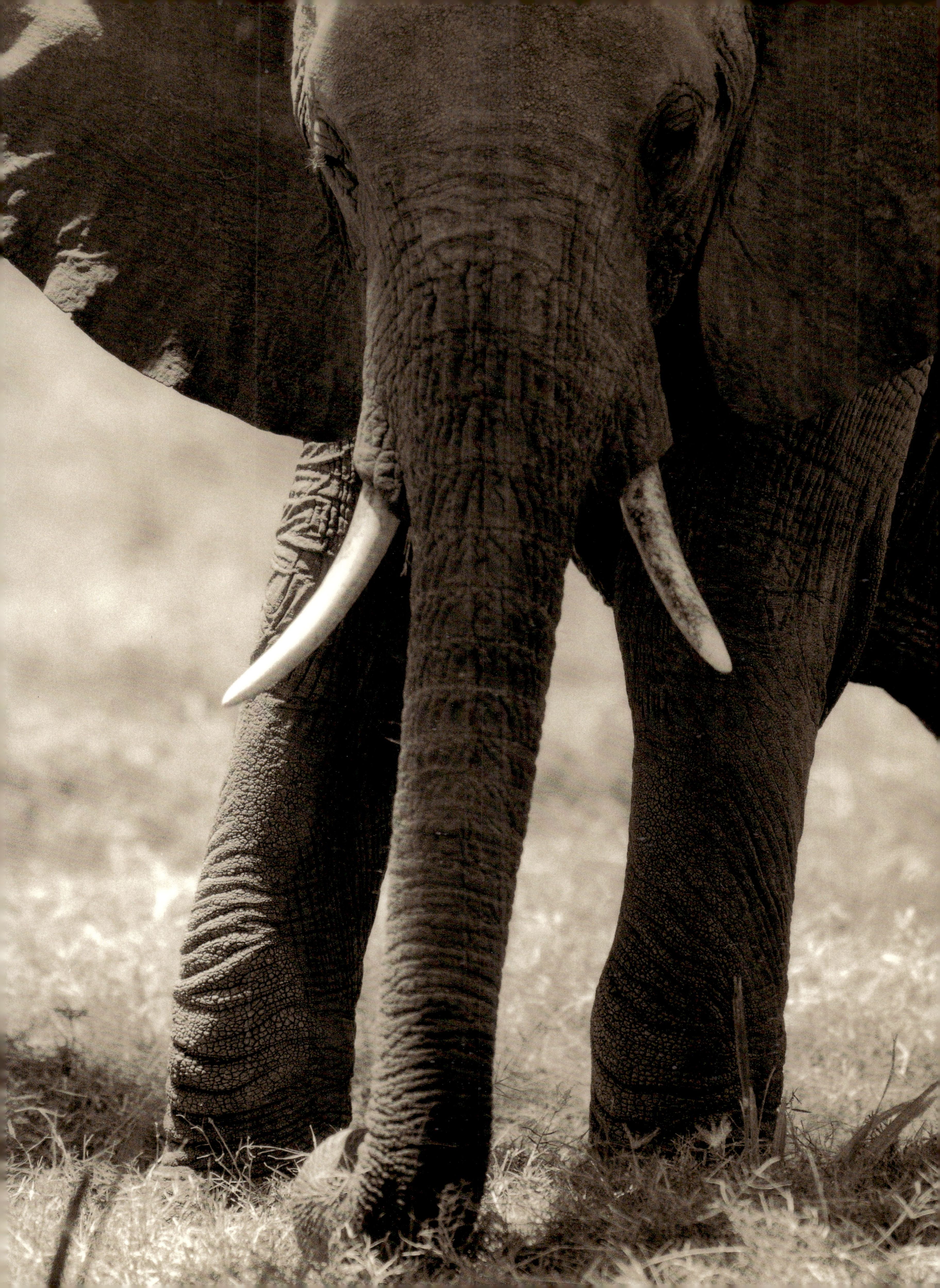

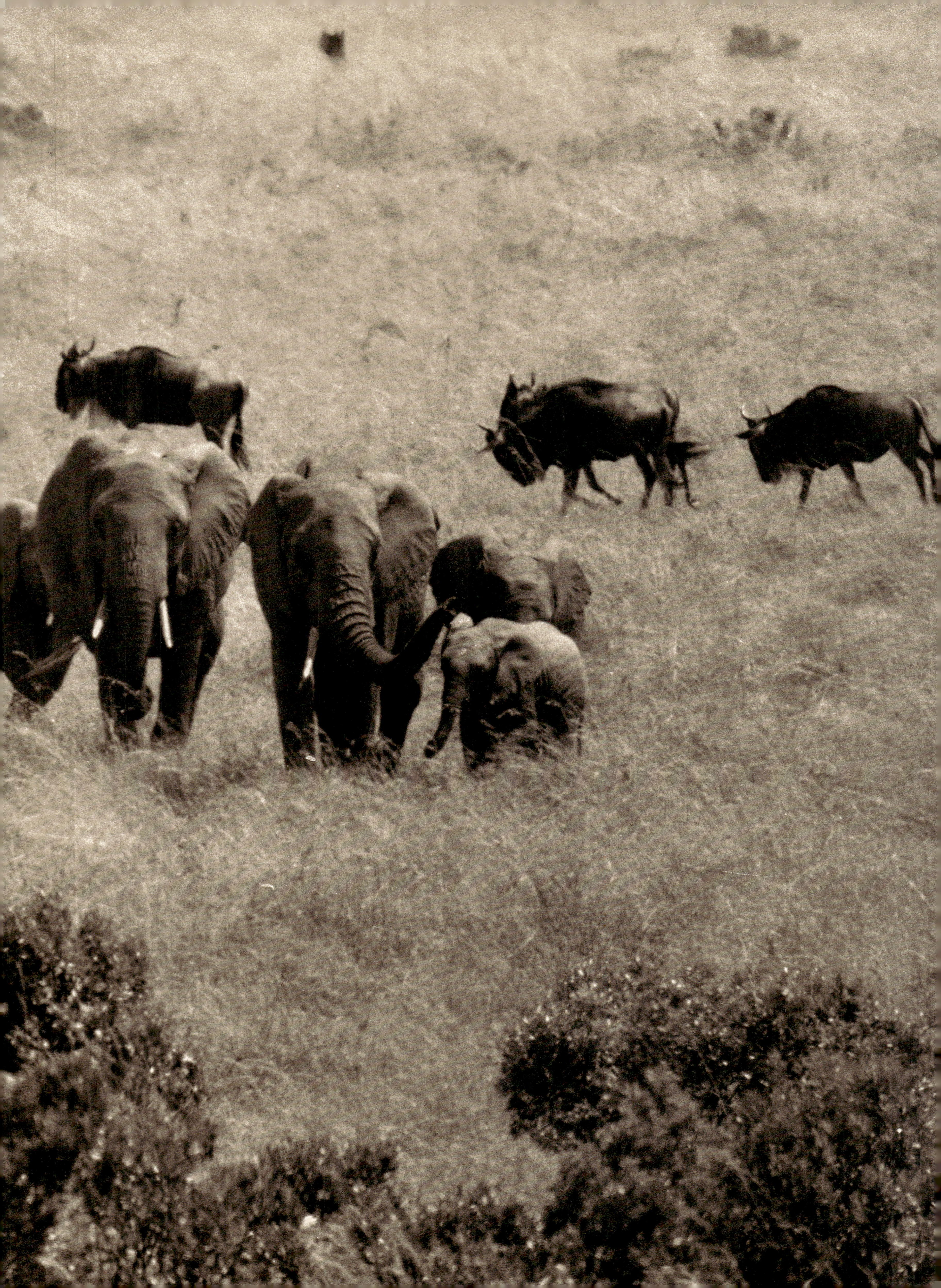

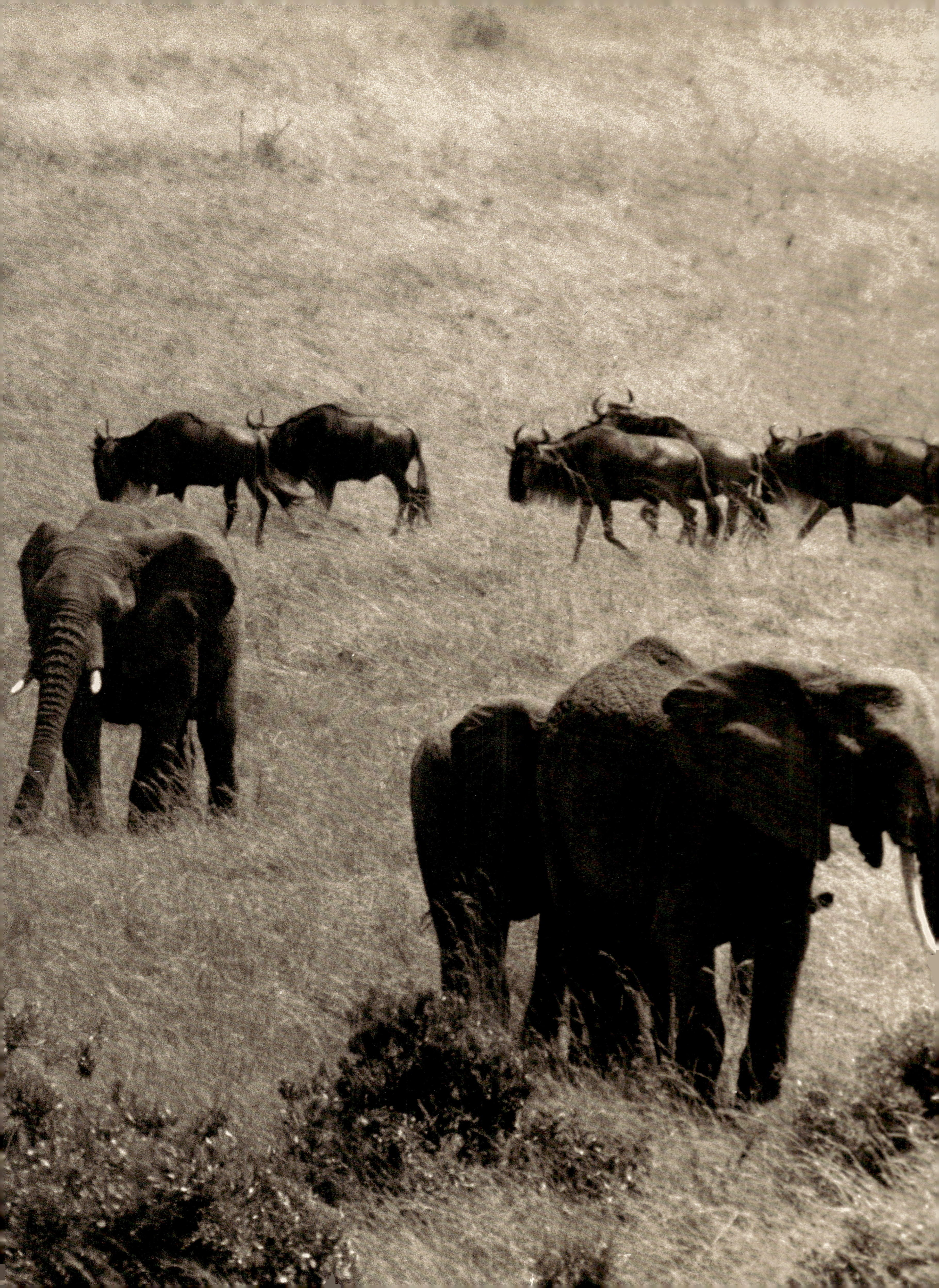

GNUS ARE FOOD TO NEARLY ALL PREDATORS, LARGE AND SMALL, AS WELL AS TO THE SCAVENGERS. THEY ALSO PROVIDE NOURISHMENT TO THIRD PARTIES SUCH AS THE RED-BILLED OXPECKER – OR TICK-BIRD – WHO HELPS KEEP THE PELAGE OF THE GNU, GIRAFFE, ZEBRA, RHINO AND BUFFALO FREE OF BLOOD-SUCKING PARASITES. THE PERMANENT PRESENCE OF PREDATORS ALONG THE MIGRATION ROUTE ENCOURAGES THE BEASTS TO KEEP ON TREKKING. YET EVEN THE LARGEST GUILD OF PREDATORS ON DRY LAND MAKES ONLY A TINY DENT IN THE MOST IMPRESSIVE OFFERING OF HERBIVORE FLESH ON EARTH.

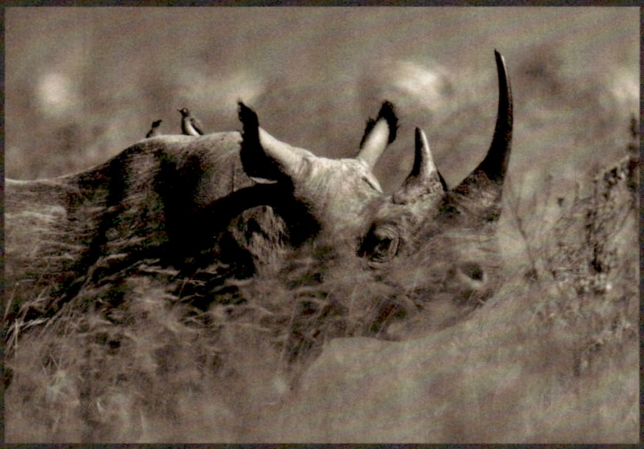

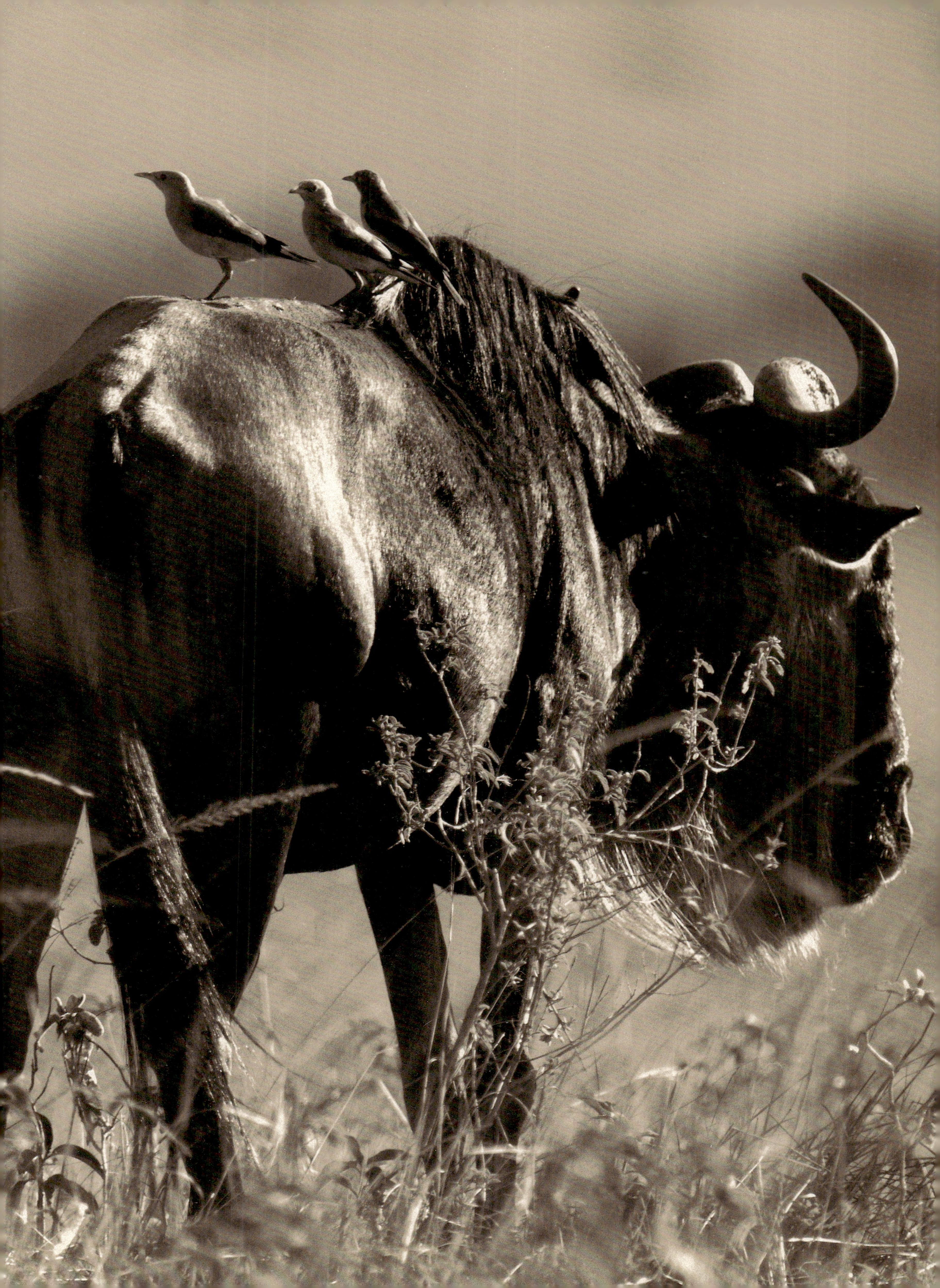

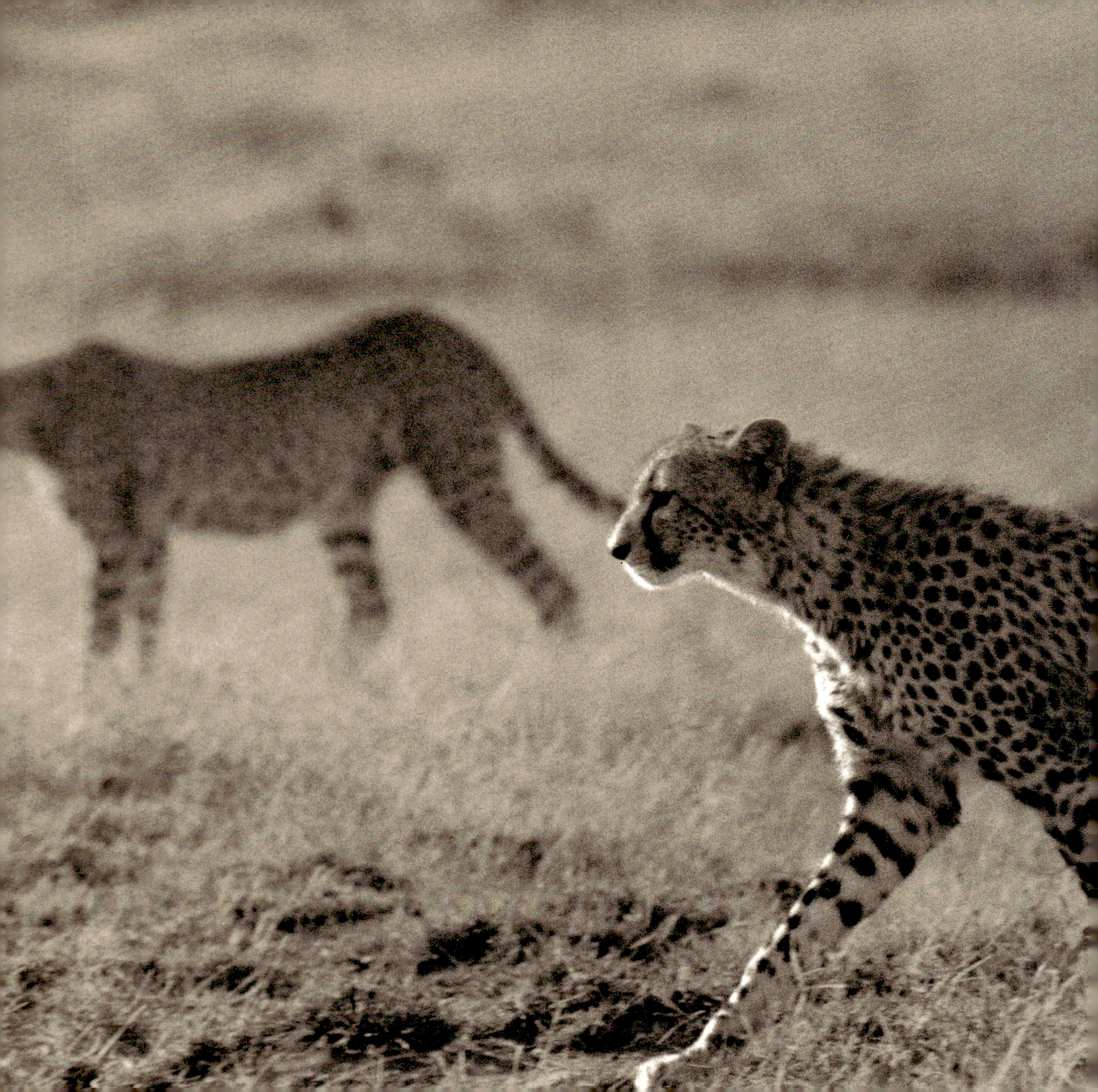

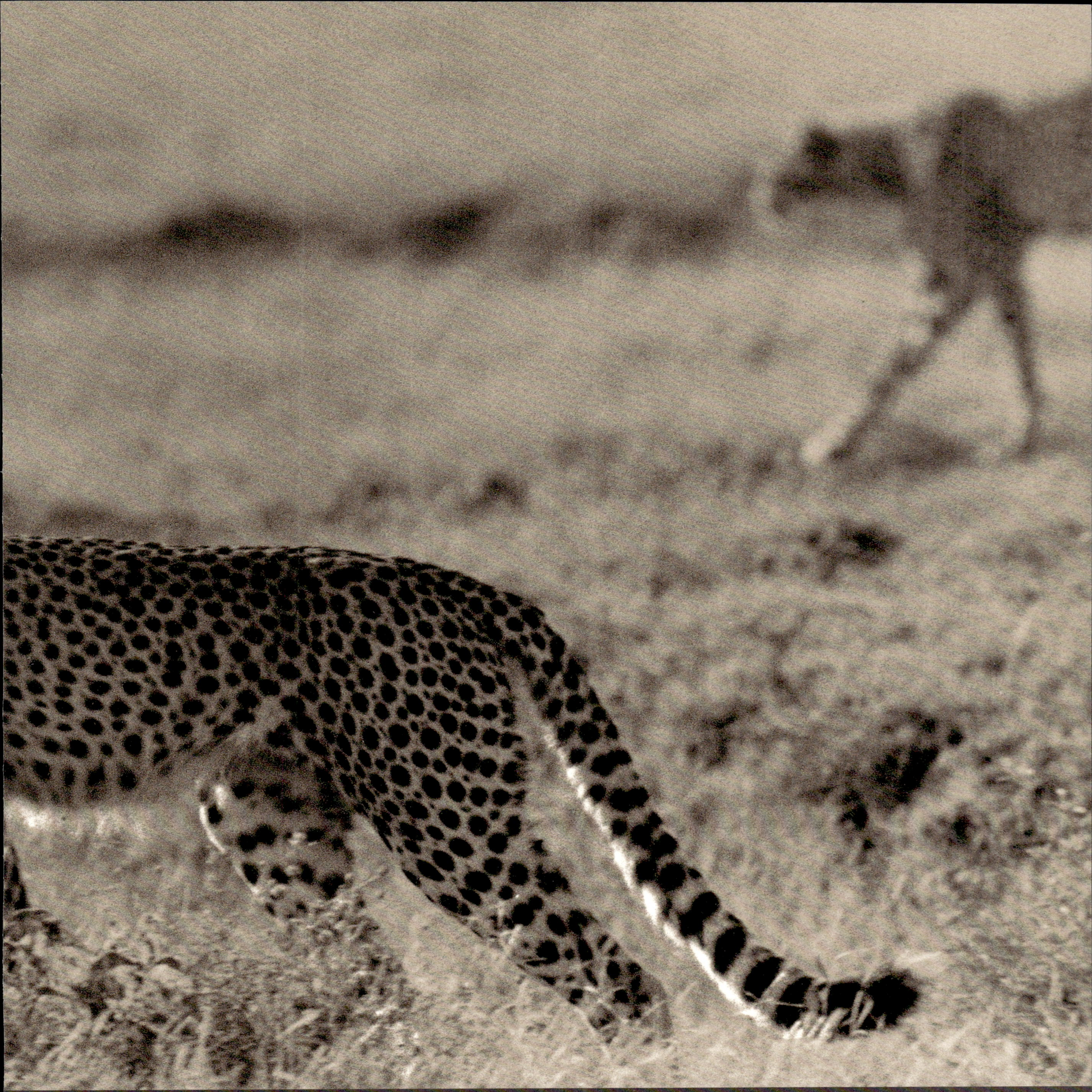

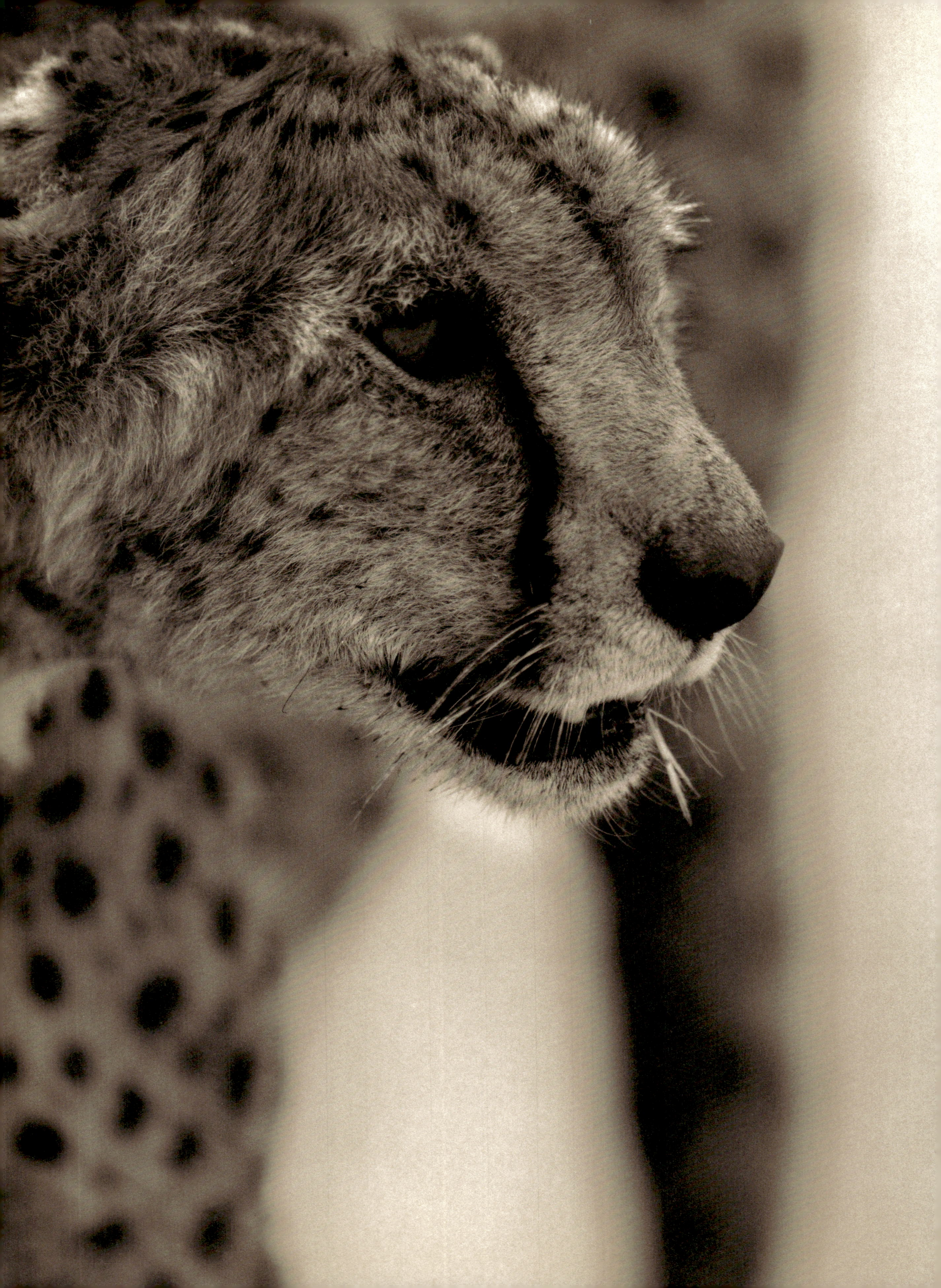

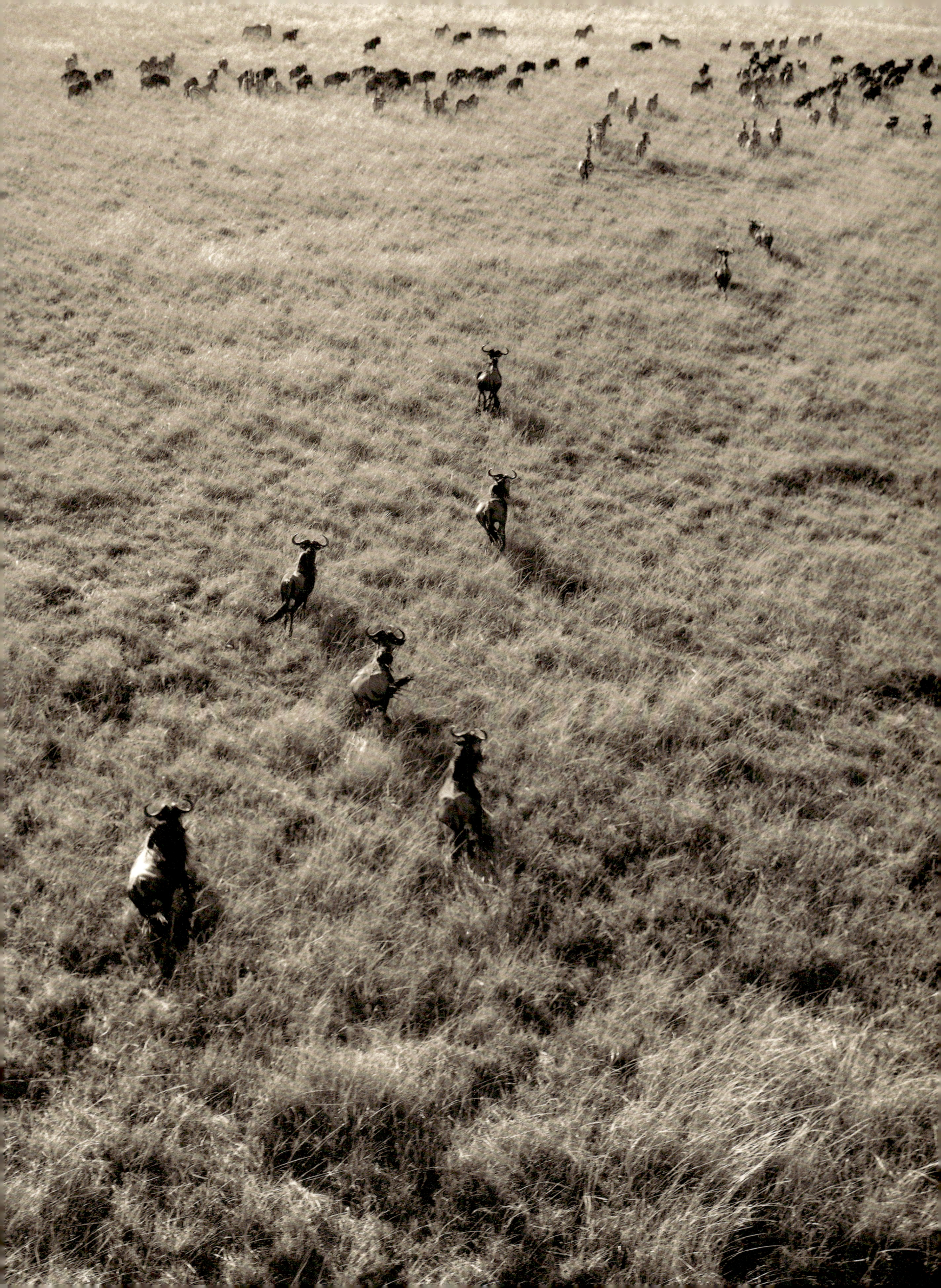

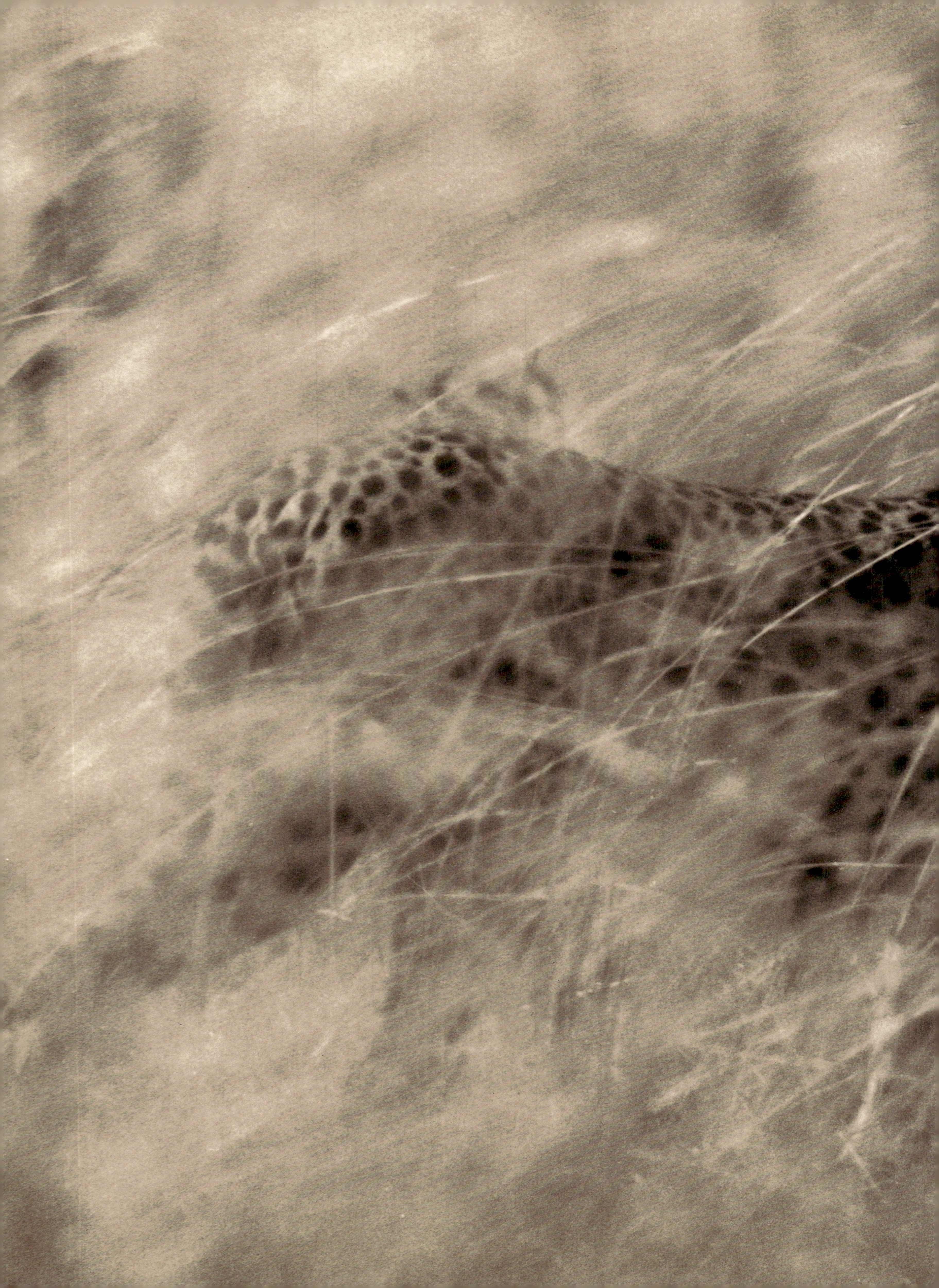

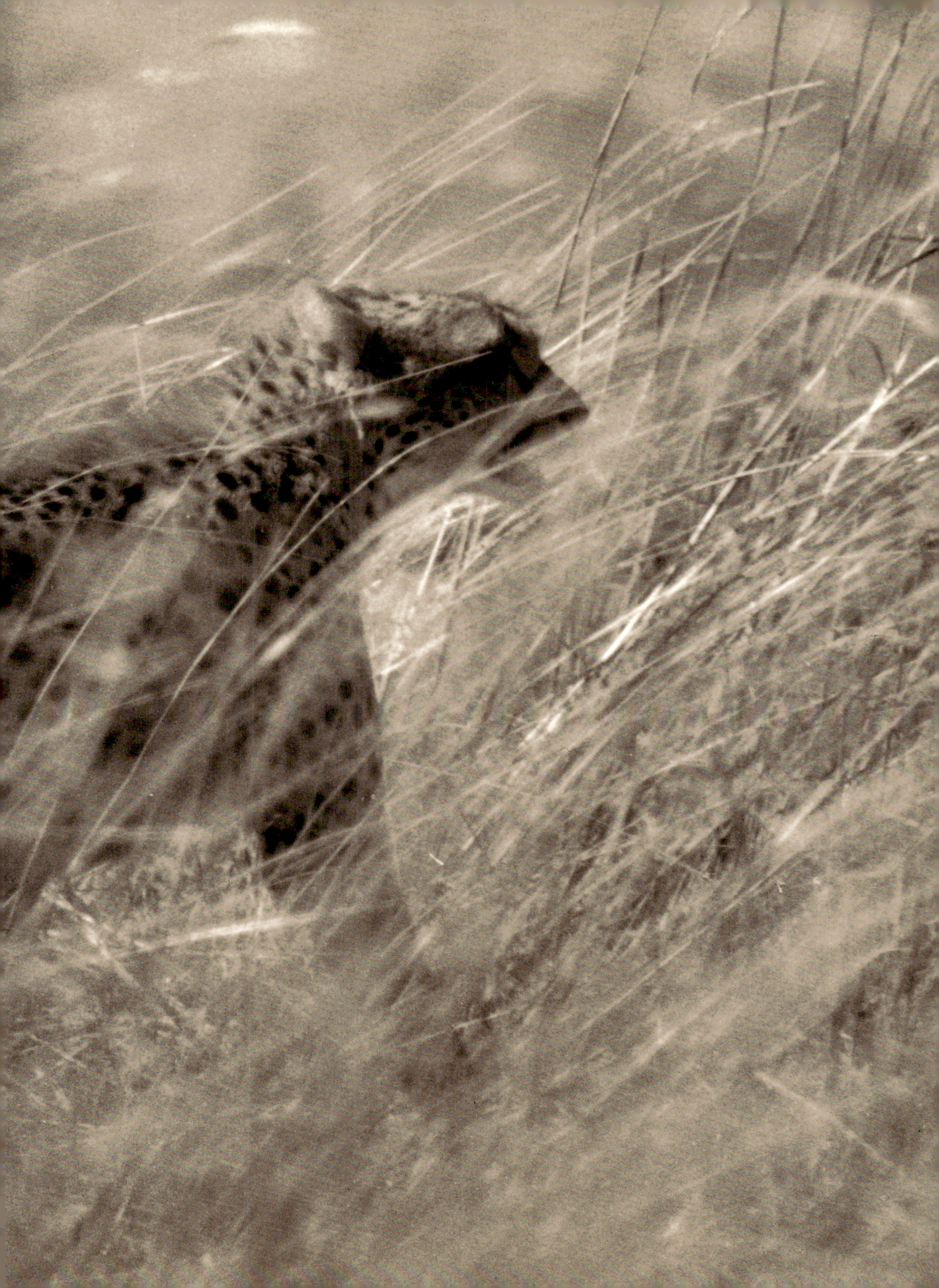

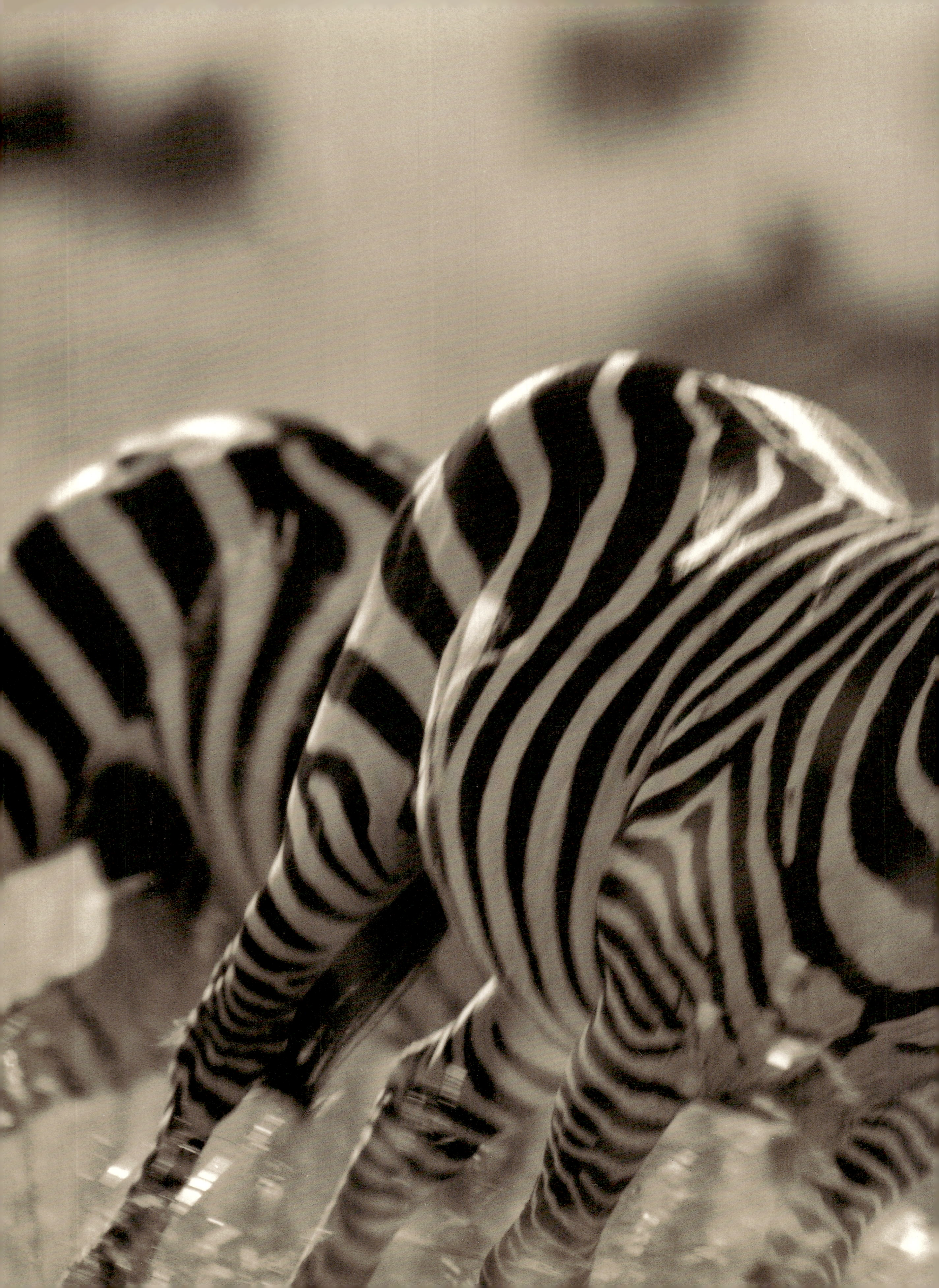

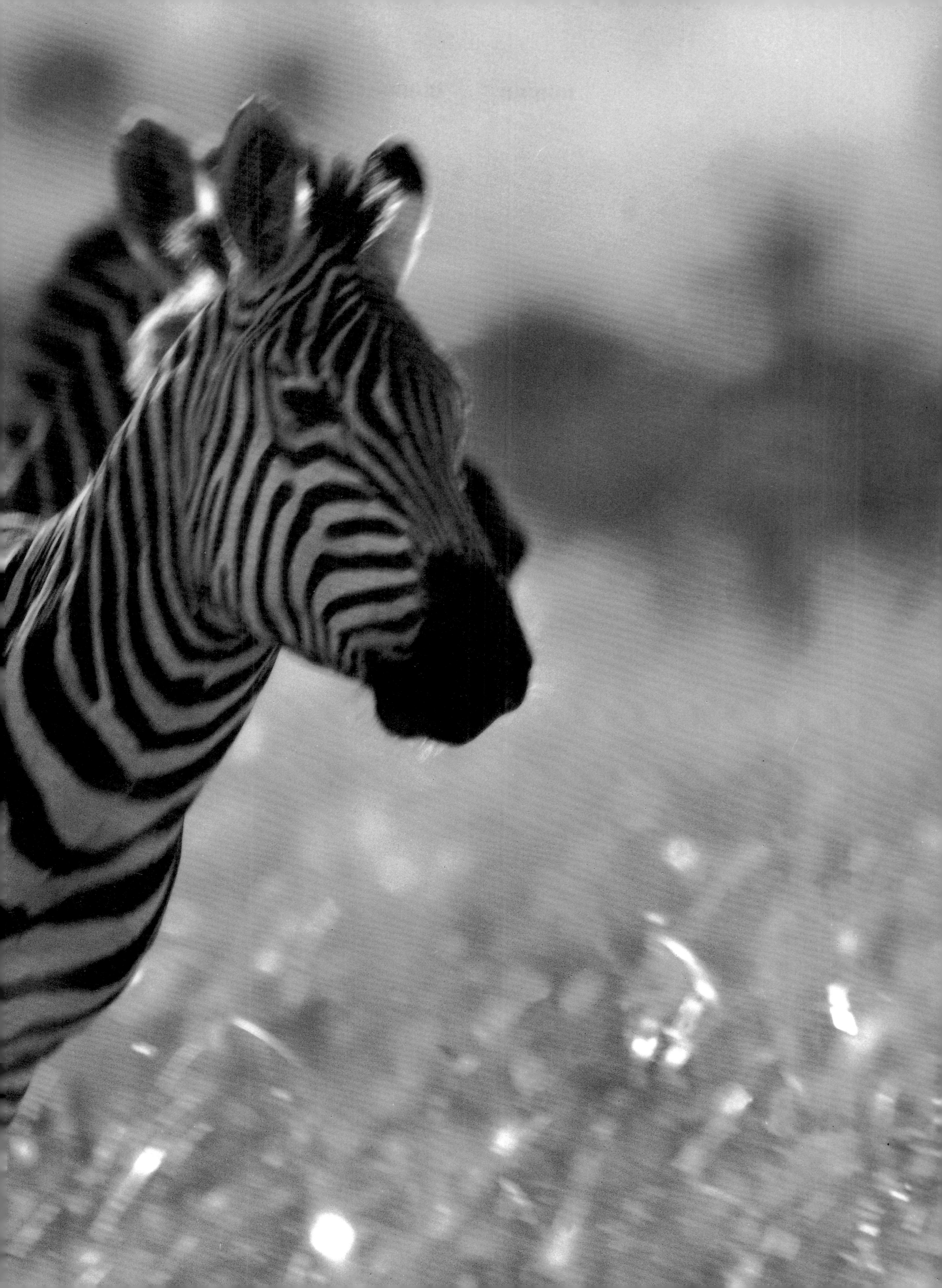

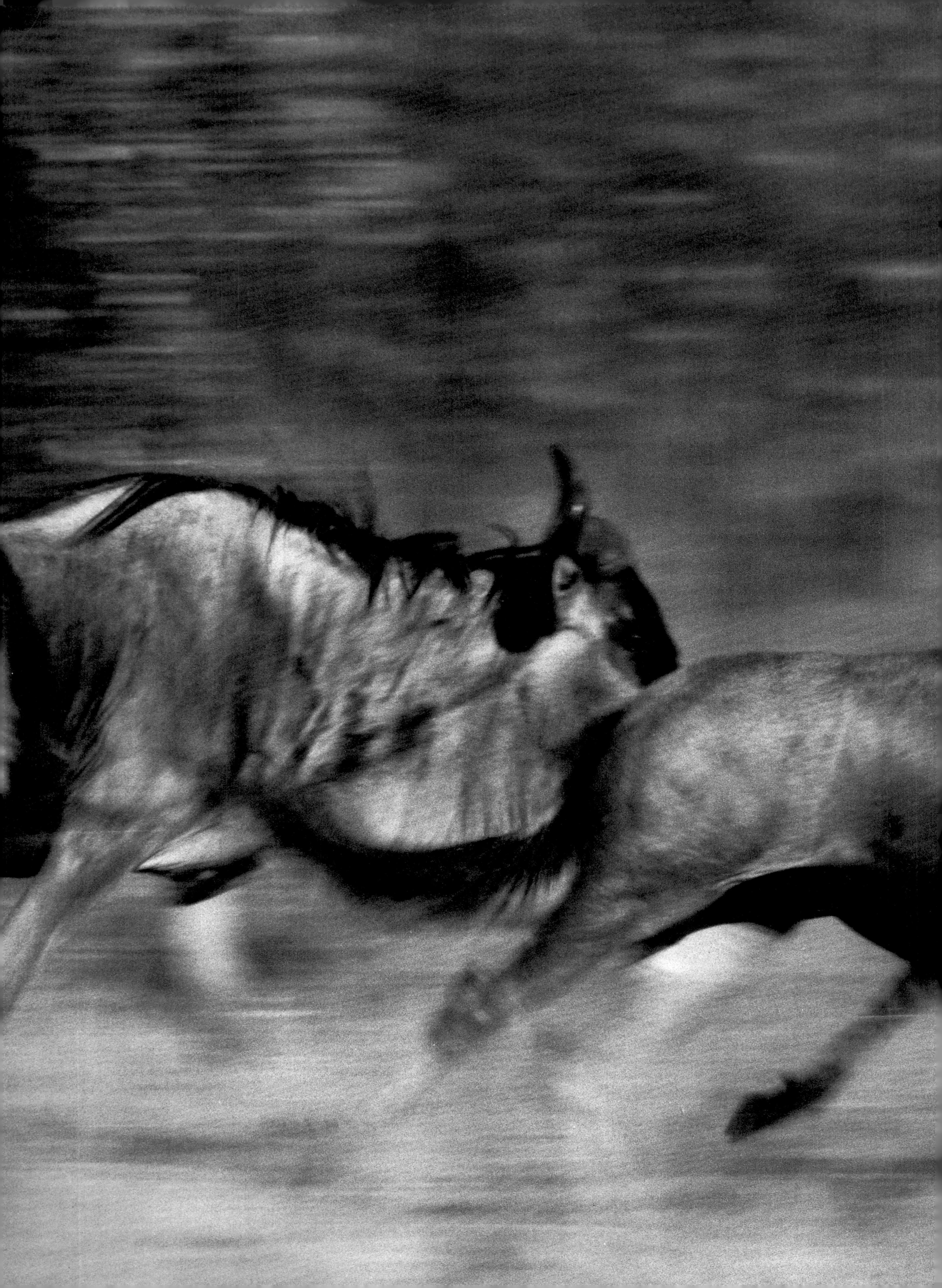

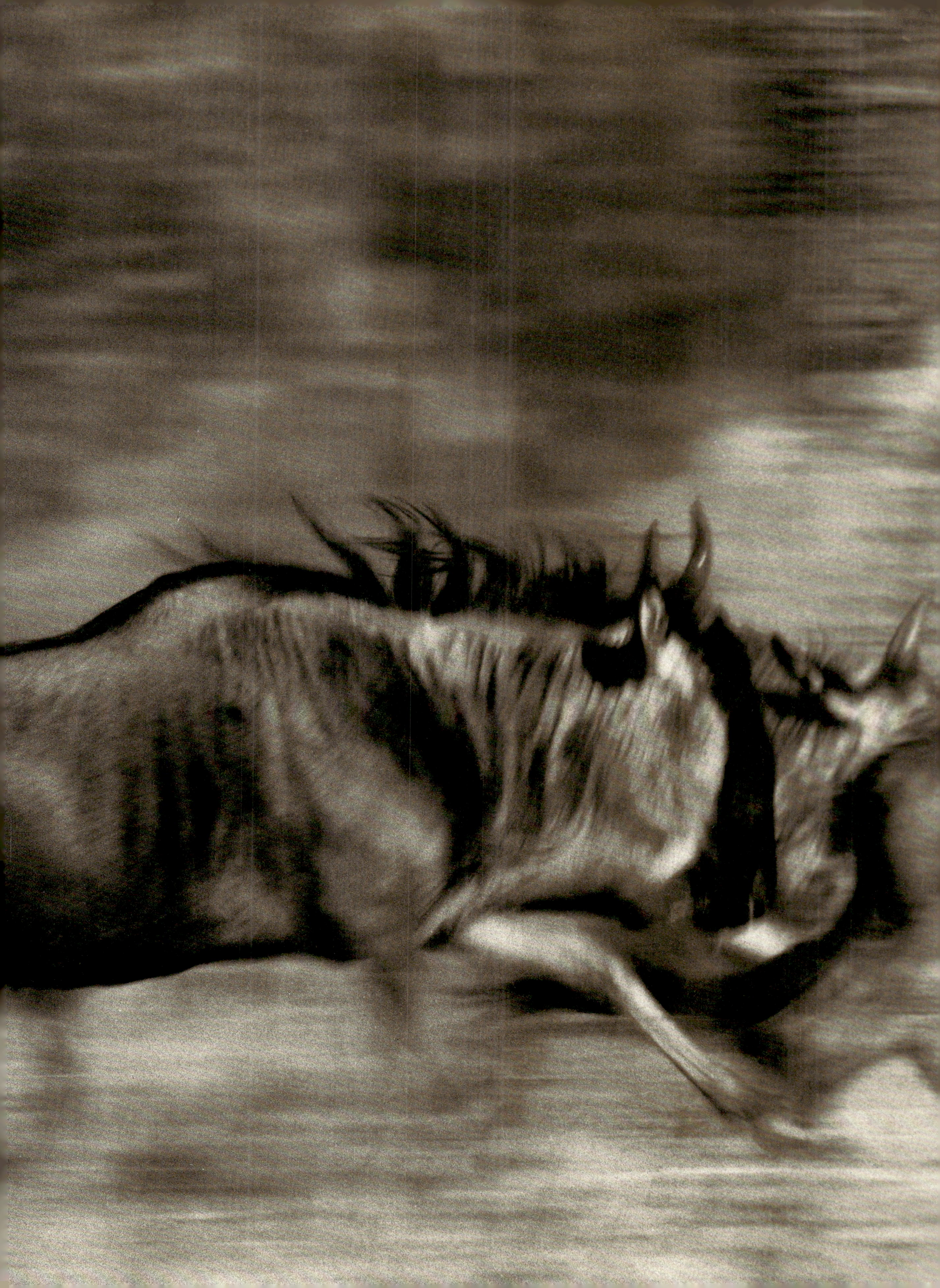

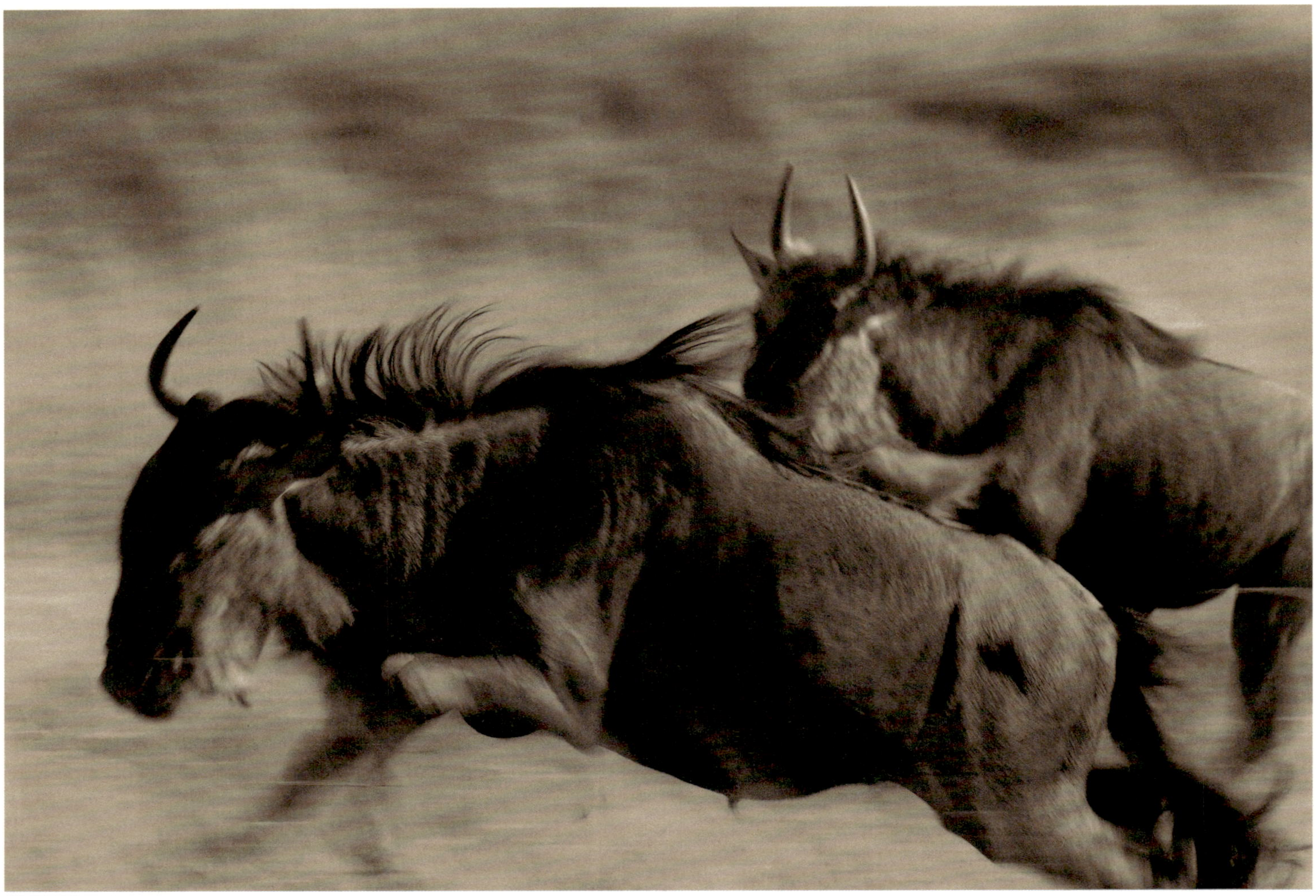

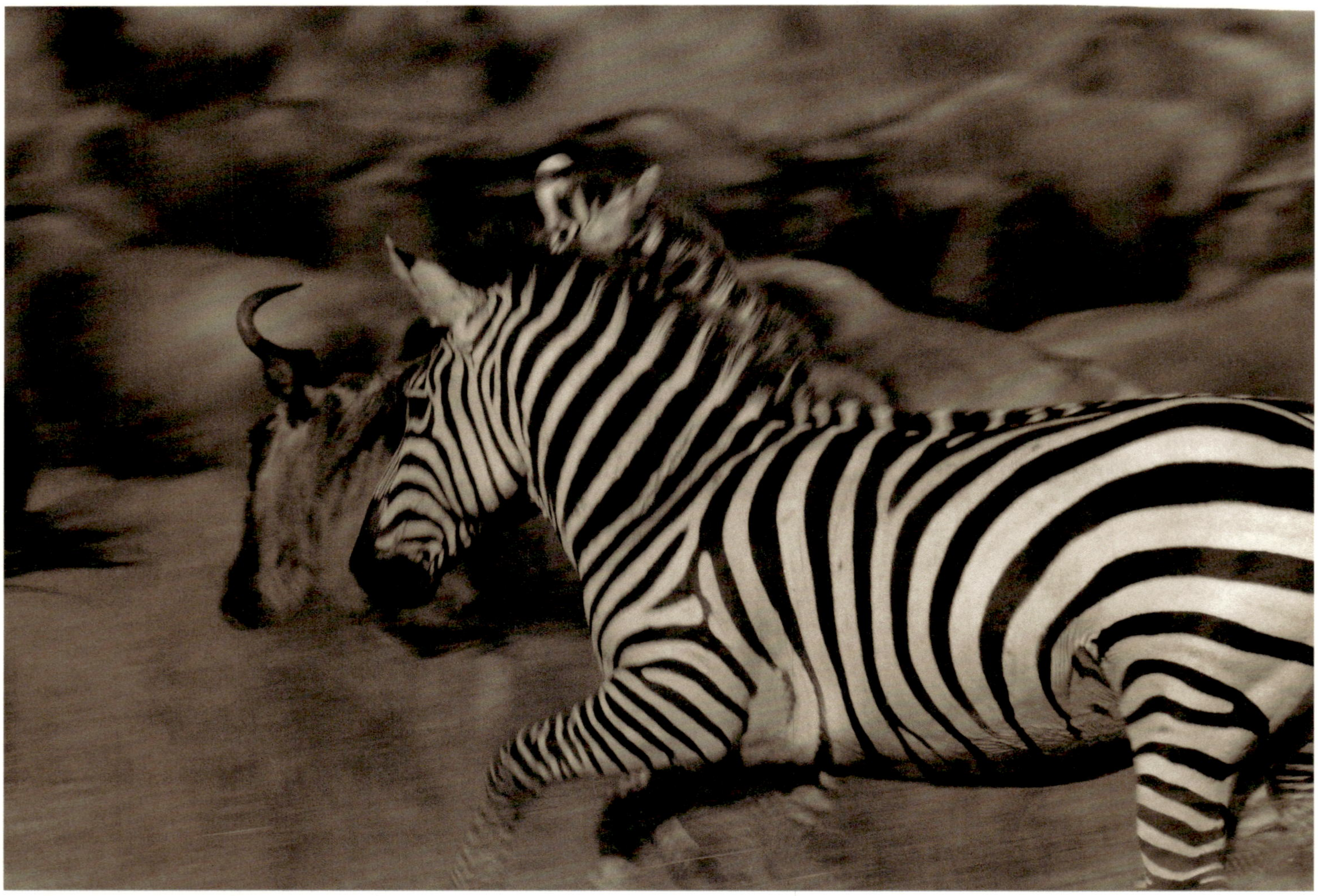

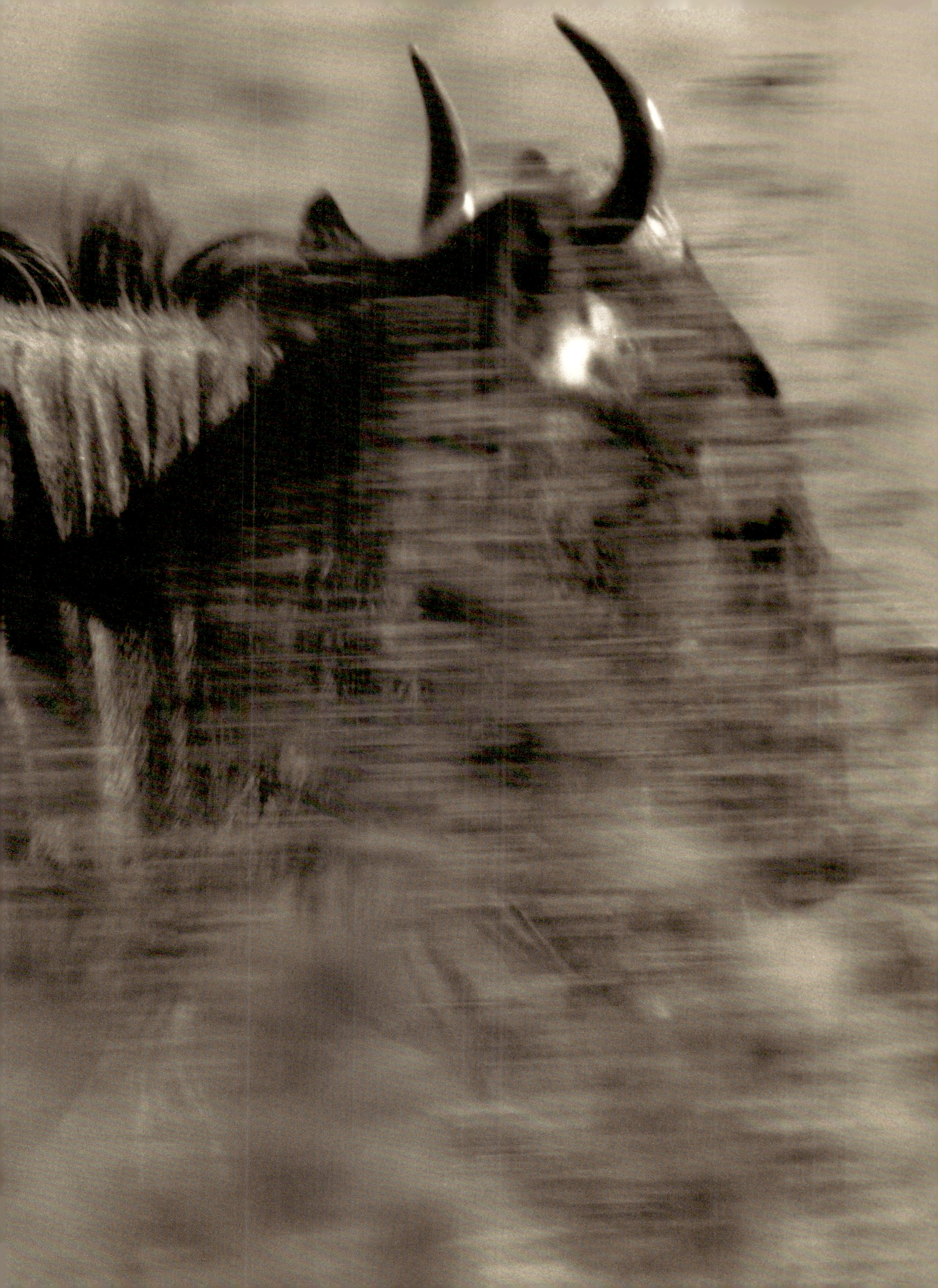

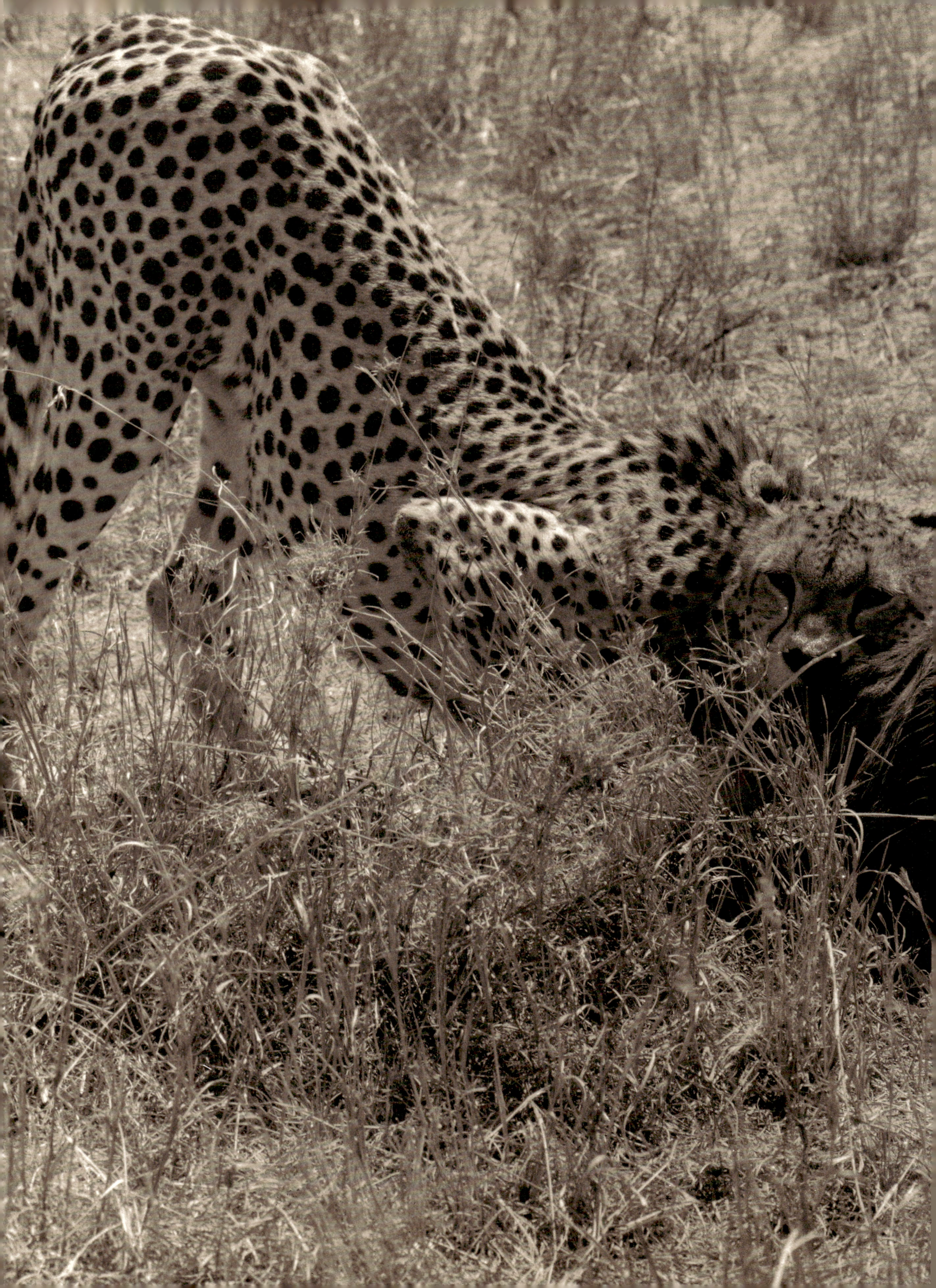

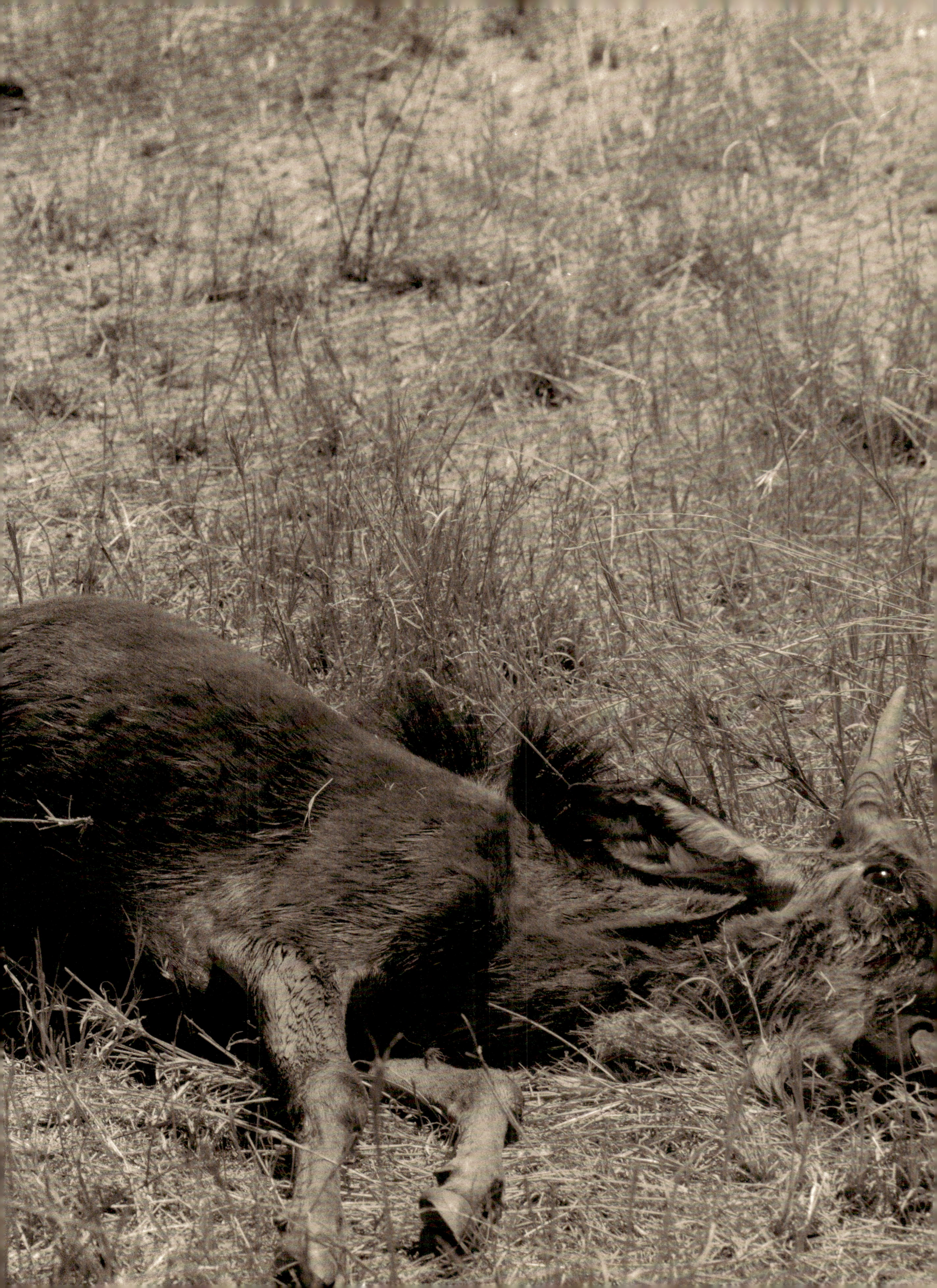

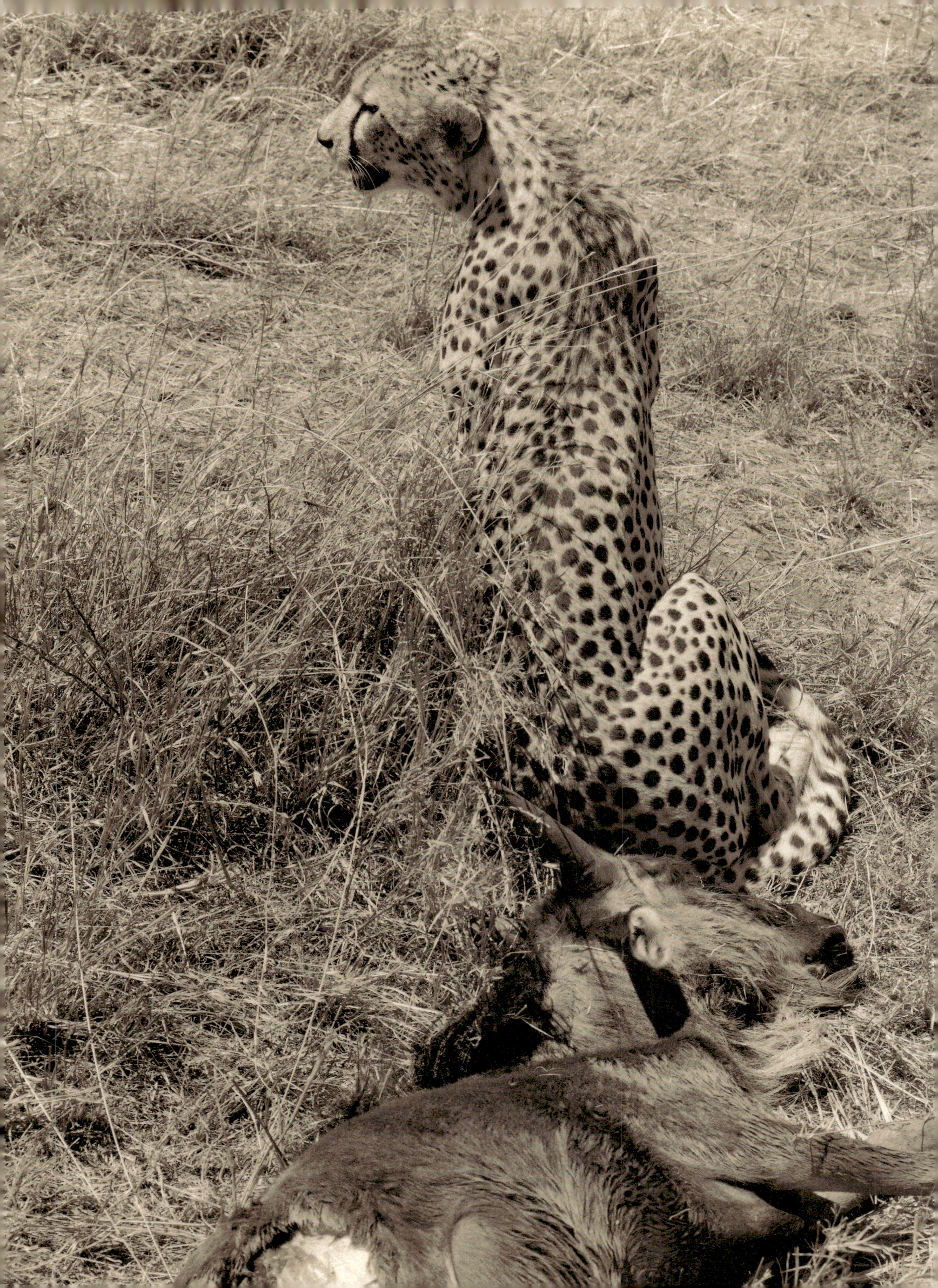

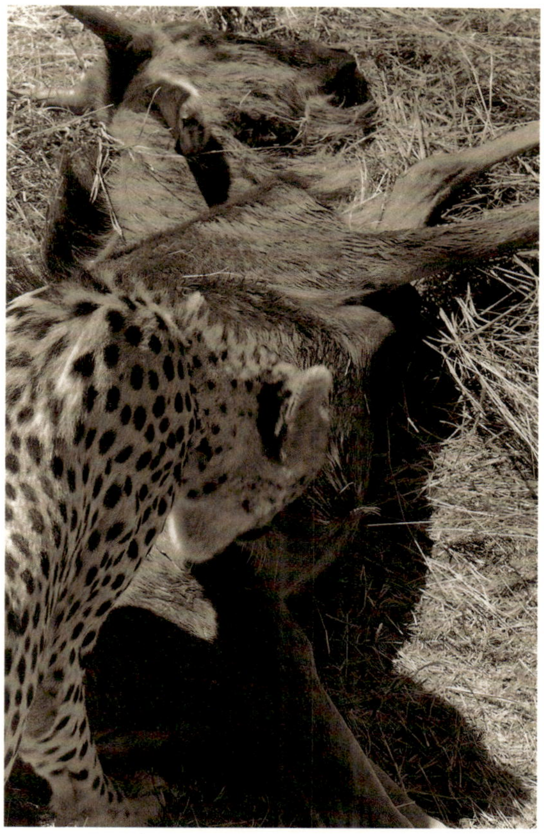

AN AGILE ATHLETE LIKE THE CHEETAH REQUIRES LIGHT BUT NOURISHING FARE, SUCH AS RUMP STEAKS OF THOMSON'S GAZELLE AND FILET OF WILDEBEEST FAWN. CONVENTIONAL WISDOM IS CORRECT ABOUT CHEETAHS BEING THE FASTEST LAND MAMMALS: UP TO SIXTY-EIGHT MILES PER HOUR IN FULL SPRINT AFTER A FLEEING GAZELLE. BUT CONVENTIONAL WISDOM IS WRONG ABOUT THEM WEEDING OUT THE SICK AND WEAK. LIKE A FINICKY SHOPPER, THEY SELECT ONLY THE BEST.

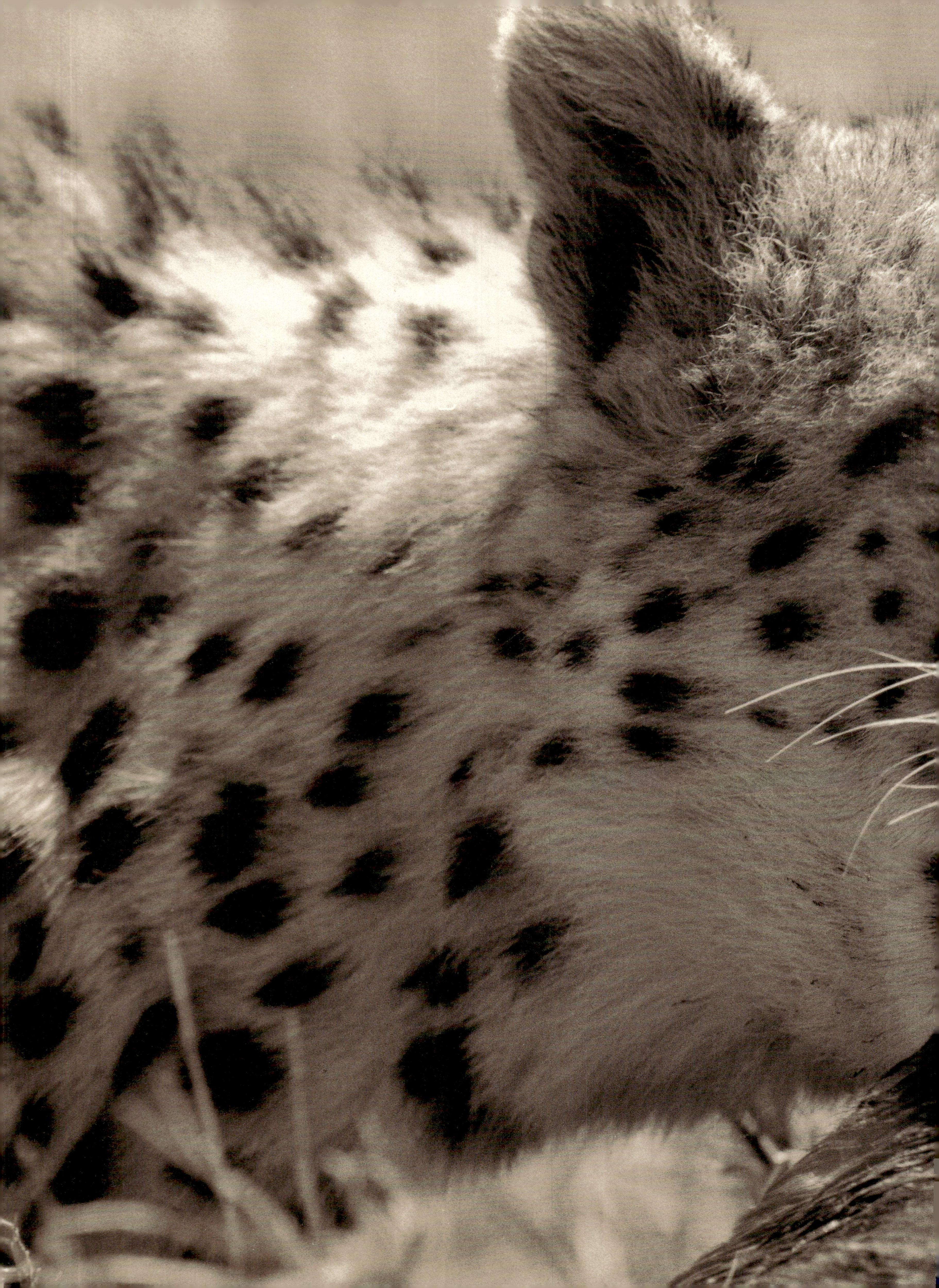

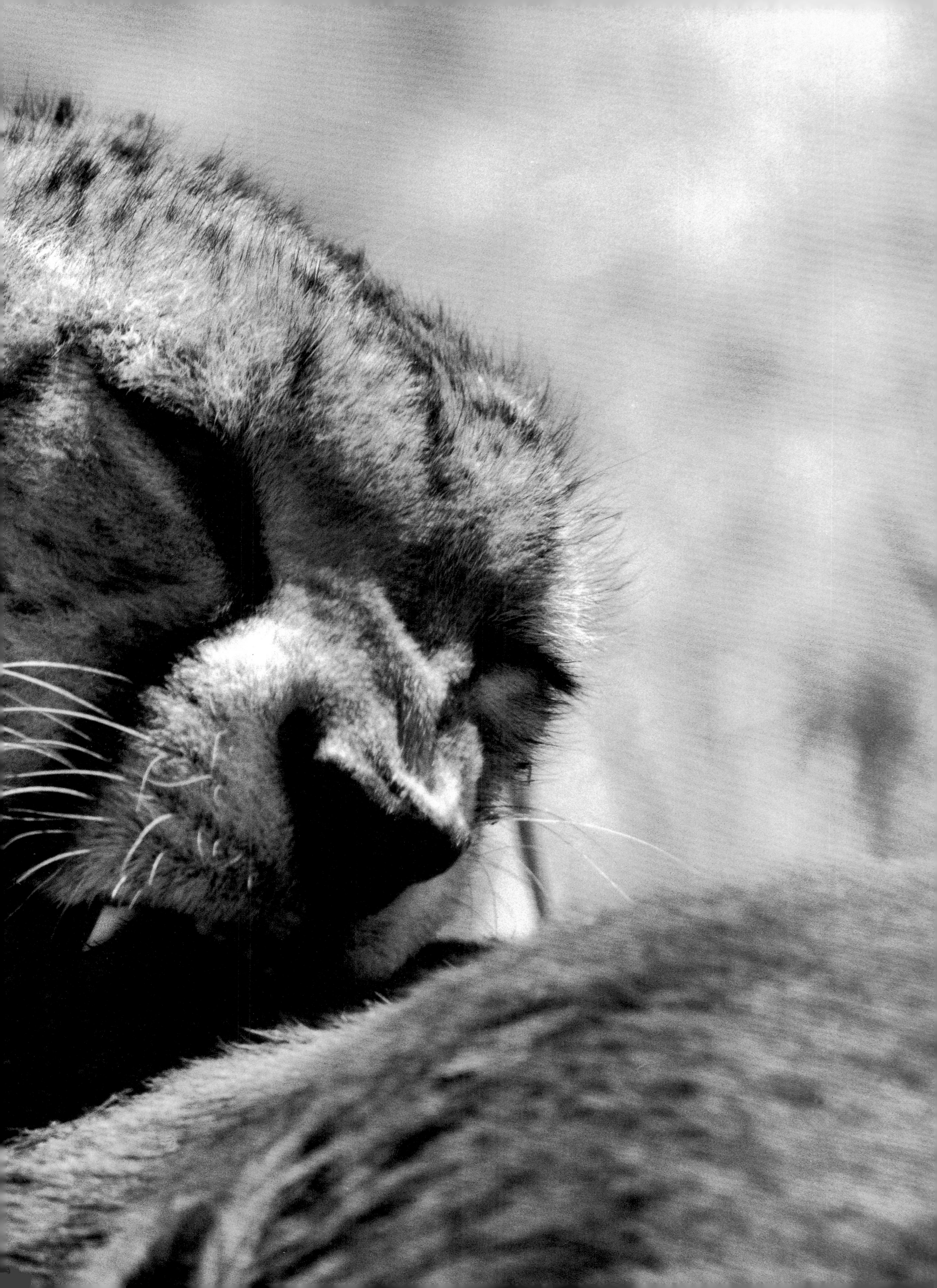

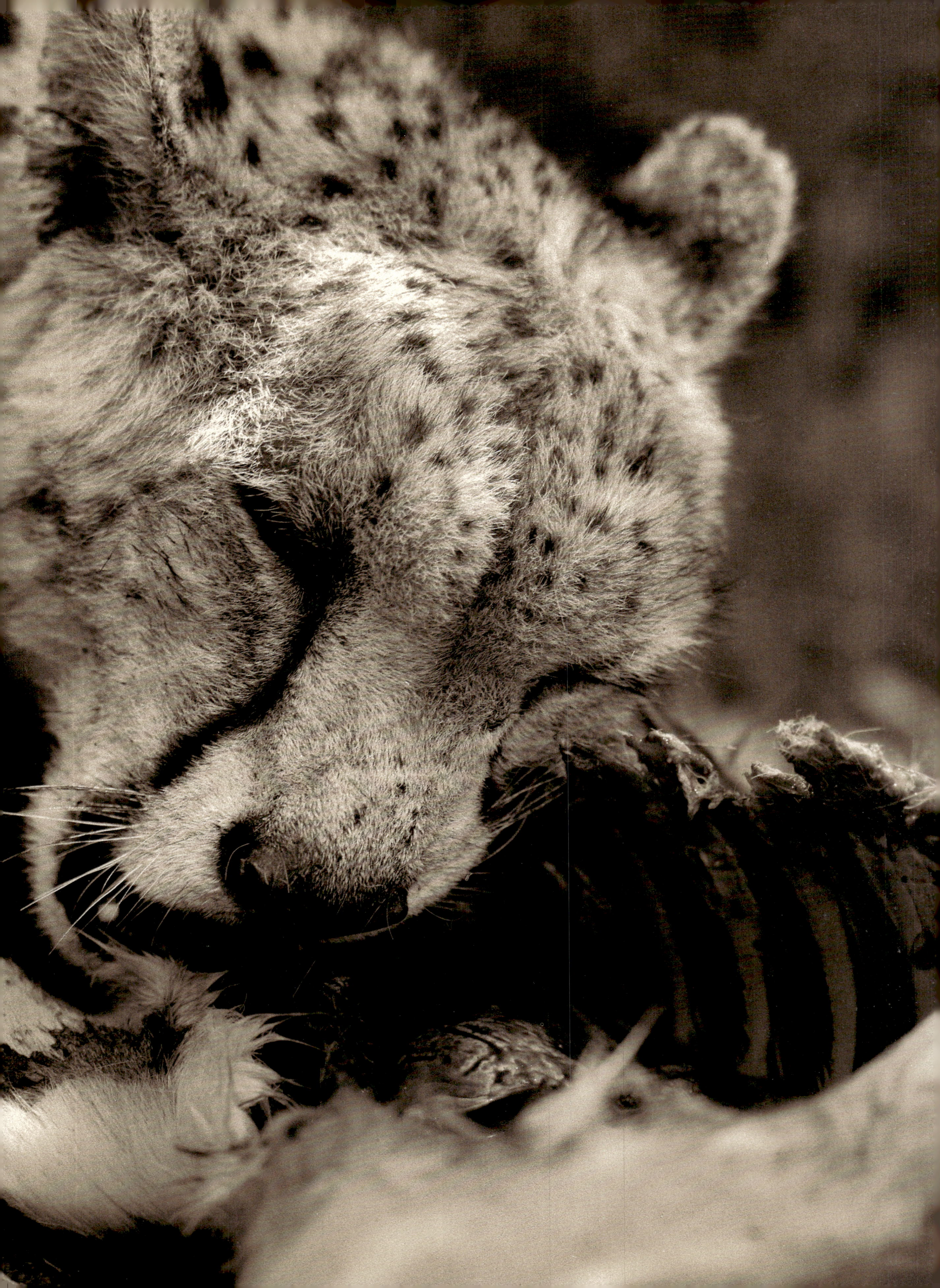

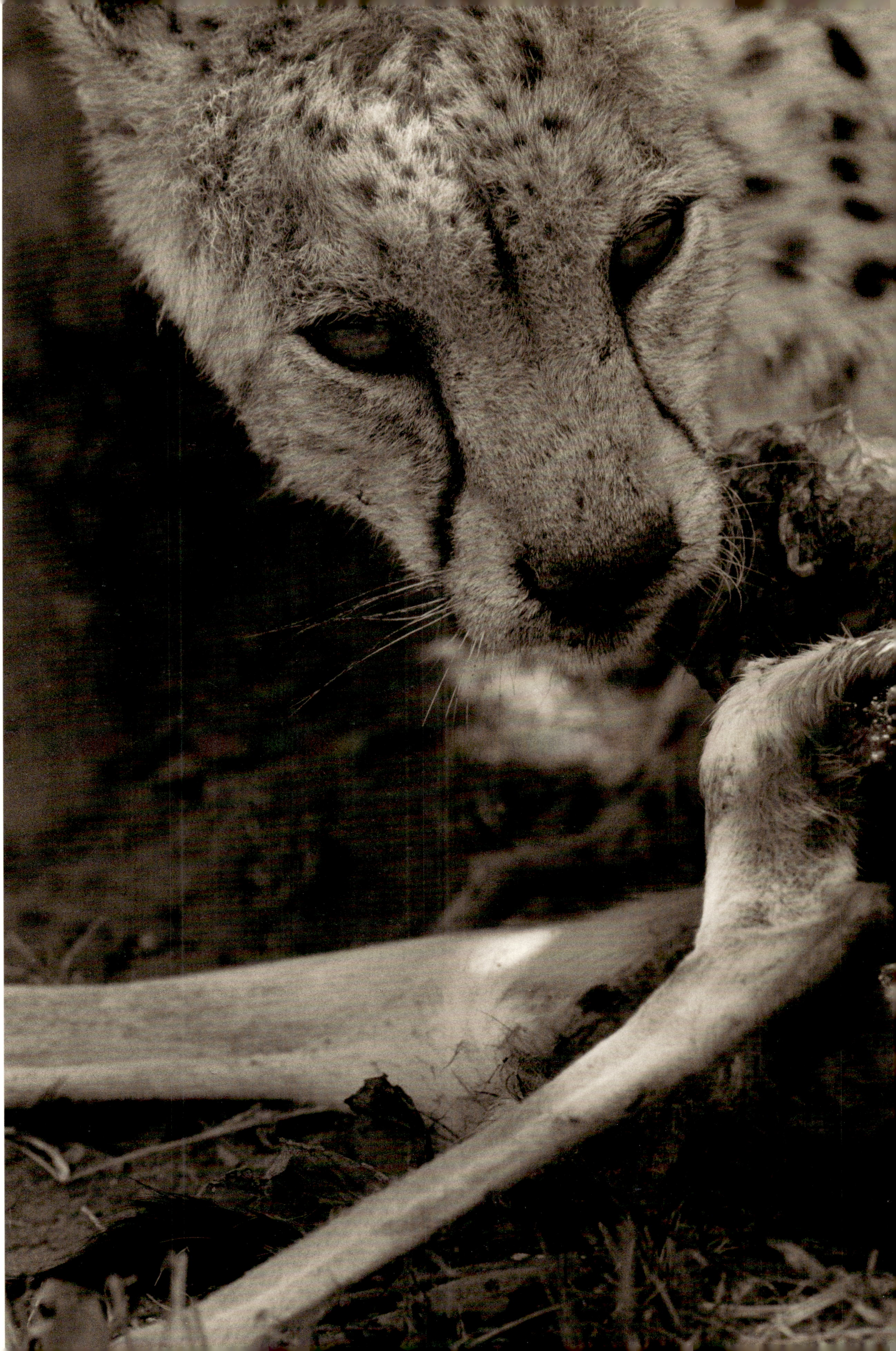

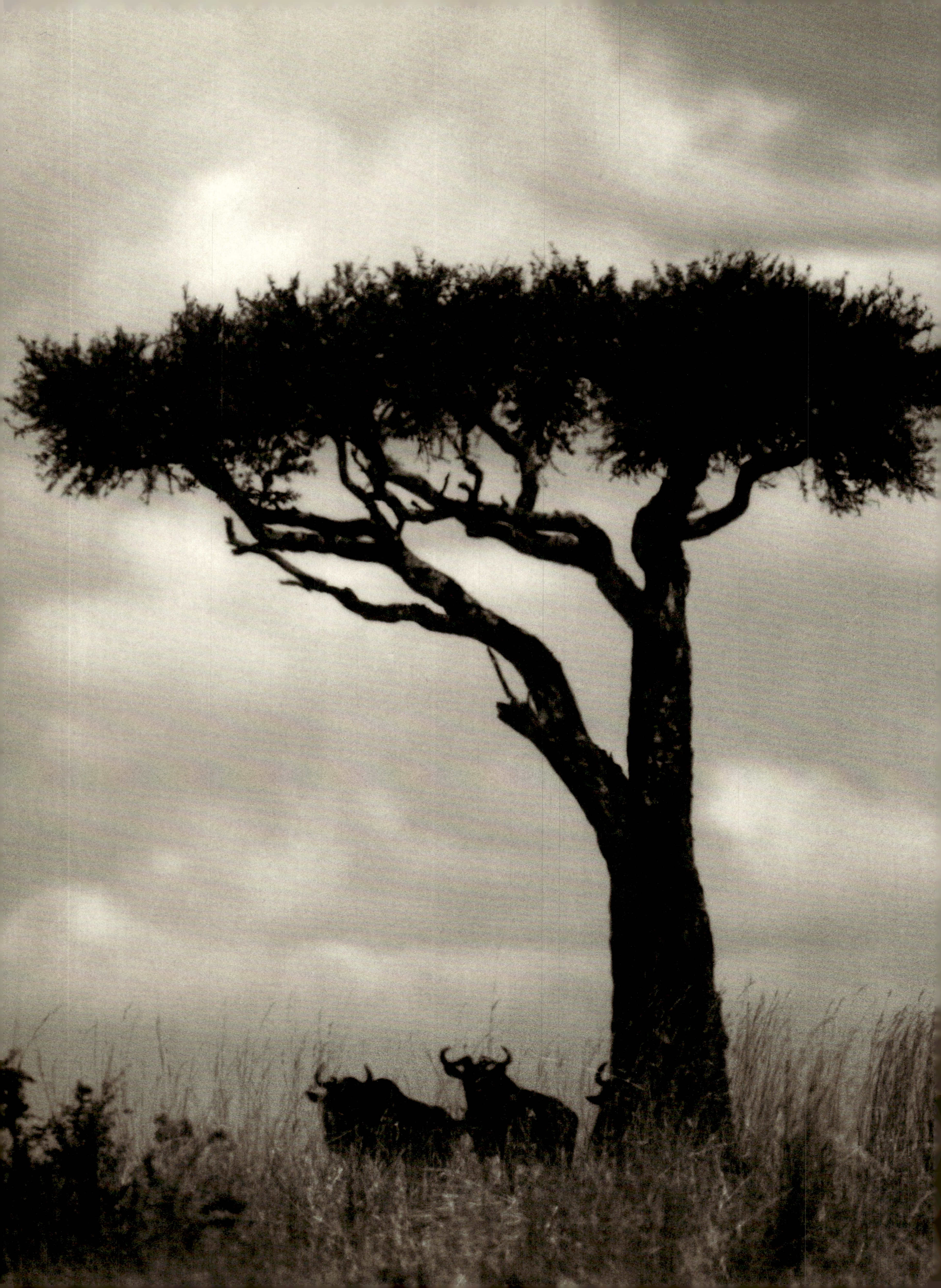

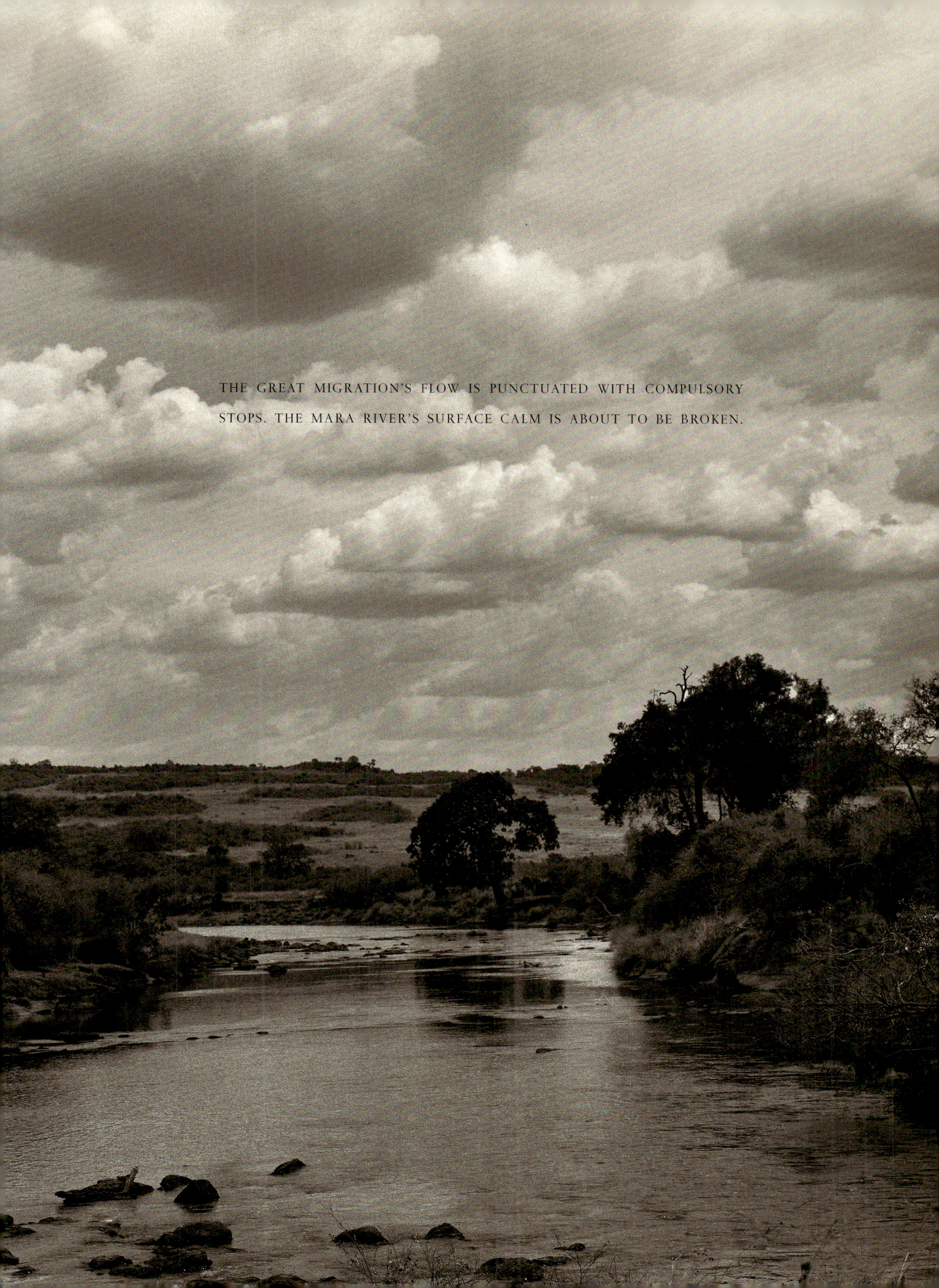

THE GREAT MIGRATION'S FLOW IS PUNCTUATED WITH COMPULSORY STOPS. THE MARA RIVER'S SURFACE CALM IS ABOUT TO BE BROKEN.

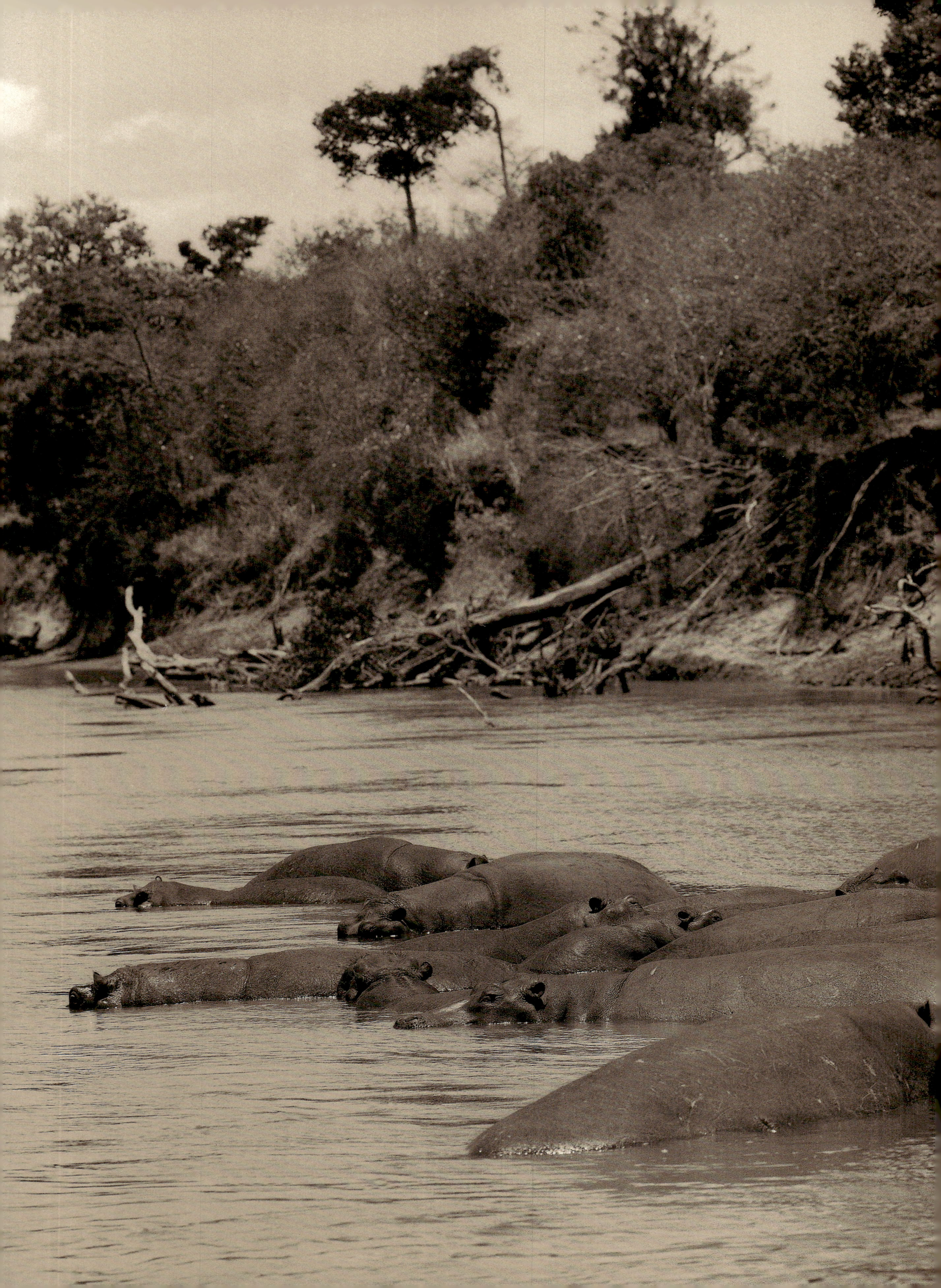

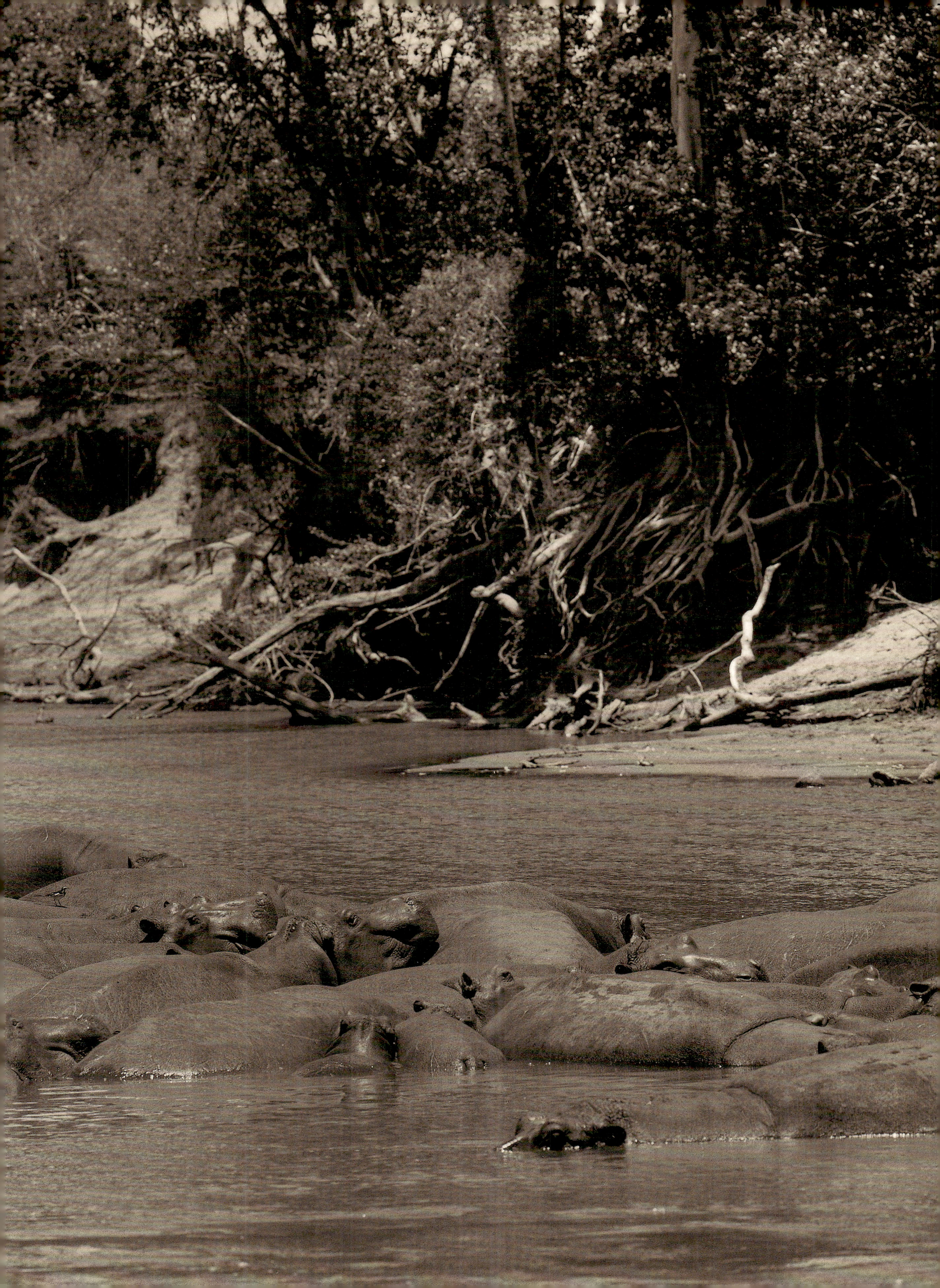

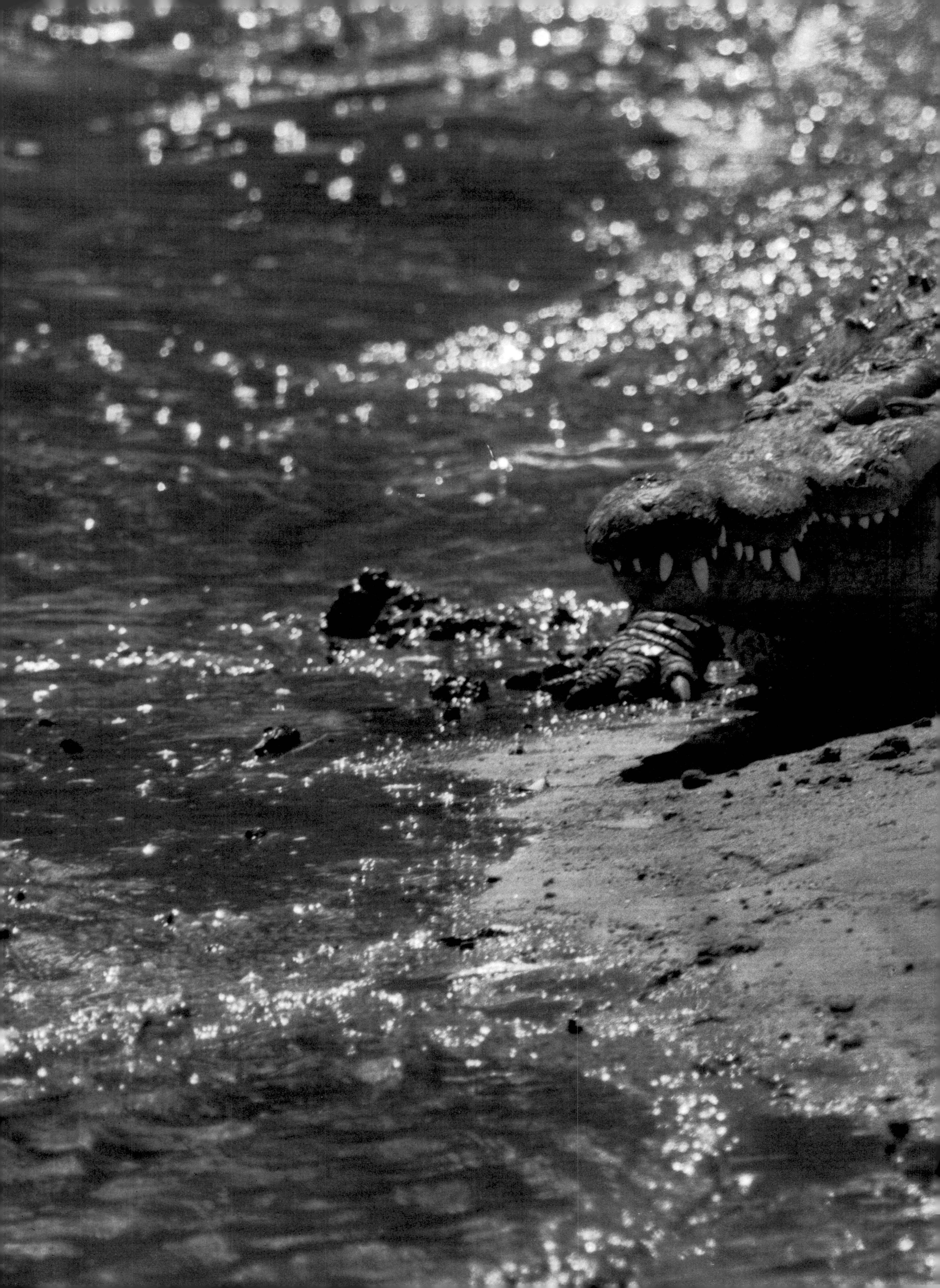

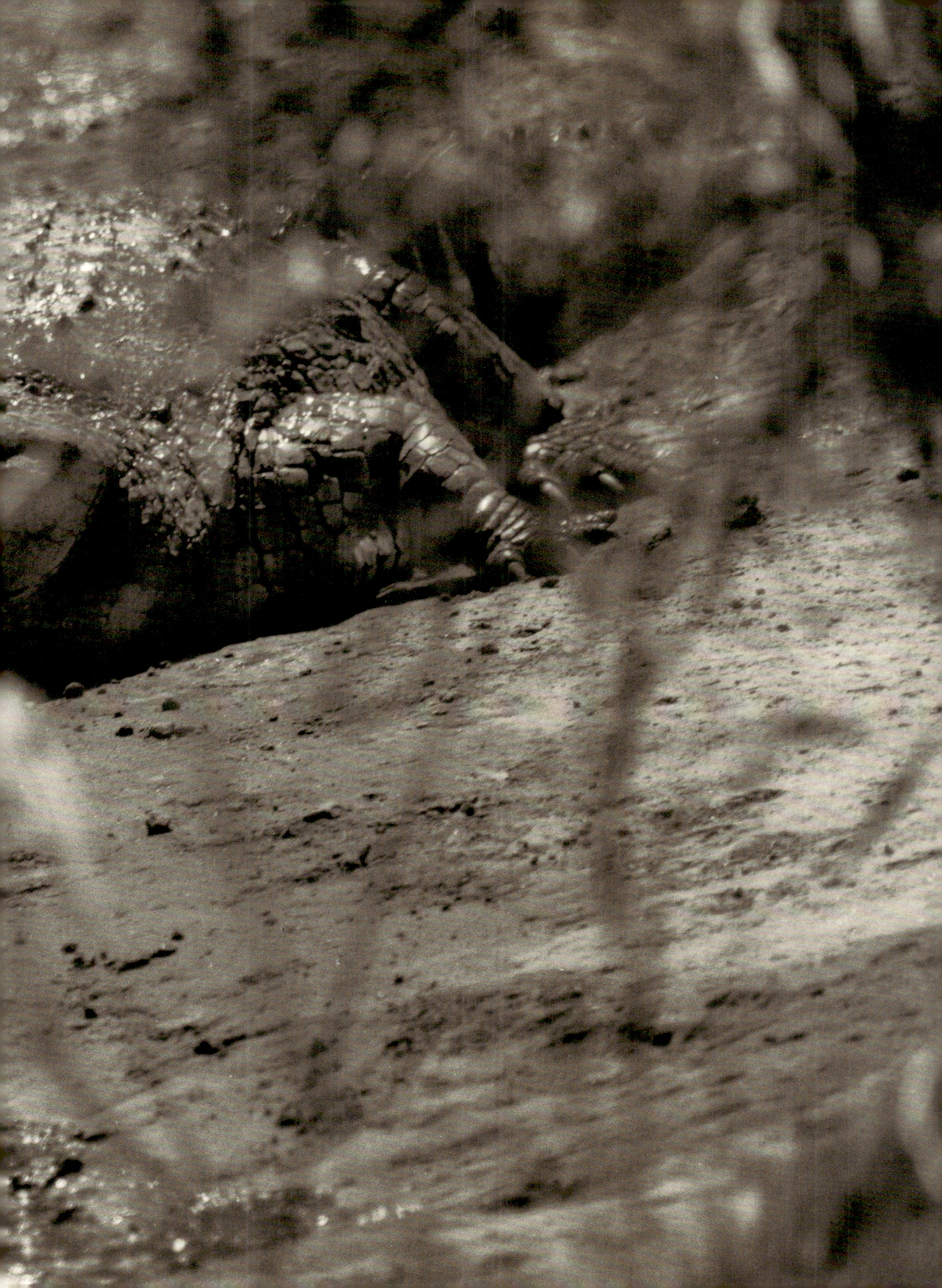

IN AN ECOSYSTEM WHERE MOST OF THE RIVERS FLOW WEST, THERE WILL INVARIABLY BE MANY CROSSINGS FOR A BEAST THAT WANTS TO GO NORTH. WATER IS A FUNNY THING: IT GIVES LIFE AND ATTRACTS, BUT IT THREATENS AND THUS REPELS. THE GNUS' ARRIVAL AT THE MARA IS ACCOMPANIED BY MIXED FEELINGS. THEY ARE HOT AND THIRSTY, EAGER TO DRINK AND GET ACROSS. ON THE OTHER HAND, THERE IS DARK RIVERINE BUSH PROBABLY BRISTLING WITH PREDATORS, AND THE WATER IS RUSHING AND DEEP AND NOT AT ALL TO THE LIKING OF A LAND MAMMAL OF LITTLE BRAIN. AND THERE ARE SO MANY LIKE-MINDED BEASTS PUSHING FROM BEHIND.

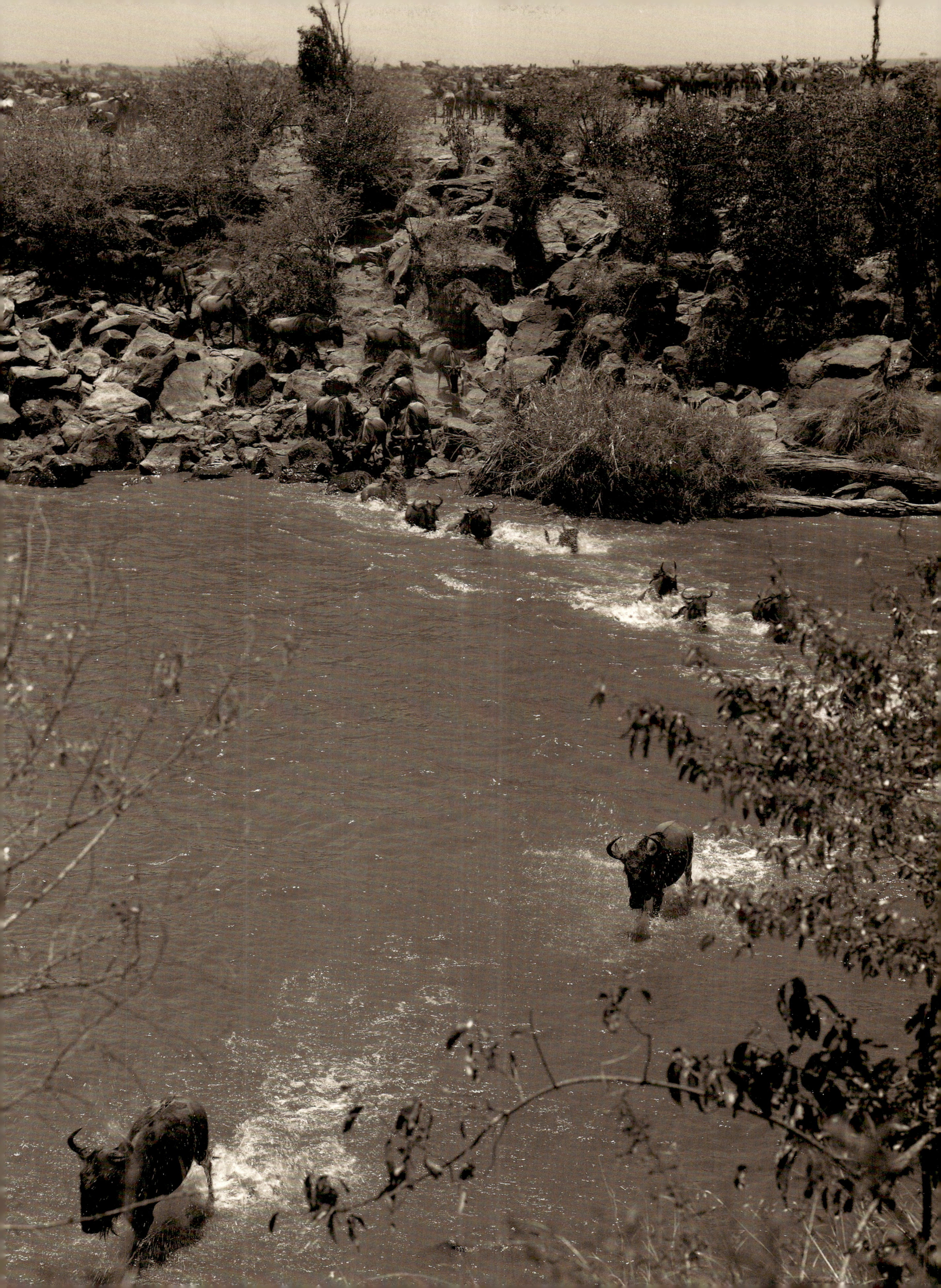

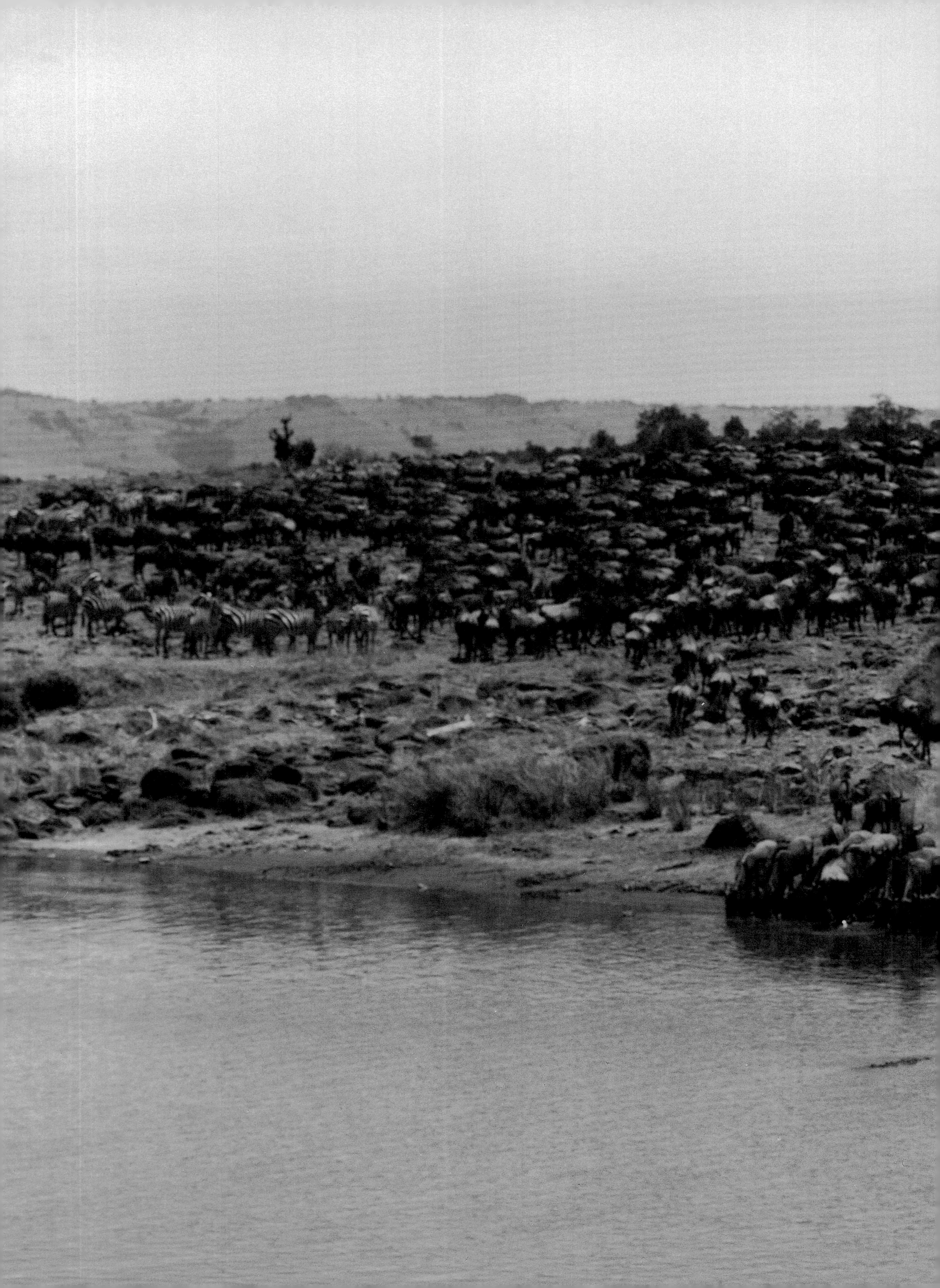

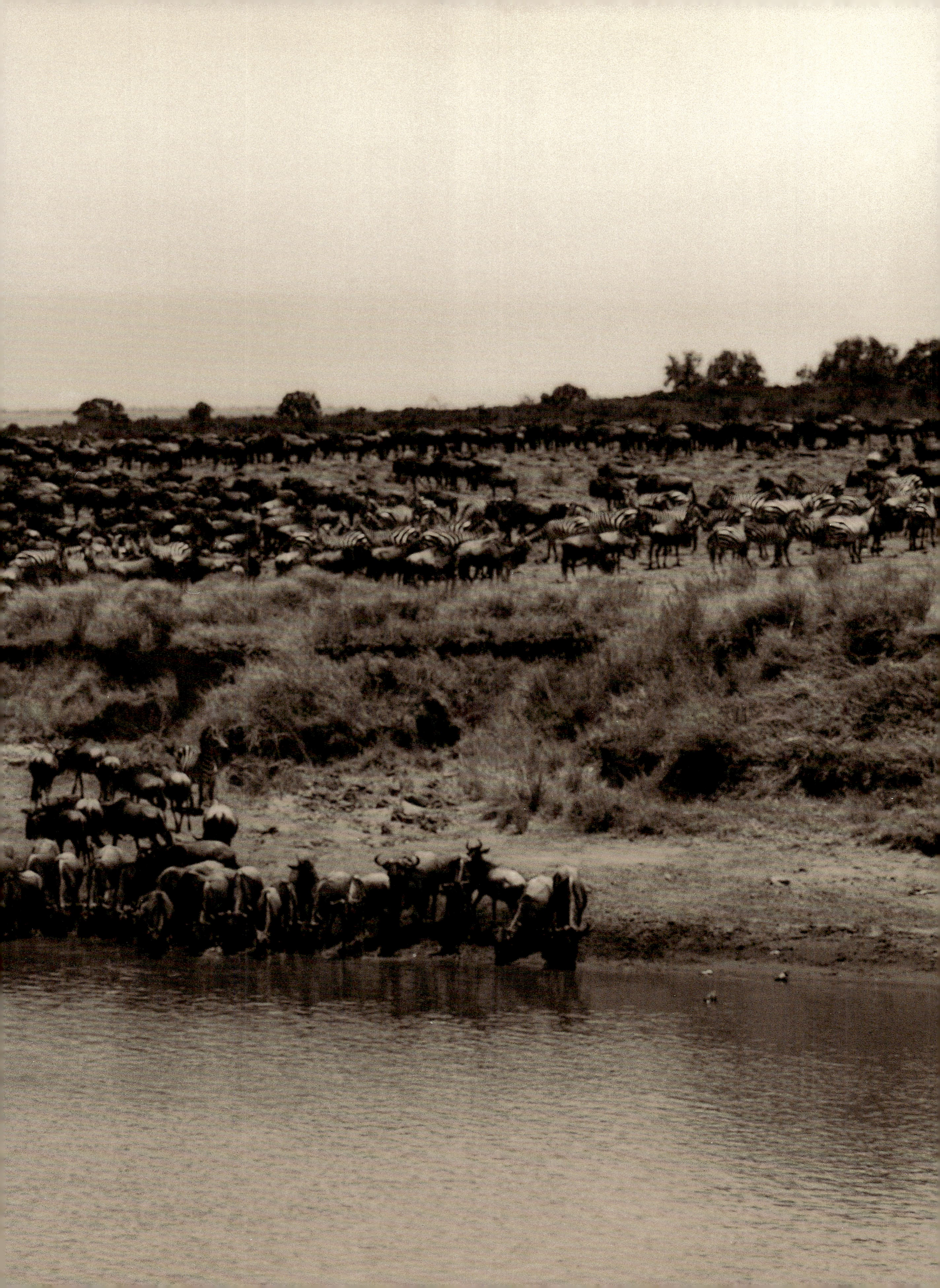

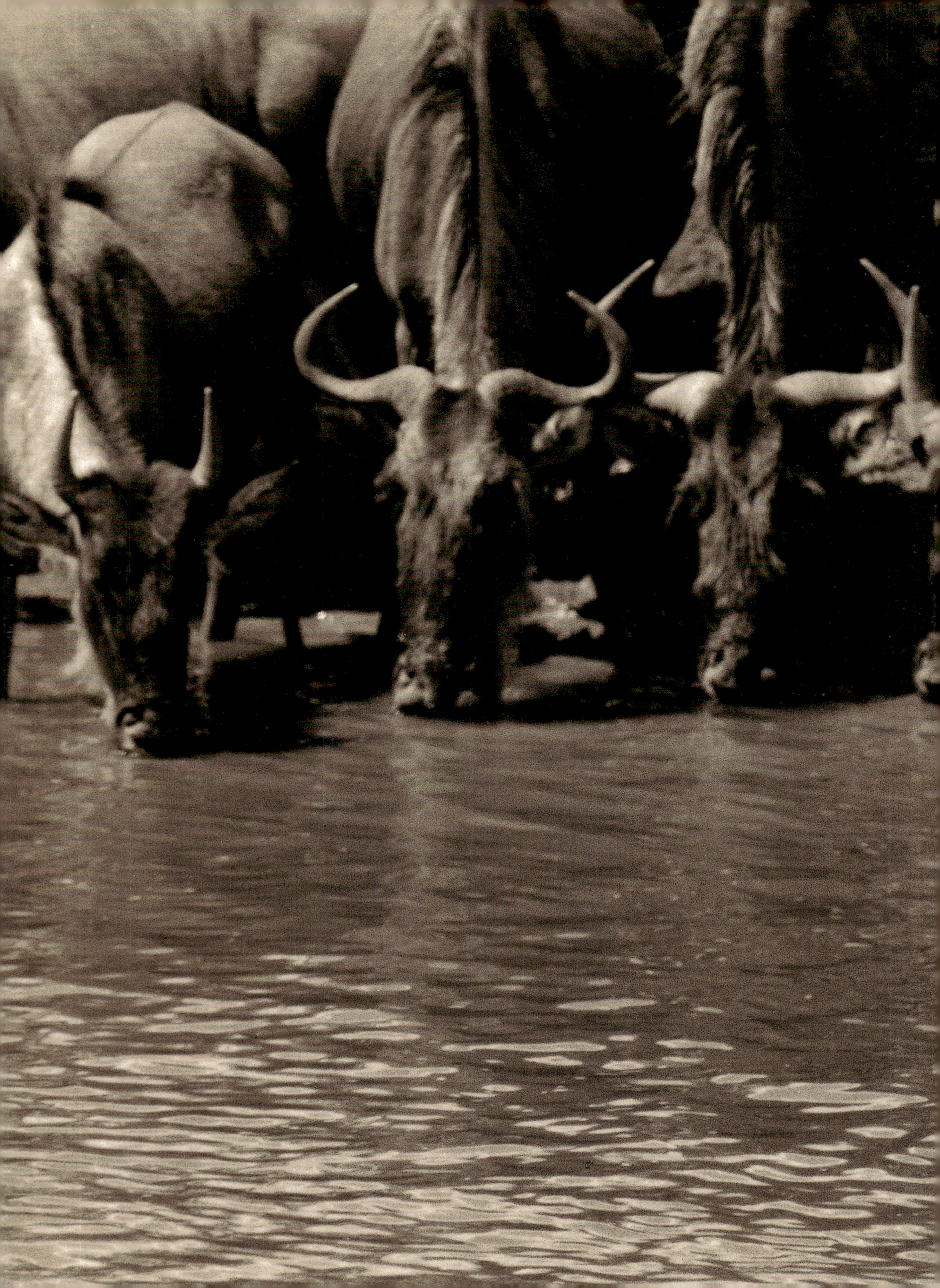

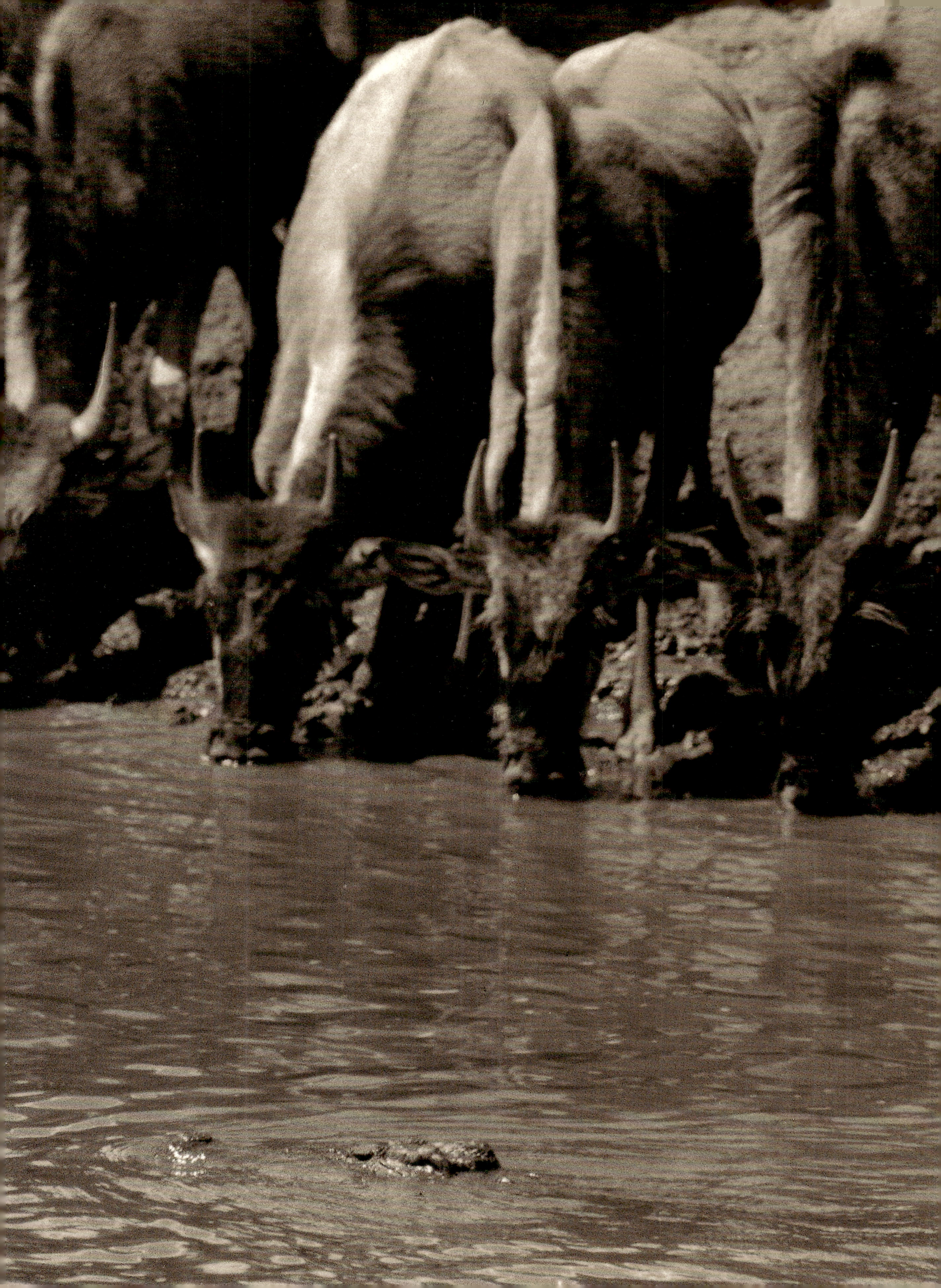

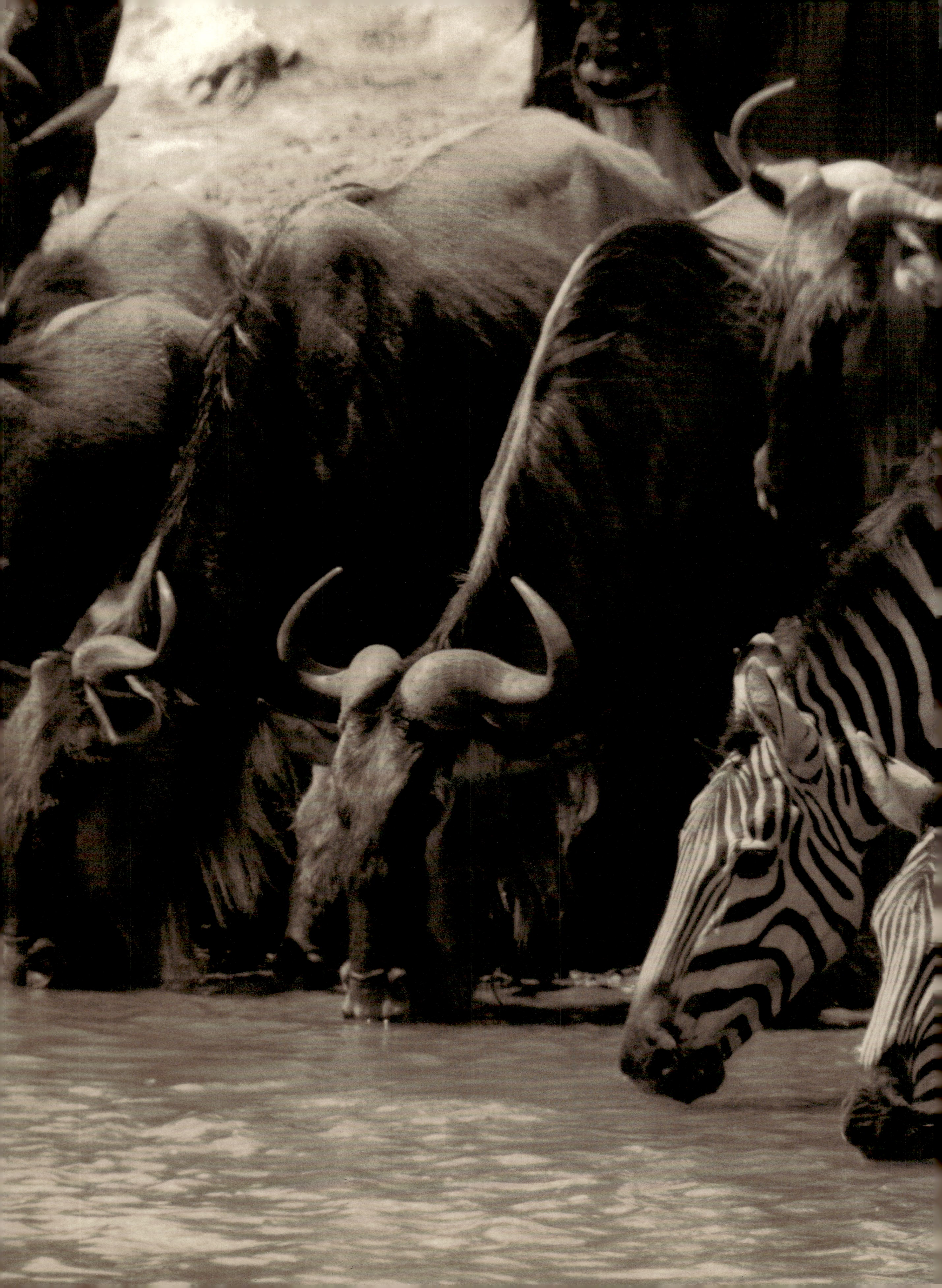

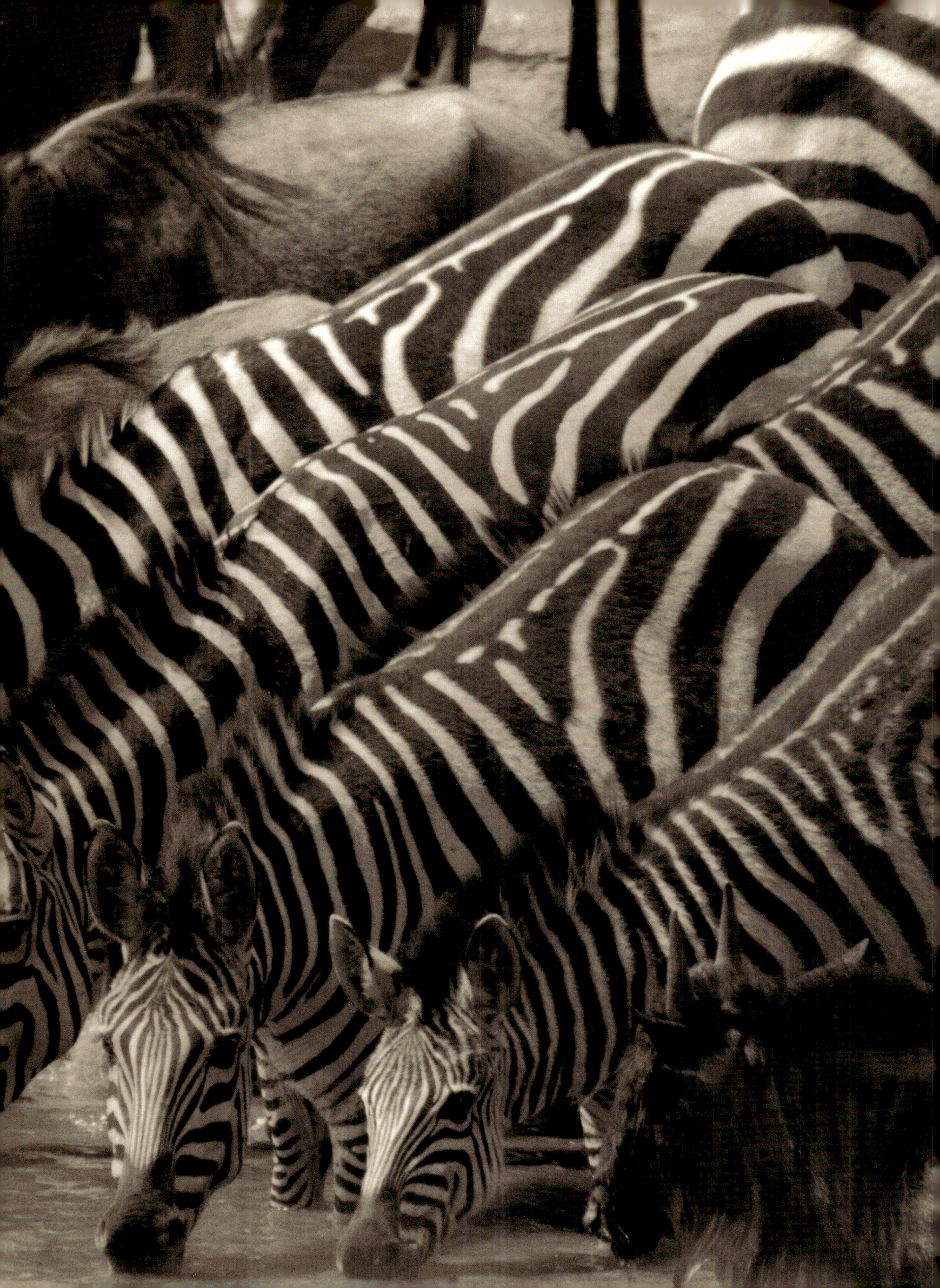

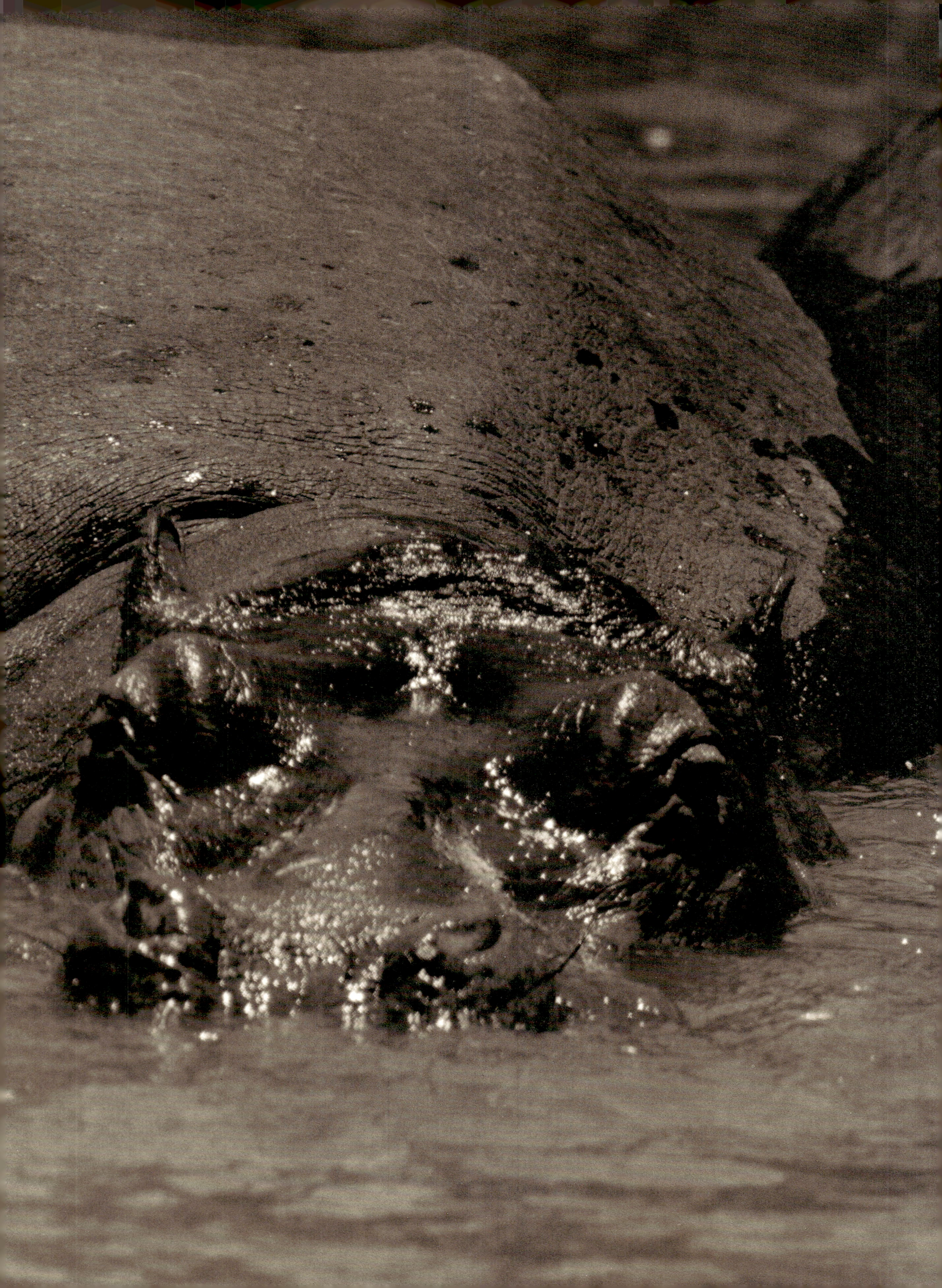

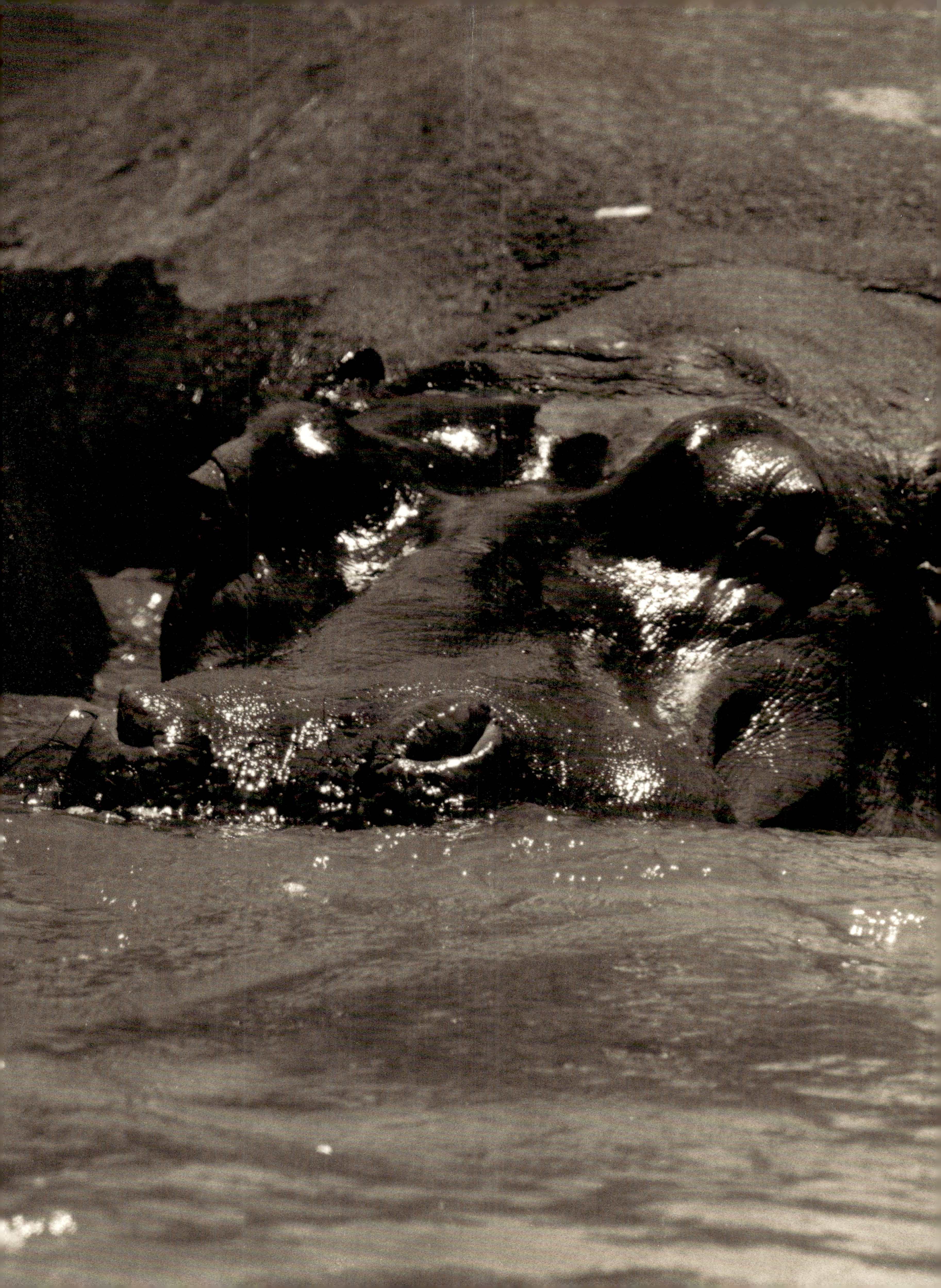

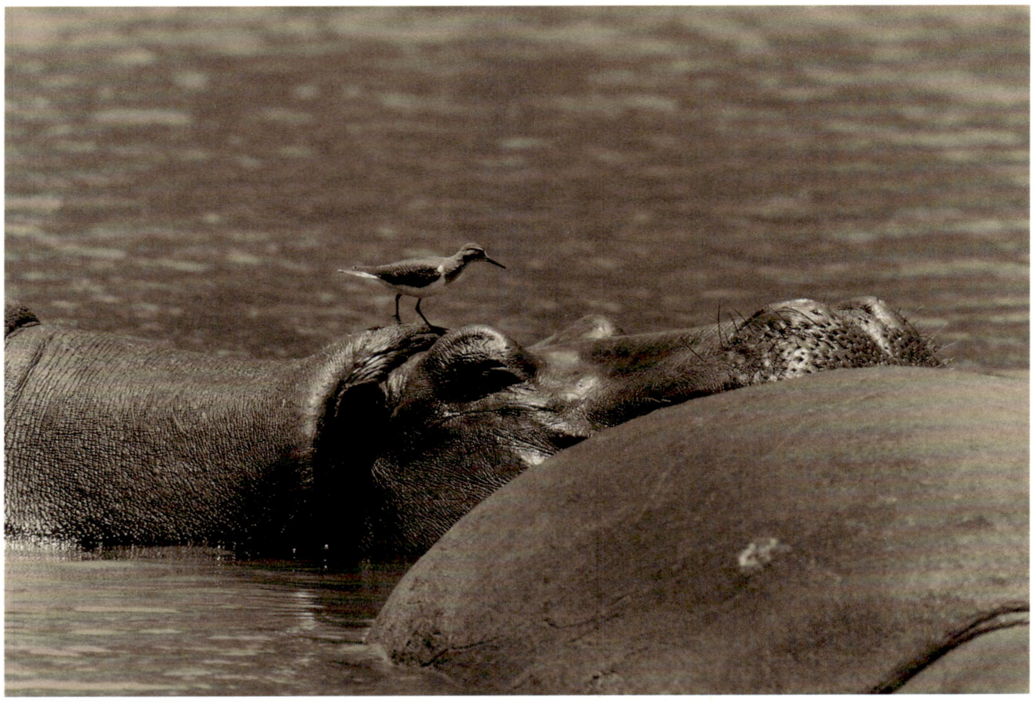

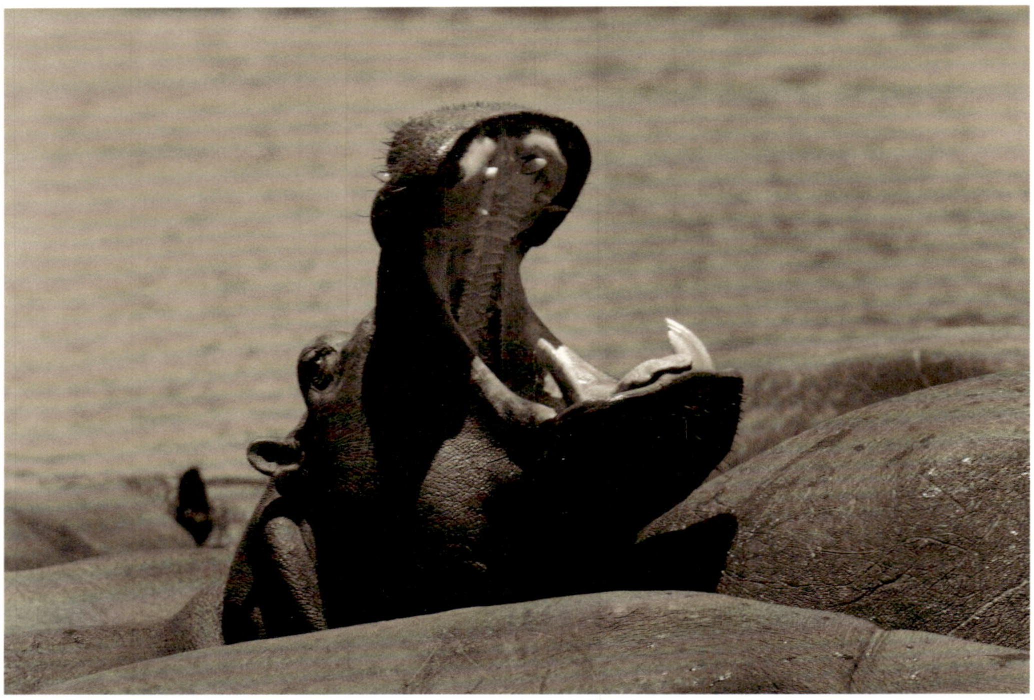

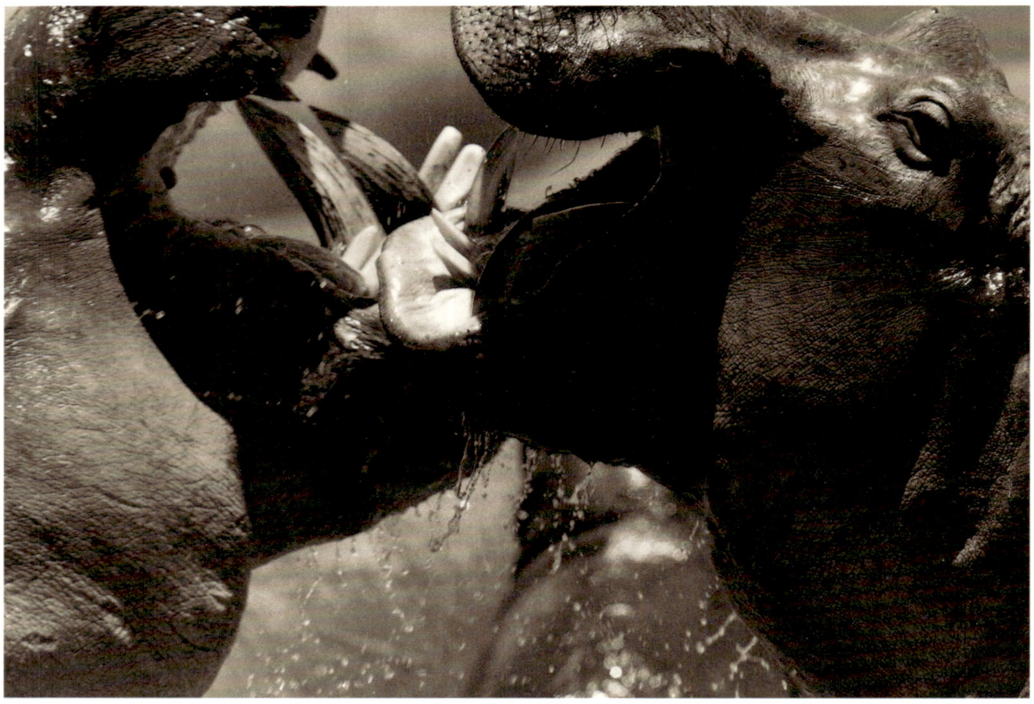

THOSE AT HOME IN THE RIVERS OR ON THEIR BANKS SUCH AS HIPPOS AND PIED WAGTAILS ARE INDIFFERENT TO THE GNU'S CROSSING PLIGHT. THEY SEEM TO WATCH, BEMUSED AS THE MIGRATION CRASHES EACH YEAR THROUGH THEIR HABITAT. WHY THE FUSS? OF GREATER CONCERN TO THE THREE-TONNE HIPPO IS THE 50 KILOS OR MORE OF FRESH GRASS IT NEEDS EACH DAY IF IT IS NOT TO STARVE. WHETHER RETURNING FROM A NIGHT OF GRAZING ON LAND OR LOLLING ABOUT IN THE COOL WATER, THE HIPPO IS TO BE AVOIDED BY WILDEBEEST AND MAN ALIKE. MORE AFRICANS ARE KILLED BY HIPPOS EACH YEAR THAN BY THE CARNIVORES THAT STALK THE GNUS.

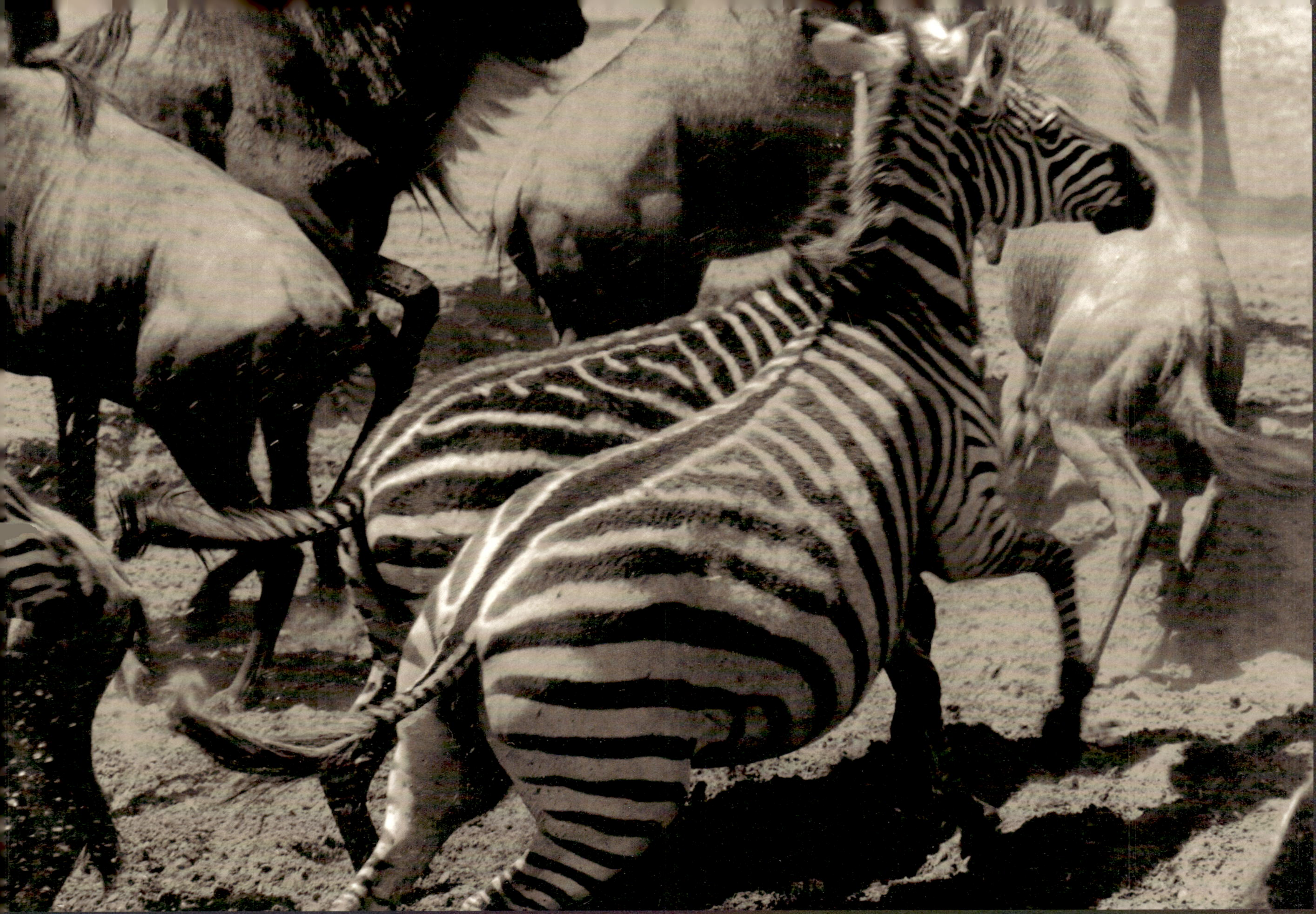

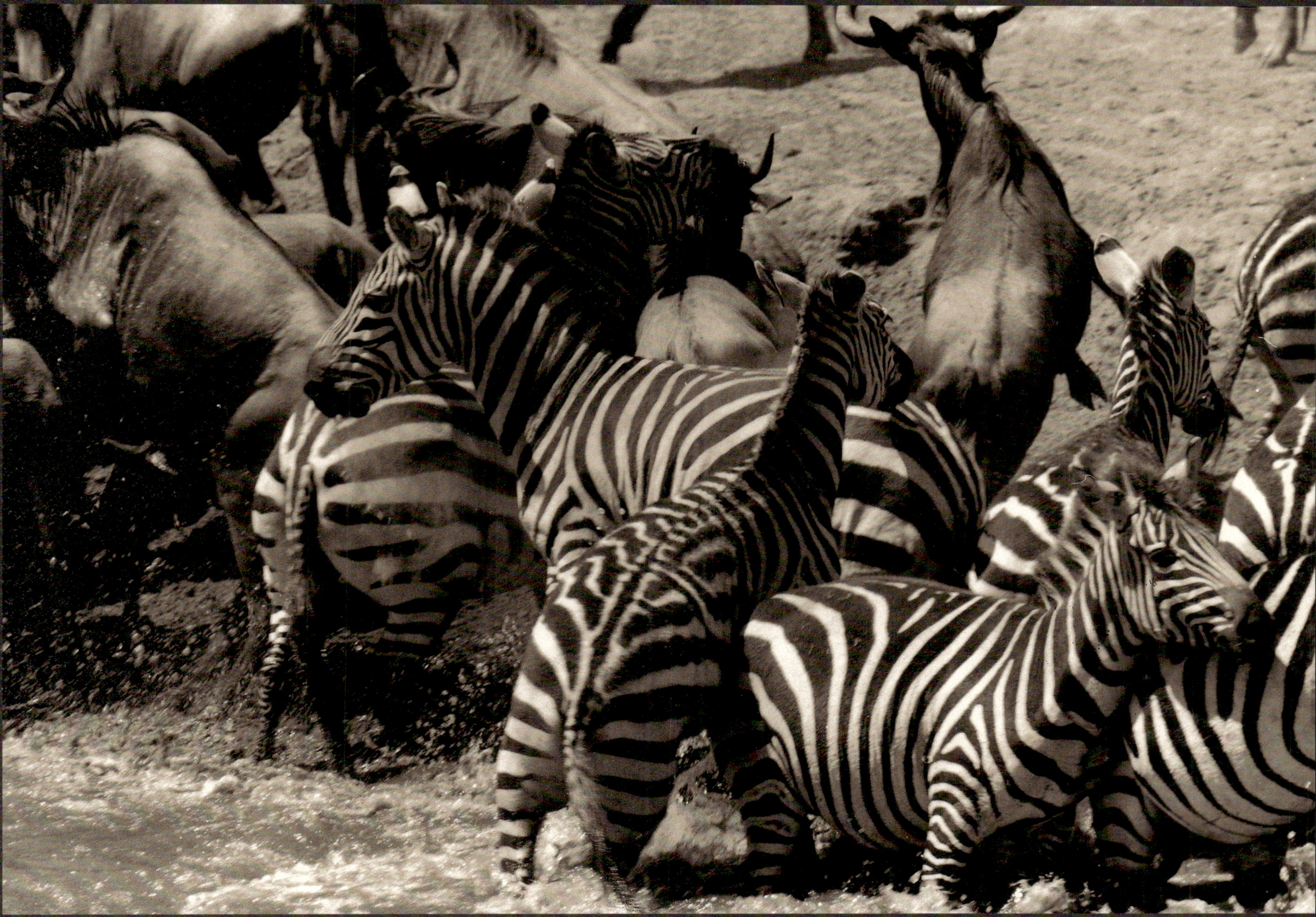

THE MOST IMPRESSIVE SCRAMBLES IN THE GNU'S WILD ROUNDABOUT ARE THOSE ACROSS THE MAASAI MARA AND THE SERENGETI ECOSYSTEM'S LESSER WATERCOURSES. IT IS NOT MERELY A LAND BEAST'S FEAR OF AN ALIEN MEDIUM THAT MAKES THE GNUS BRAKE AND HESITATE FRANTICALLY ON THE SHORES UNTIL THE VERY LAST MINUTE, AND THEN SURGE FORWARD LIKE THINGS POSSESSED TO REACH AND SCRAMBLE UP THE OPPOSITE BANK. IT IS AN IN-BUILT IMAGE OF DEATH LURKING IN THE SHADOWS AND BENEATH THE SURFACE.

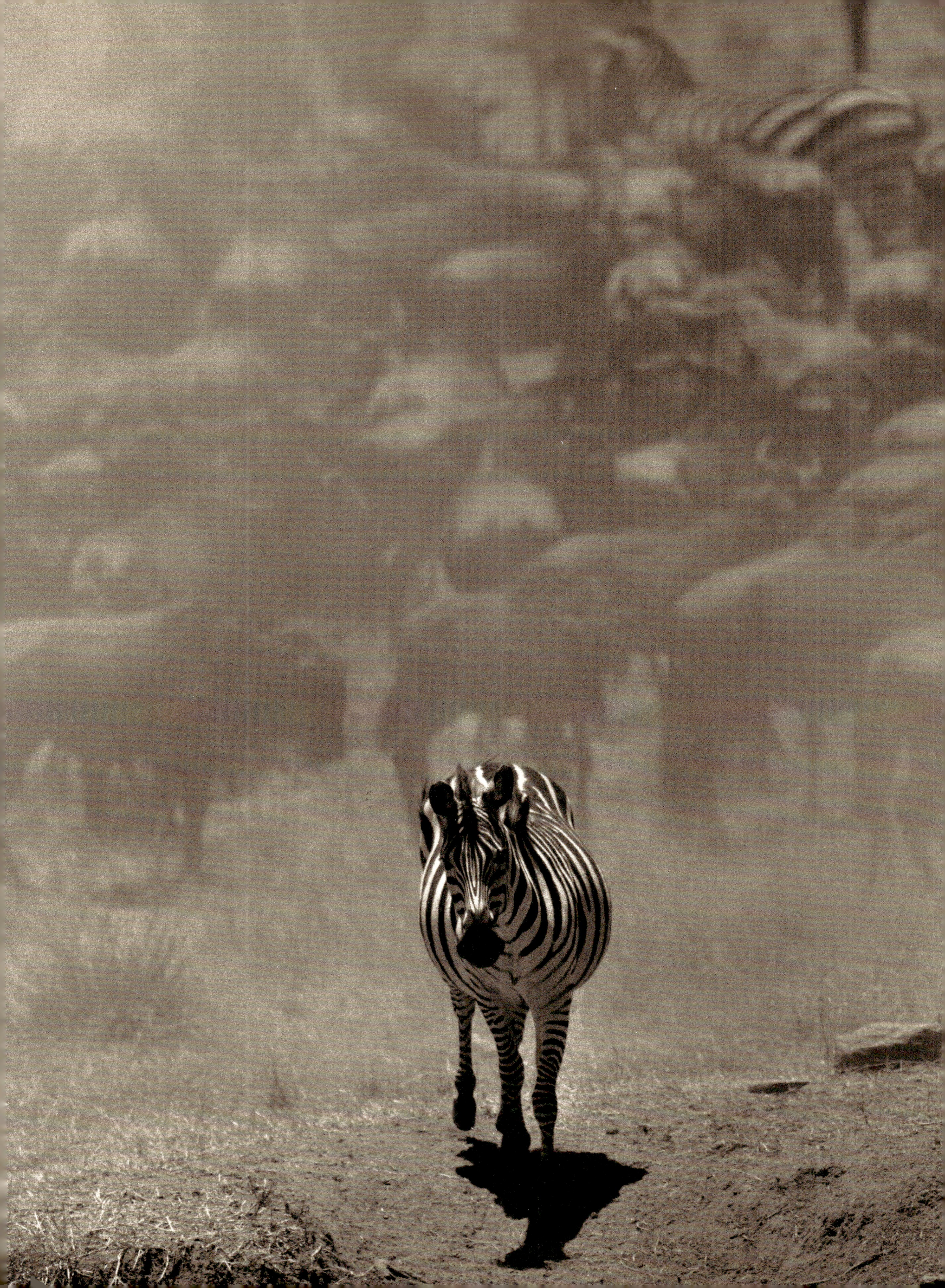

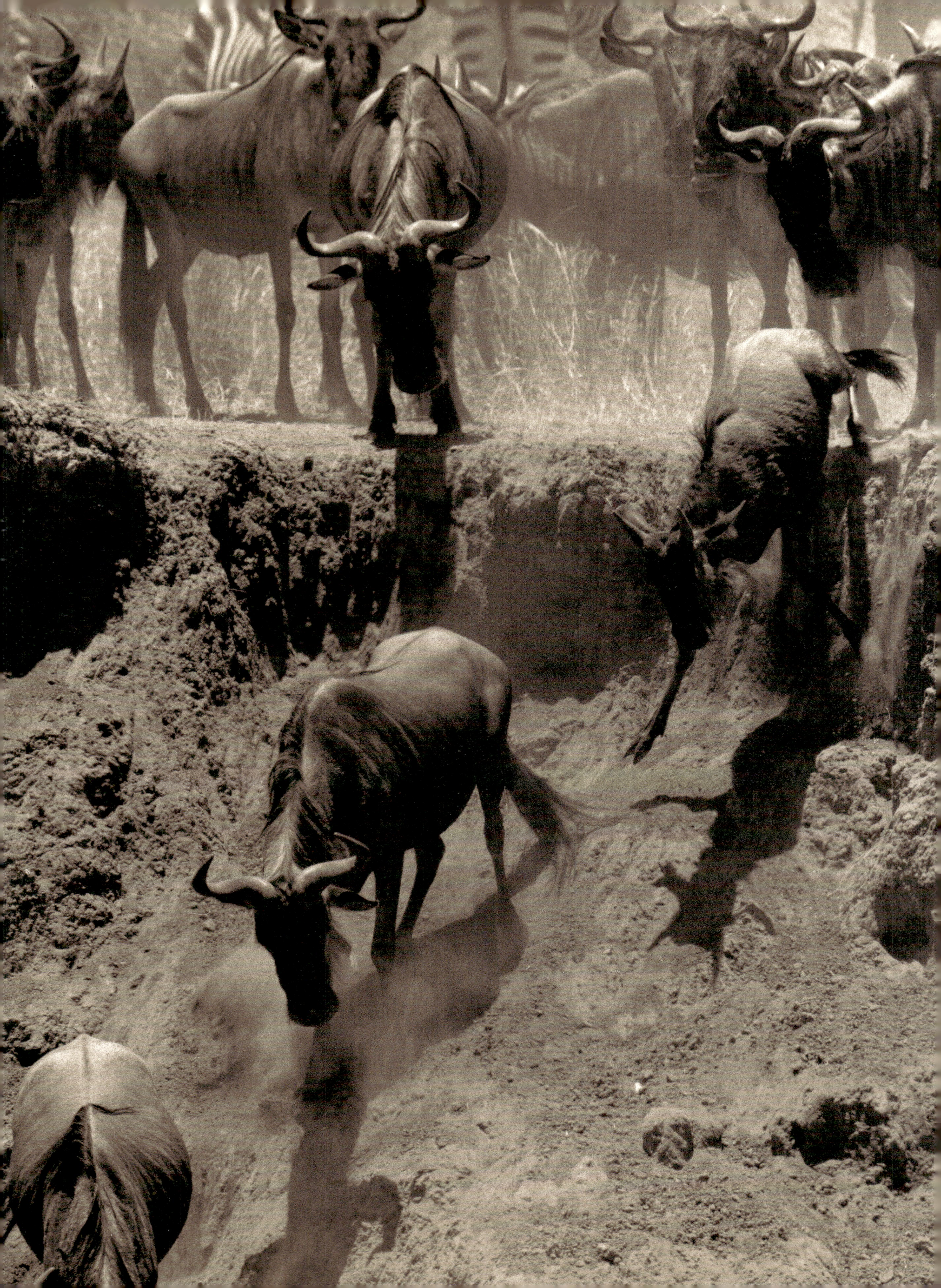

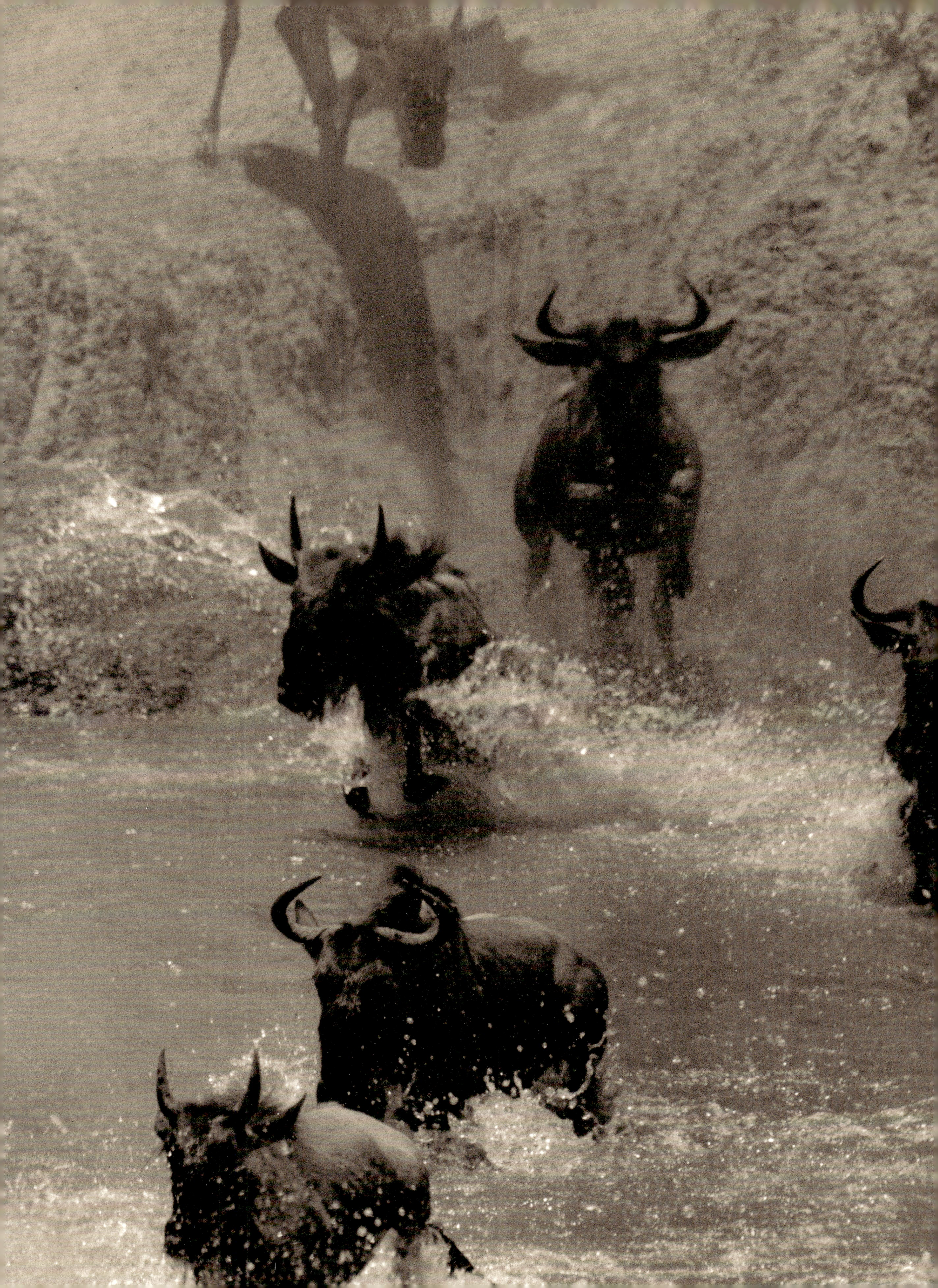

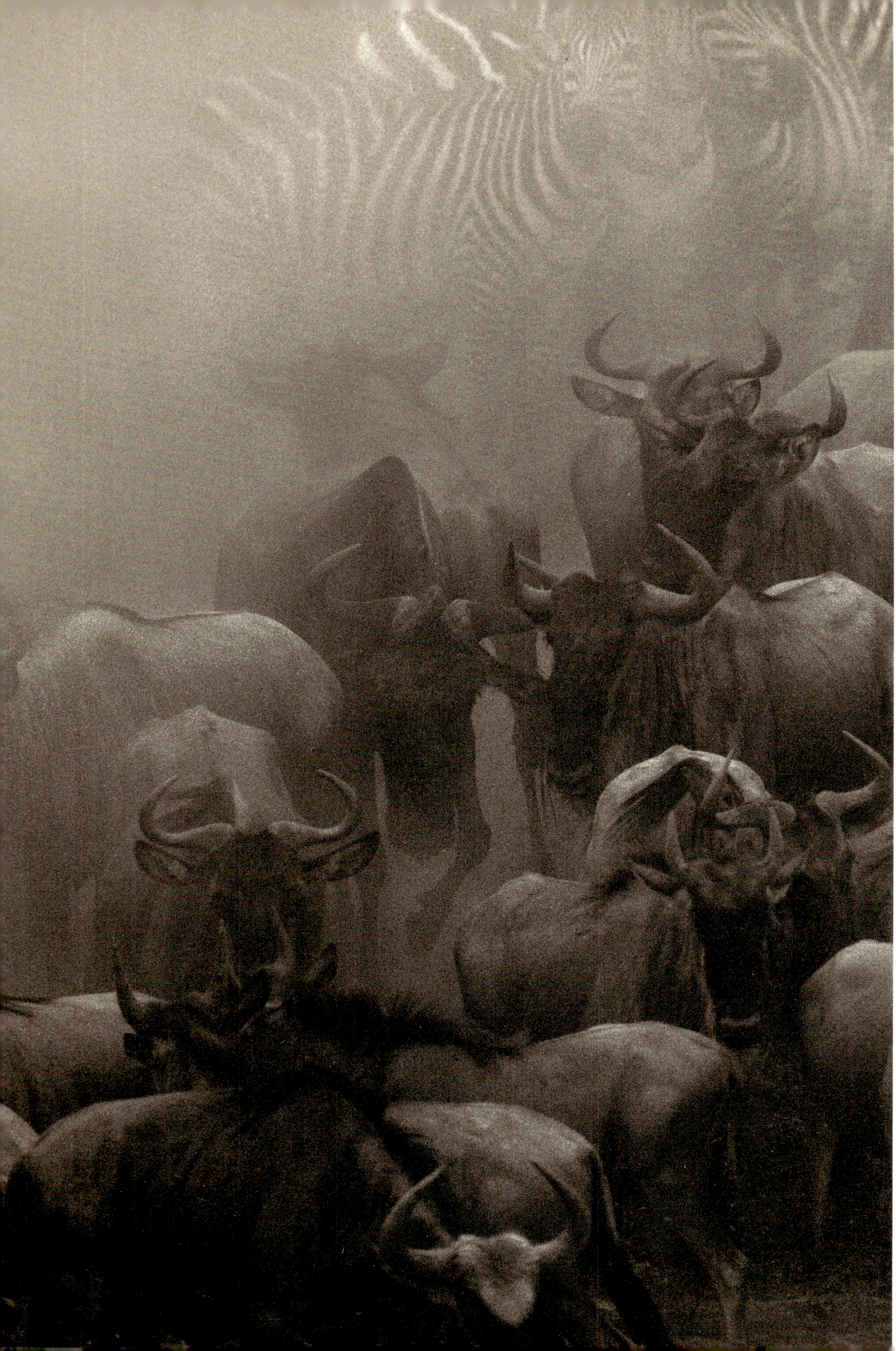

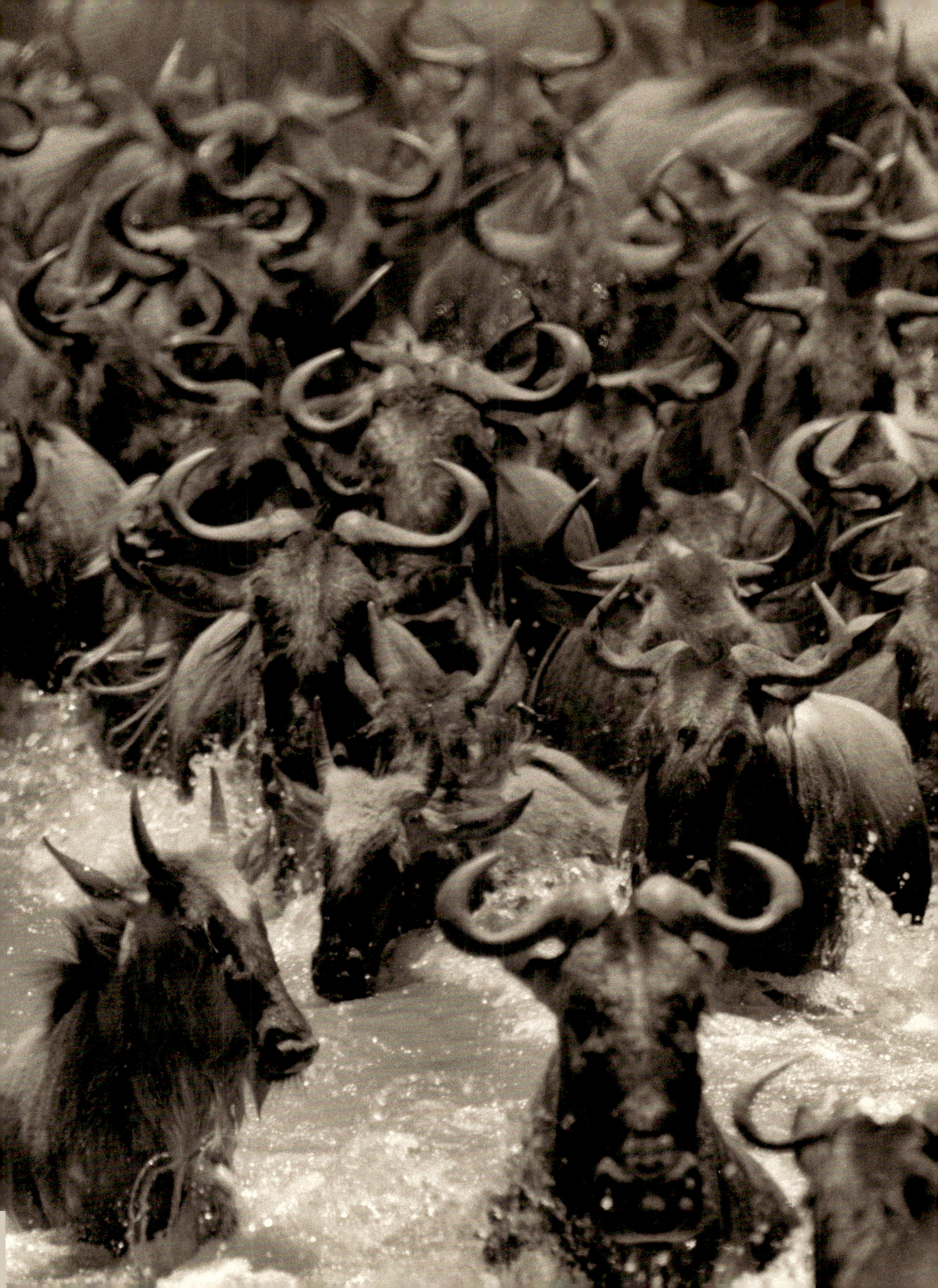

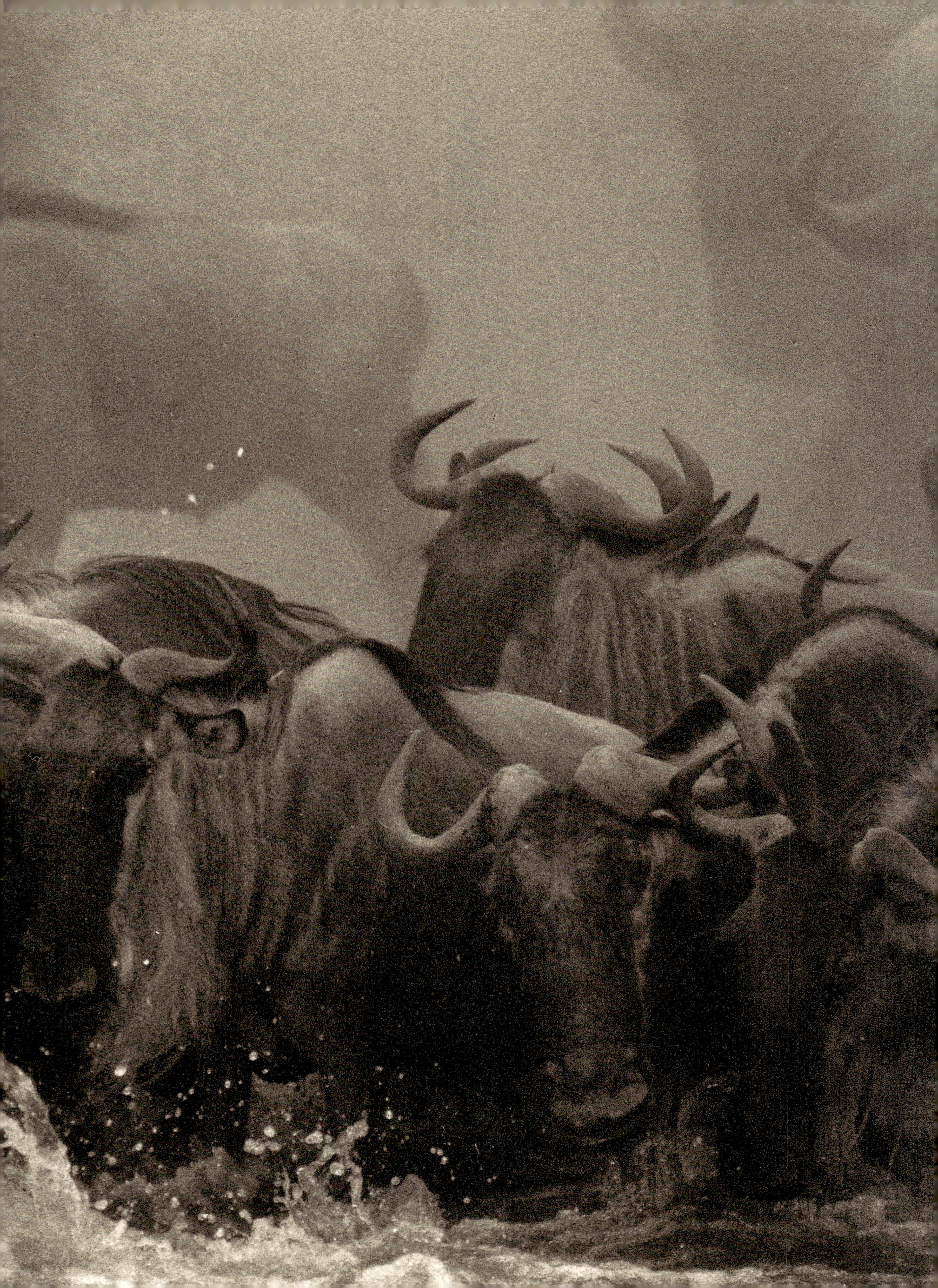

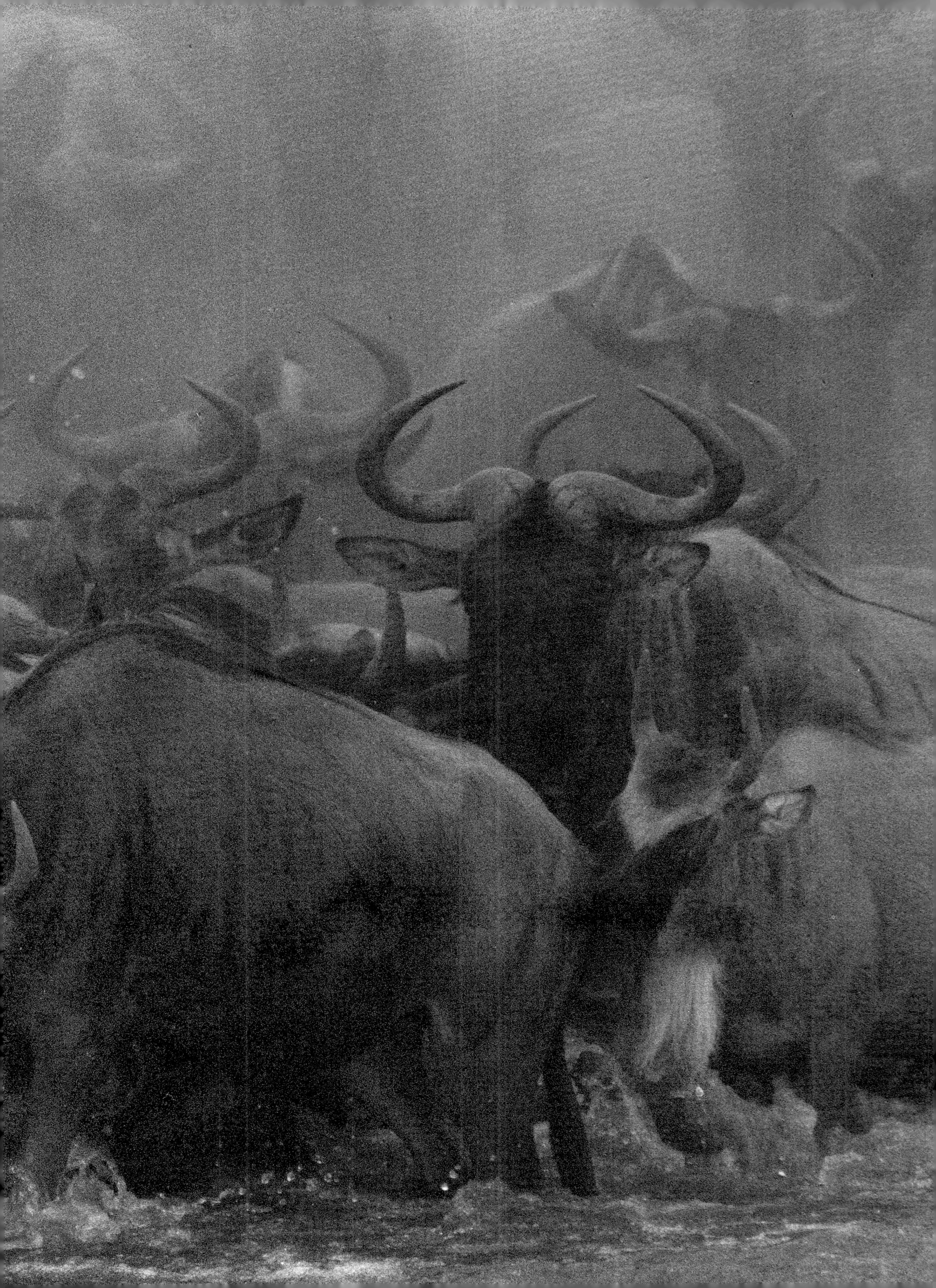

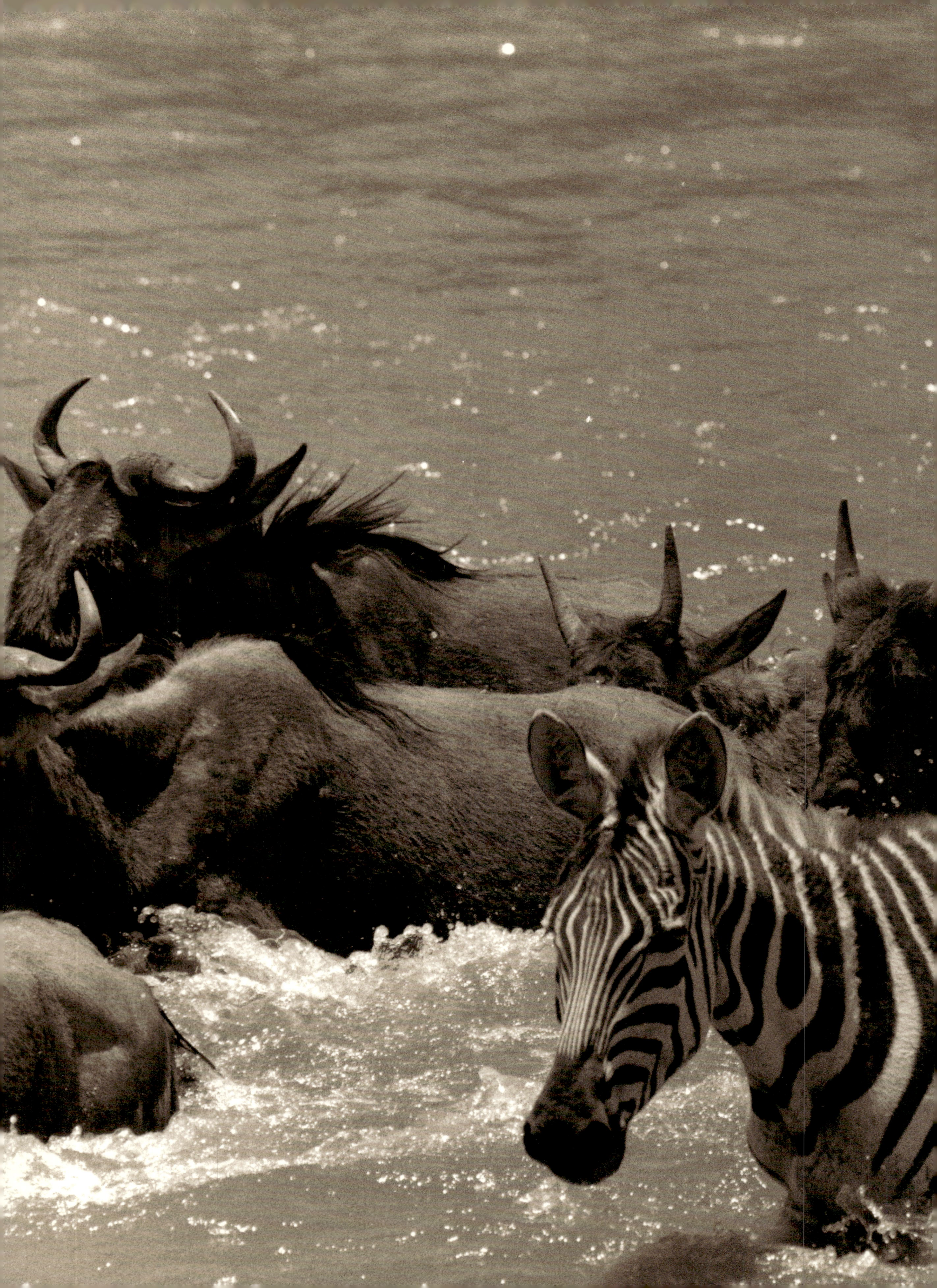

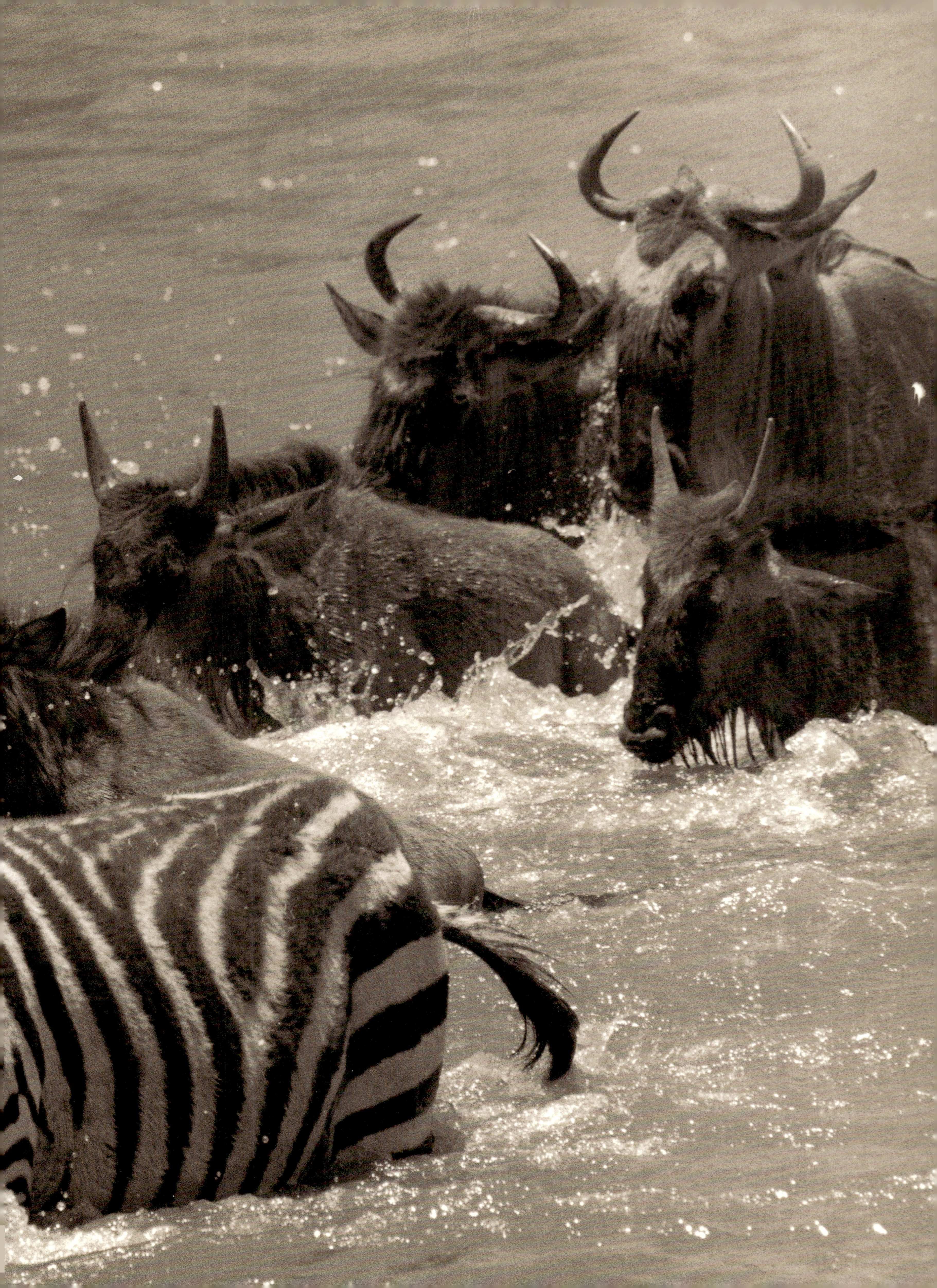

EVENTUALLY A CROSSING POINT IS CHOSEN, THE RIVER AT THE RIGHT HEIGHT AND NOT TOO FAST-FLOWING, BANKS FAIRLY FREE OF PREDATOR-CONCEALING BUSH. ZEBRA OFTEN START TO CROSS SEDATELY, BUT SOON ALL ARE CAUGHT UP IN A COLLECTIVE MADNESS. CROCODILES JUST HAVE TO BE WATCHED OR EVEN IGNORED. IN REALITY, MANY MORE WILDEBEESTE WILL DIE FROM FOOLISH PANIC THAN FROM THE JAWS OF A CROCODILE. BY THE SAME TOKEN, MORE CROCODILES WILL GROW FAT FROM GNU FLESH THAN FROM BOTTOM-DWELLING FISH.

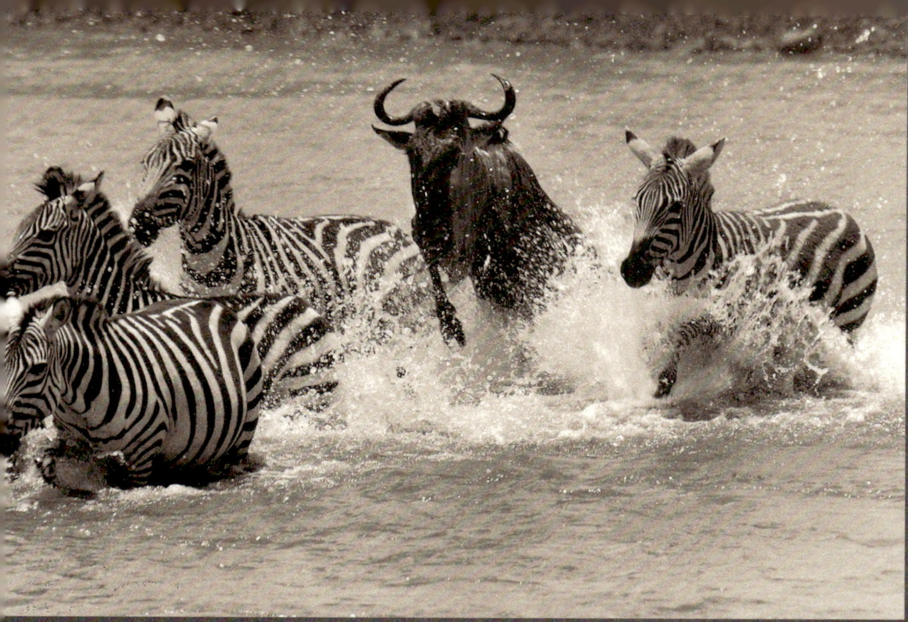
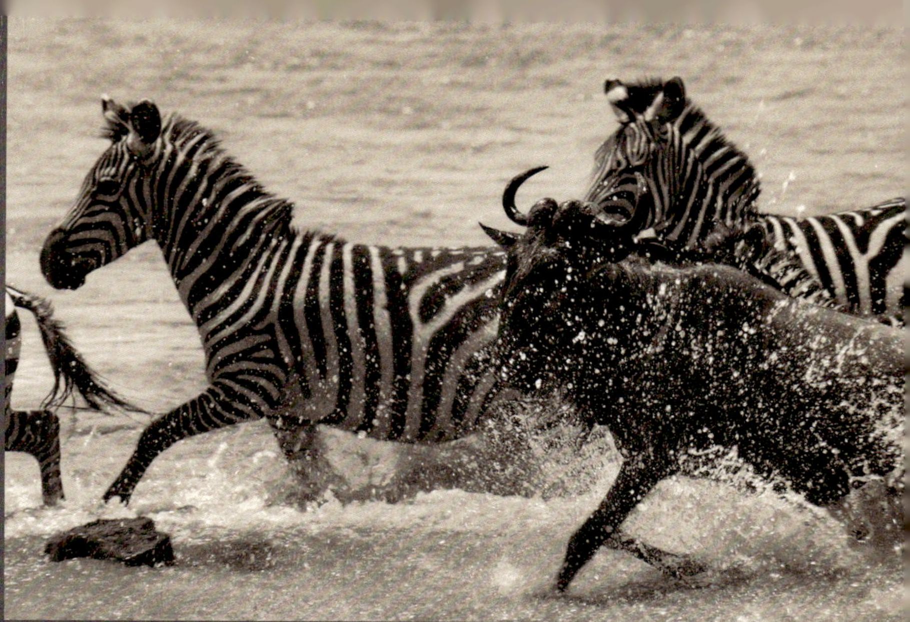
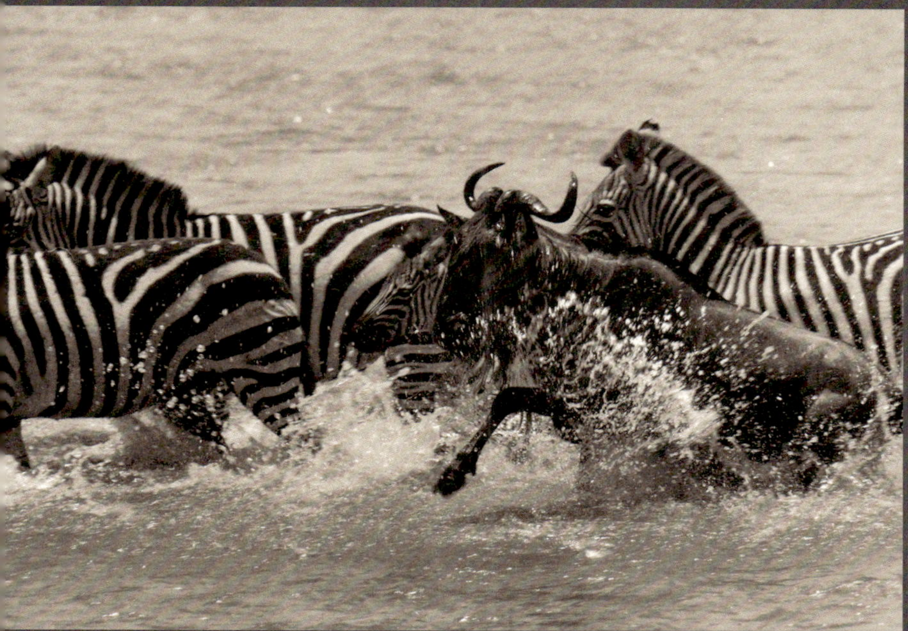
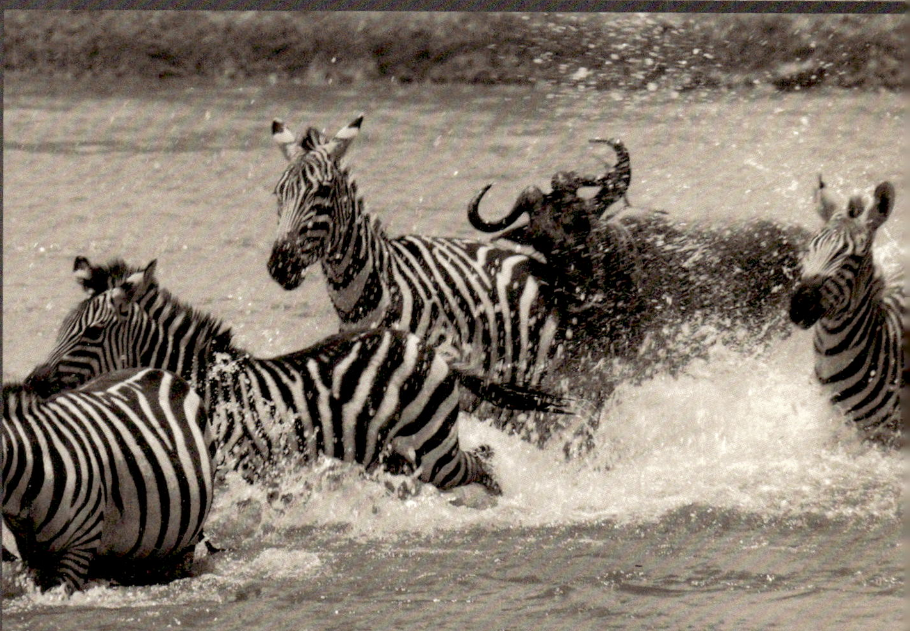
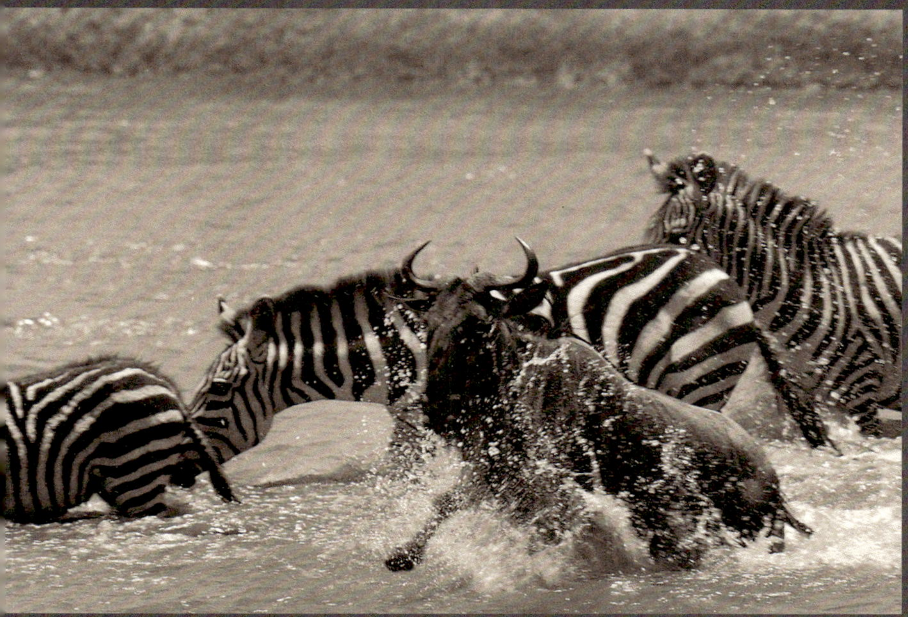
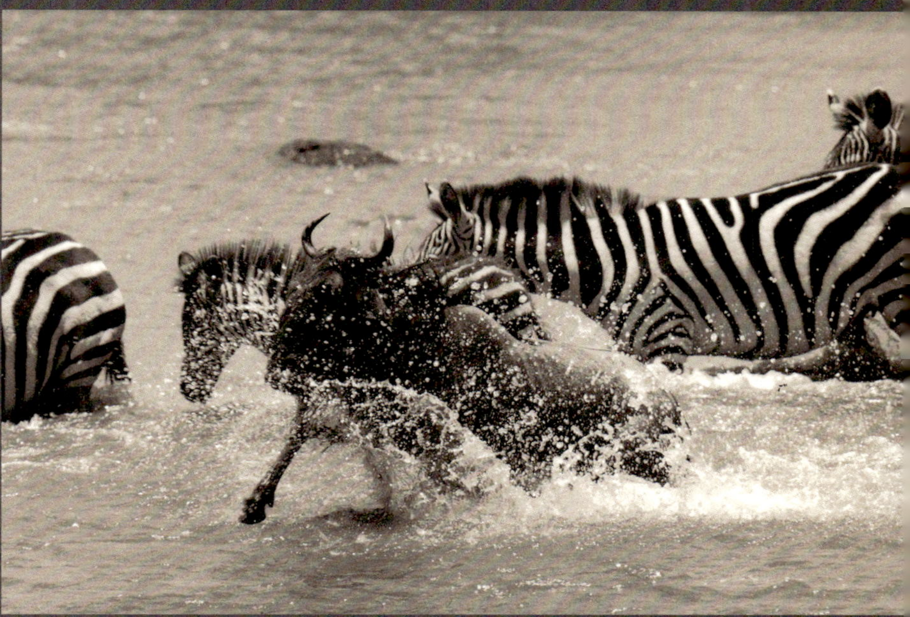
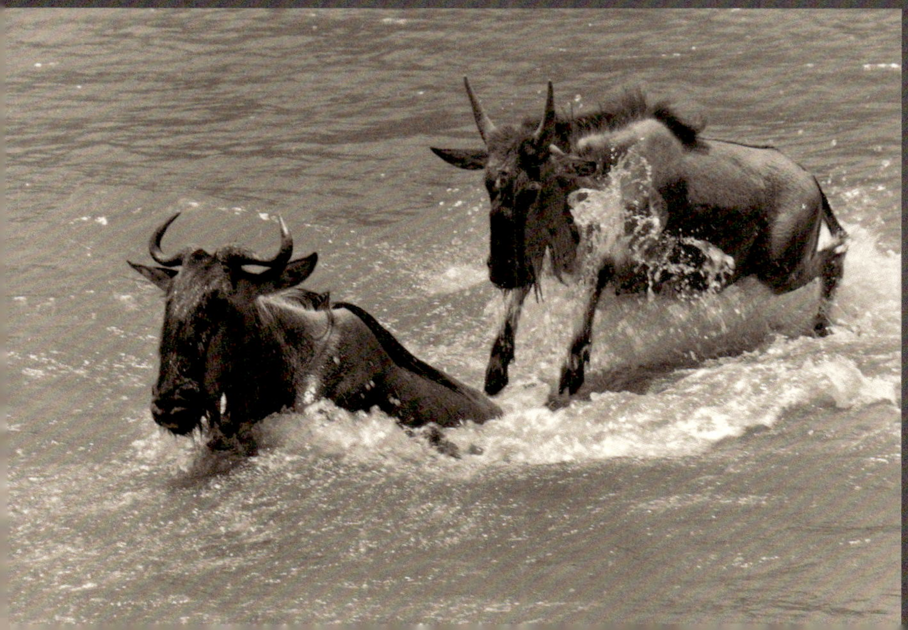
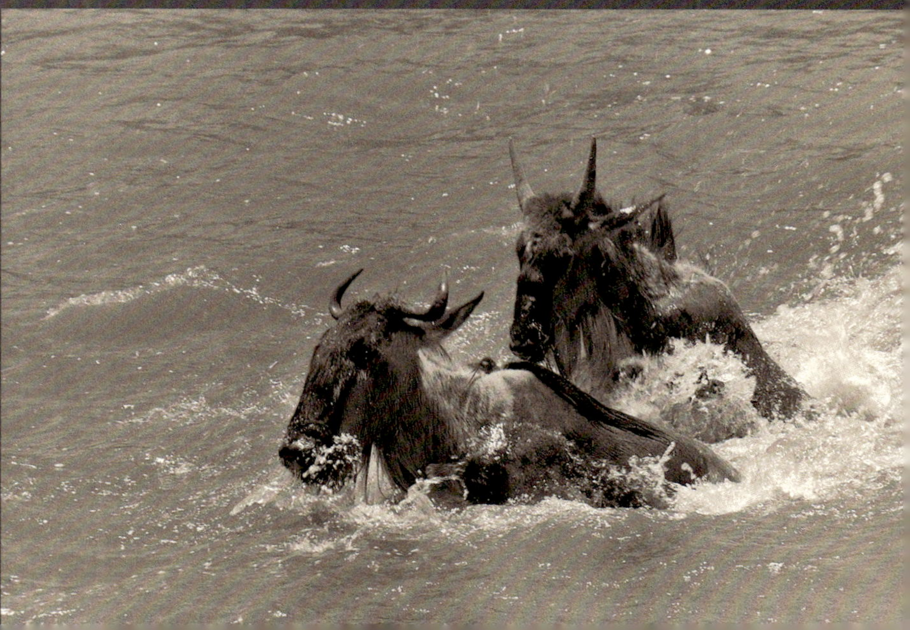

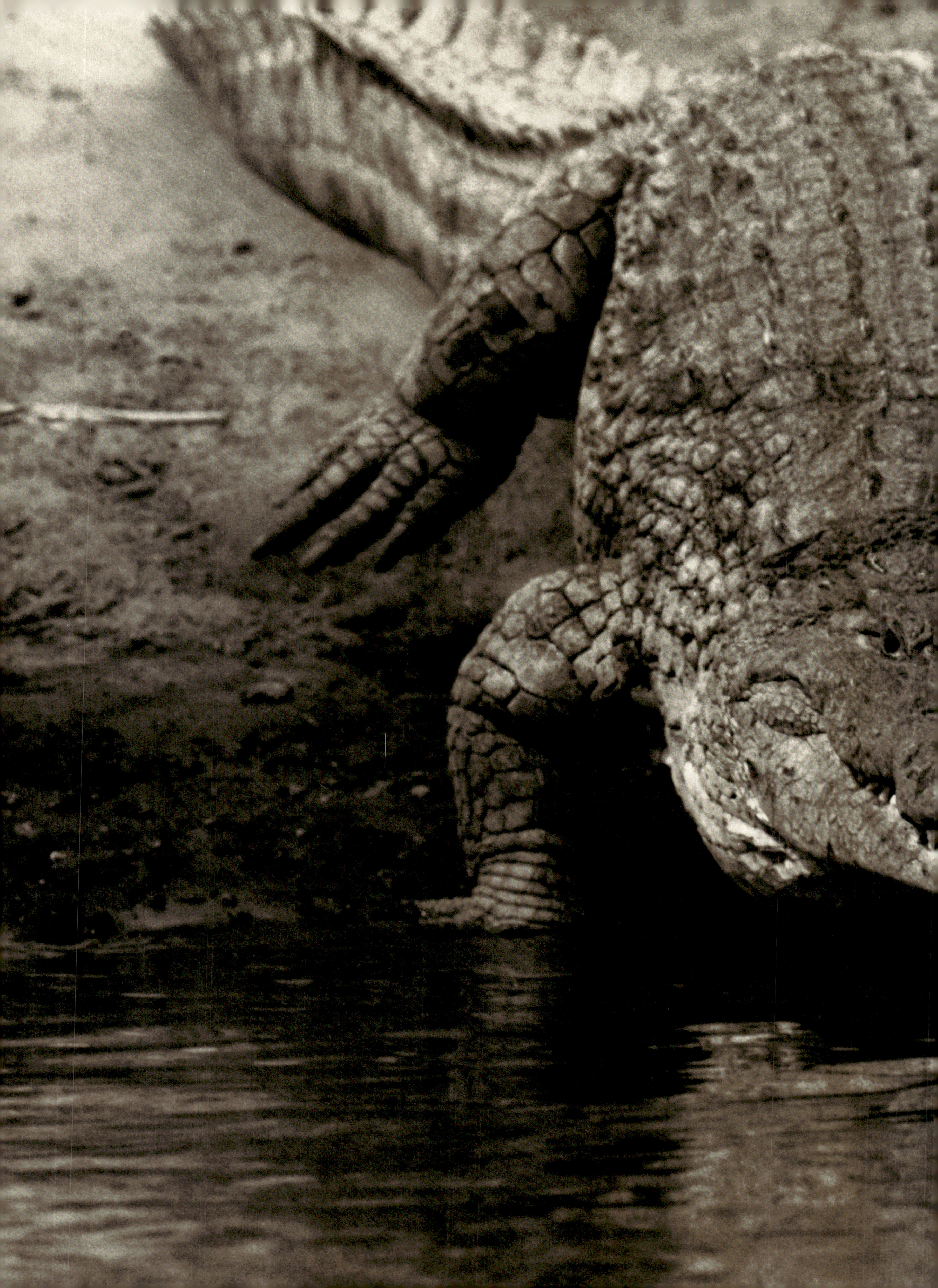

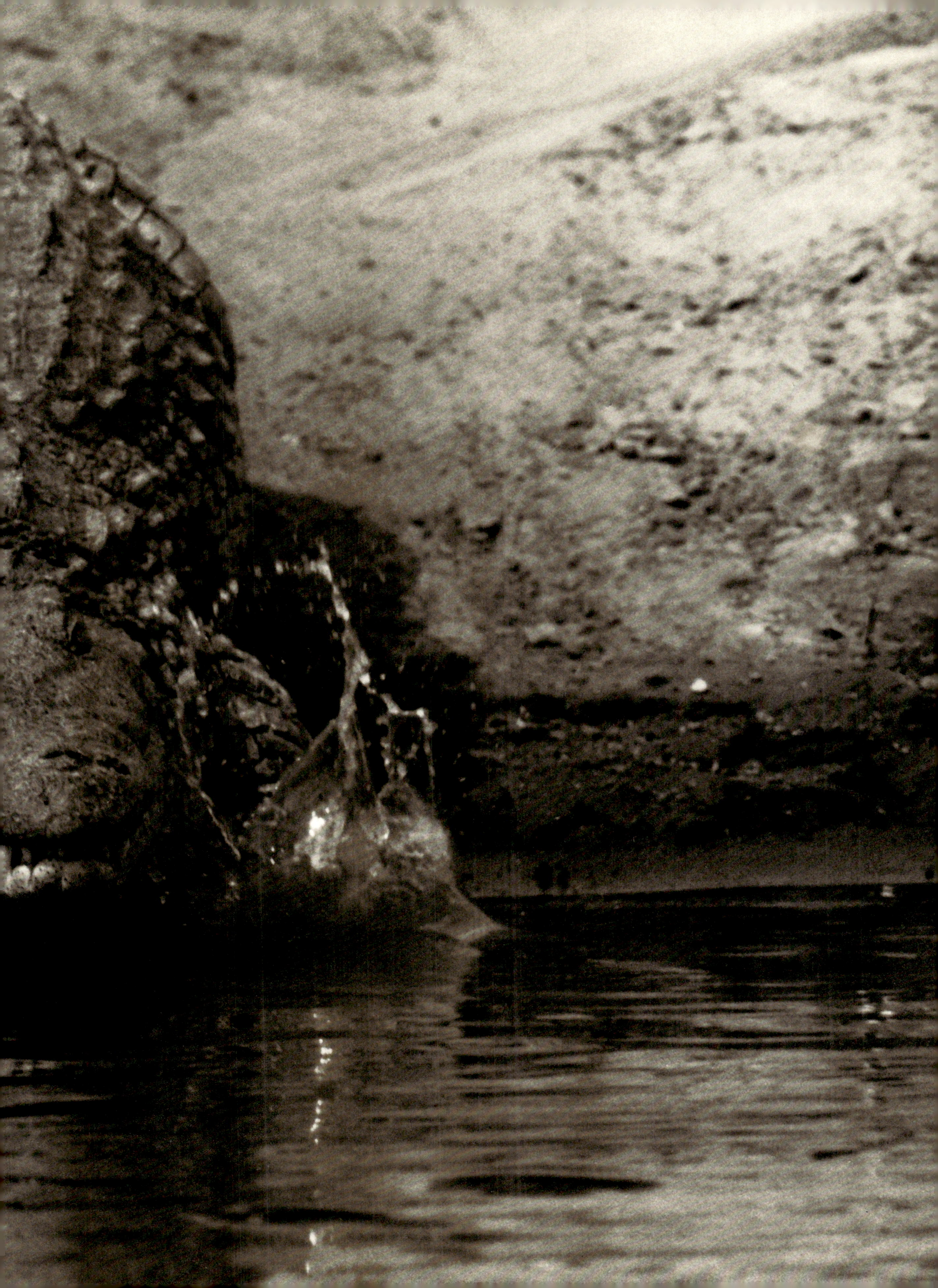

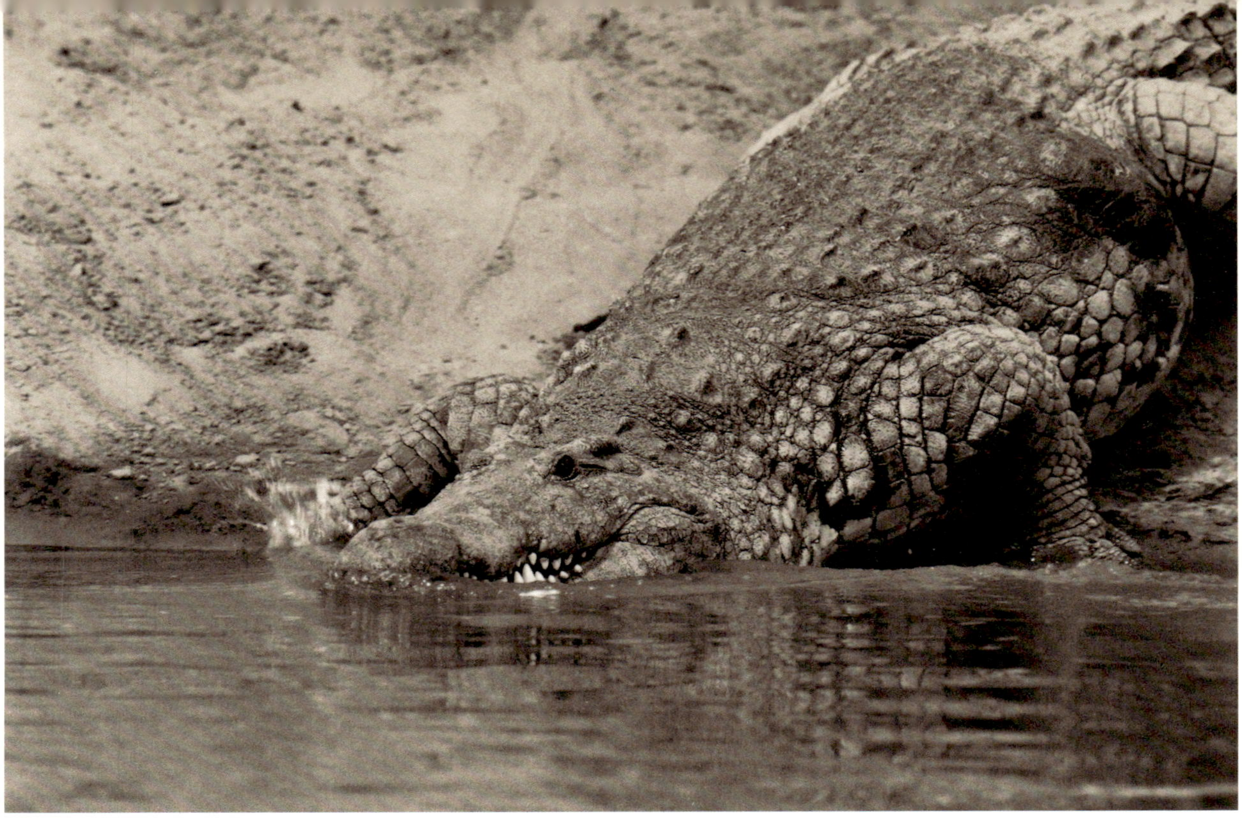
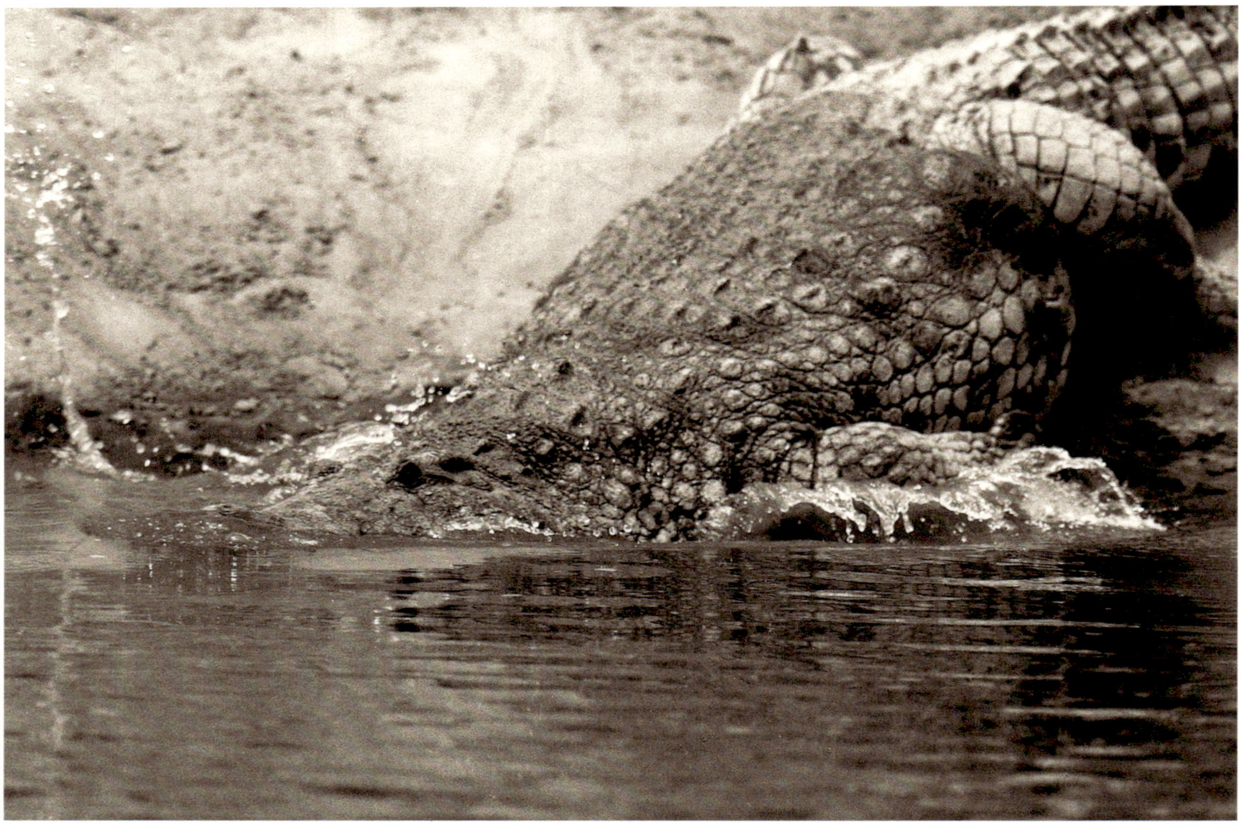
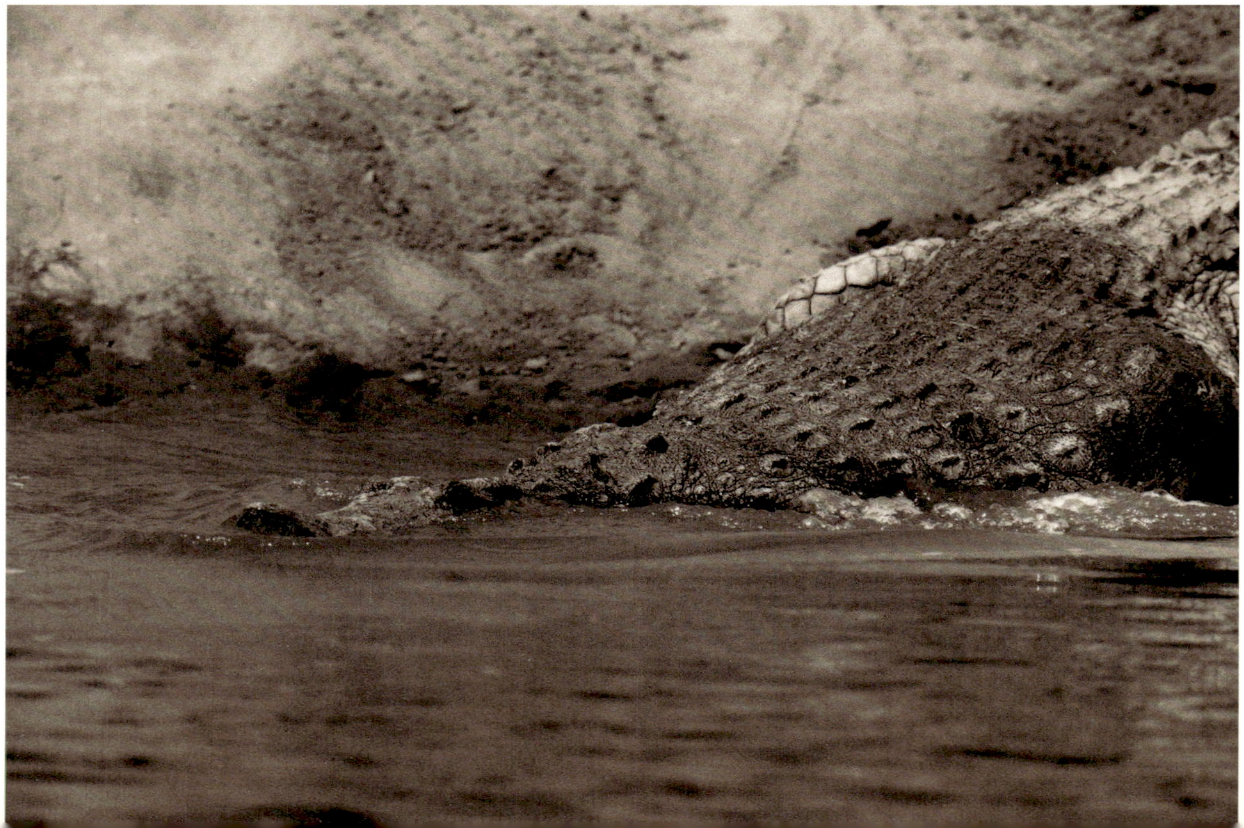

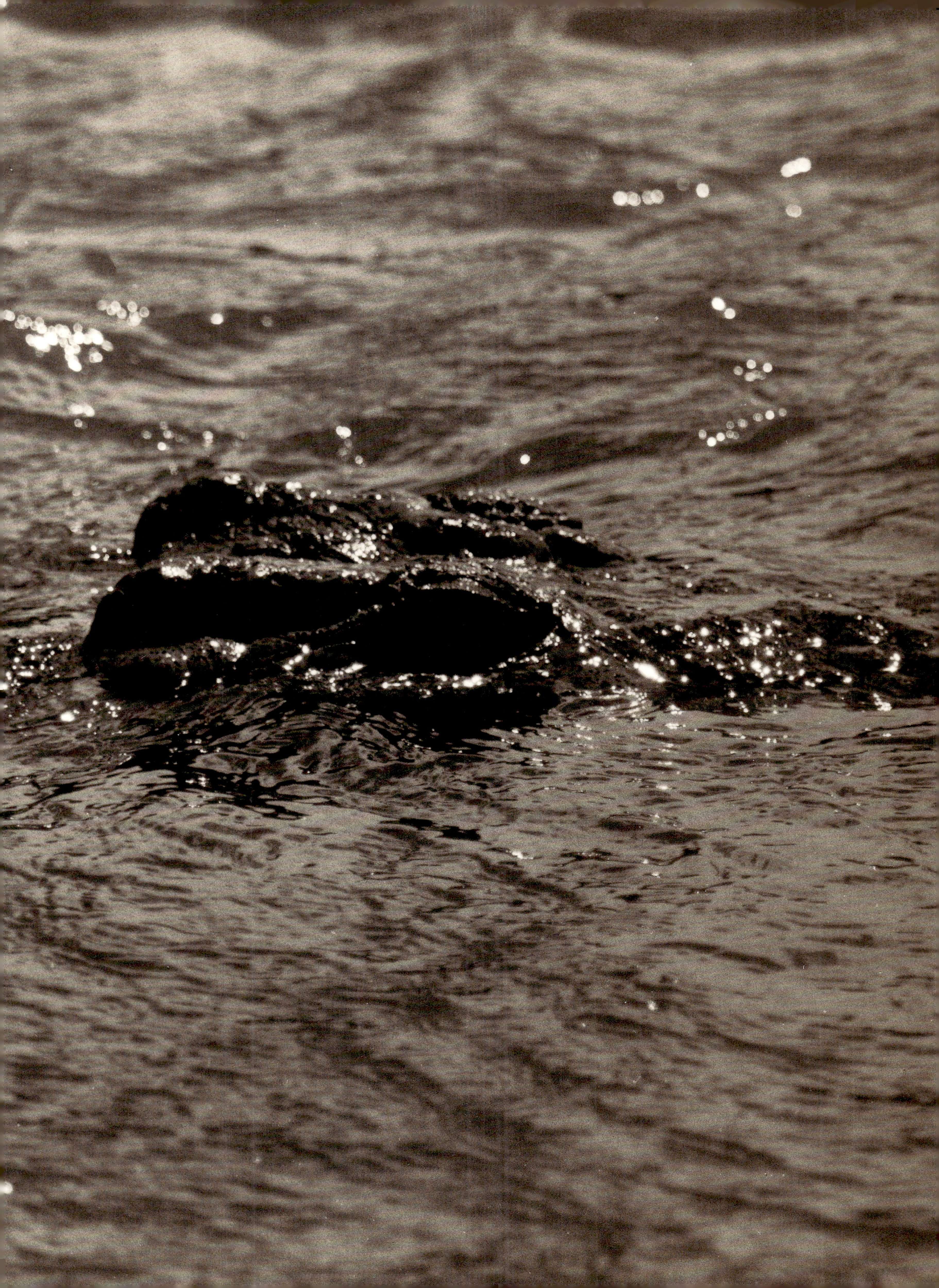

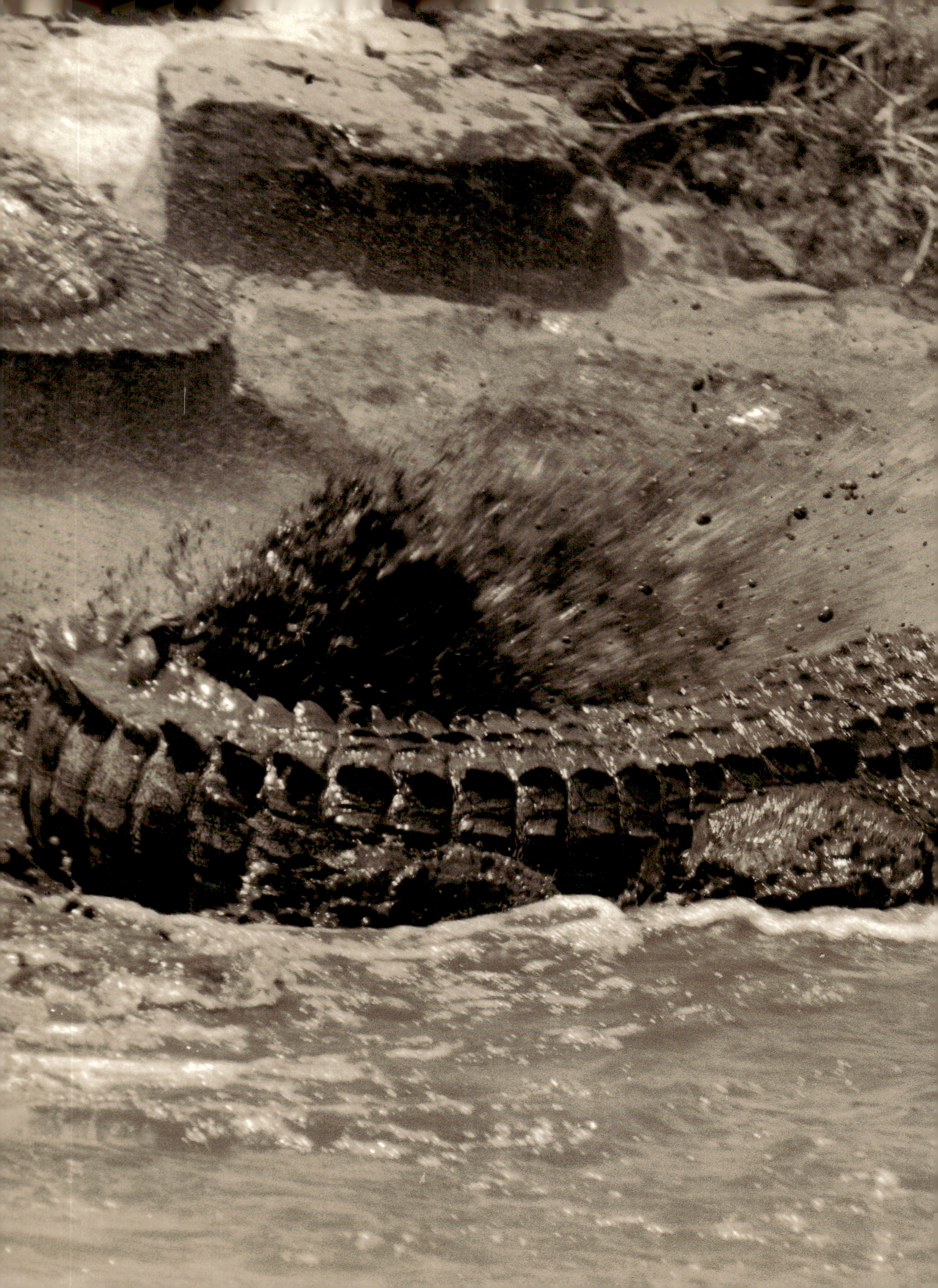

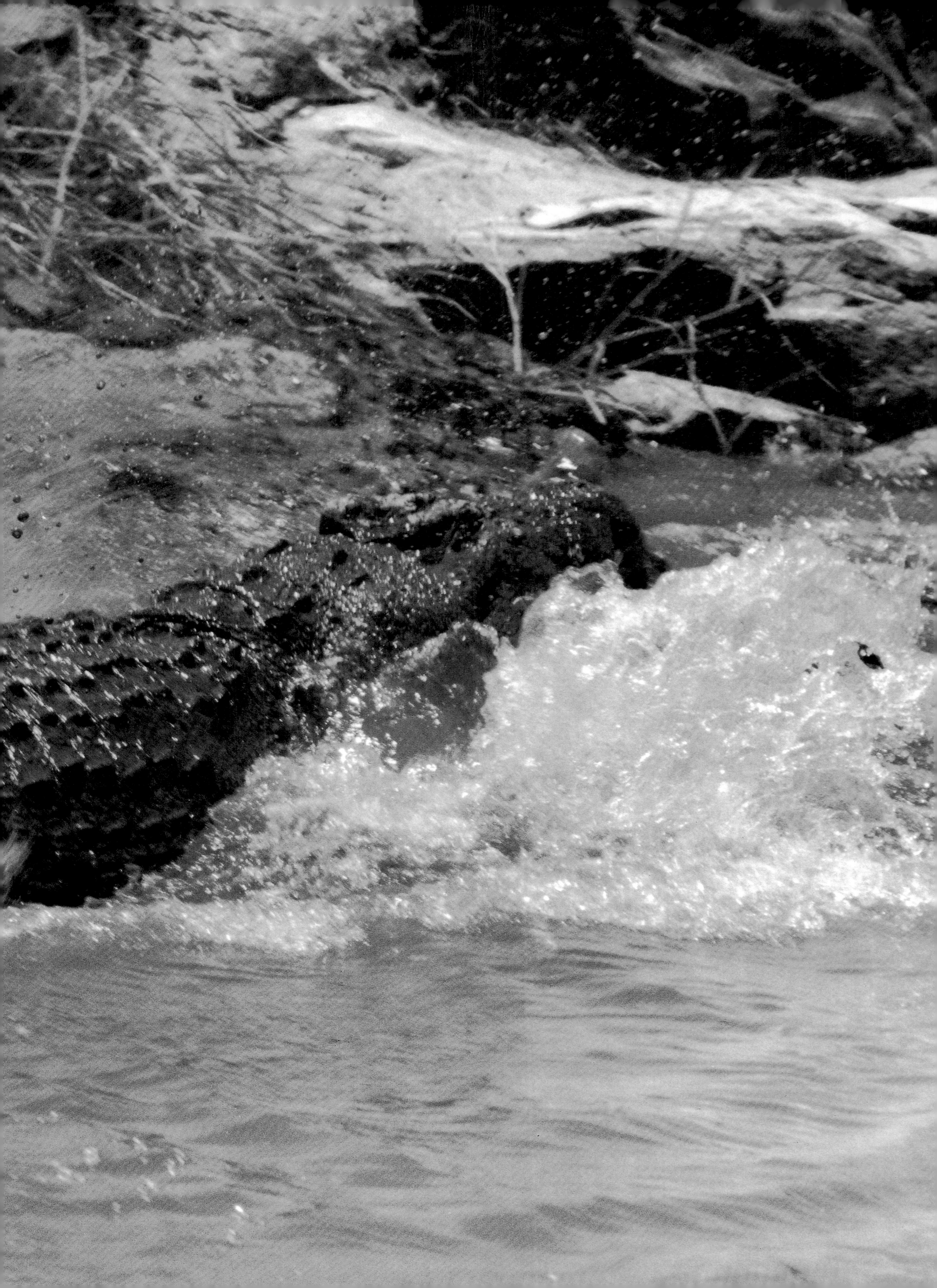

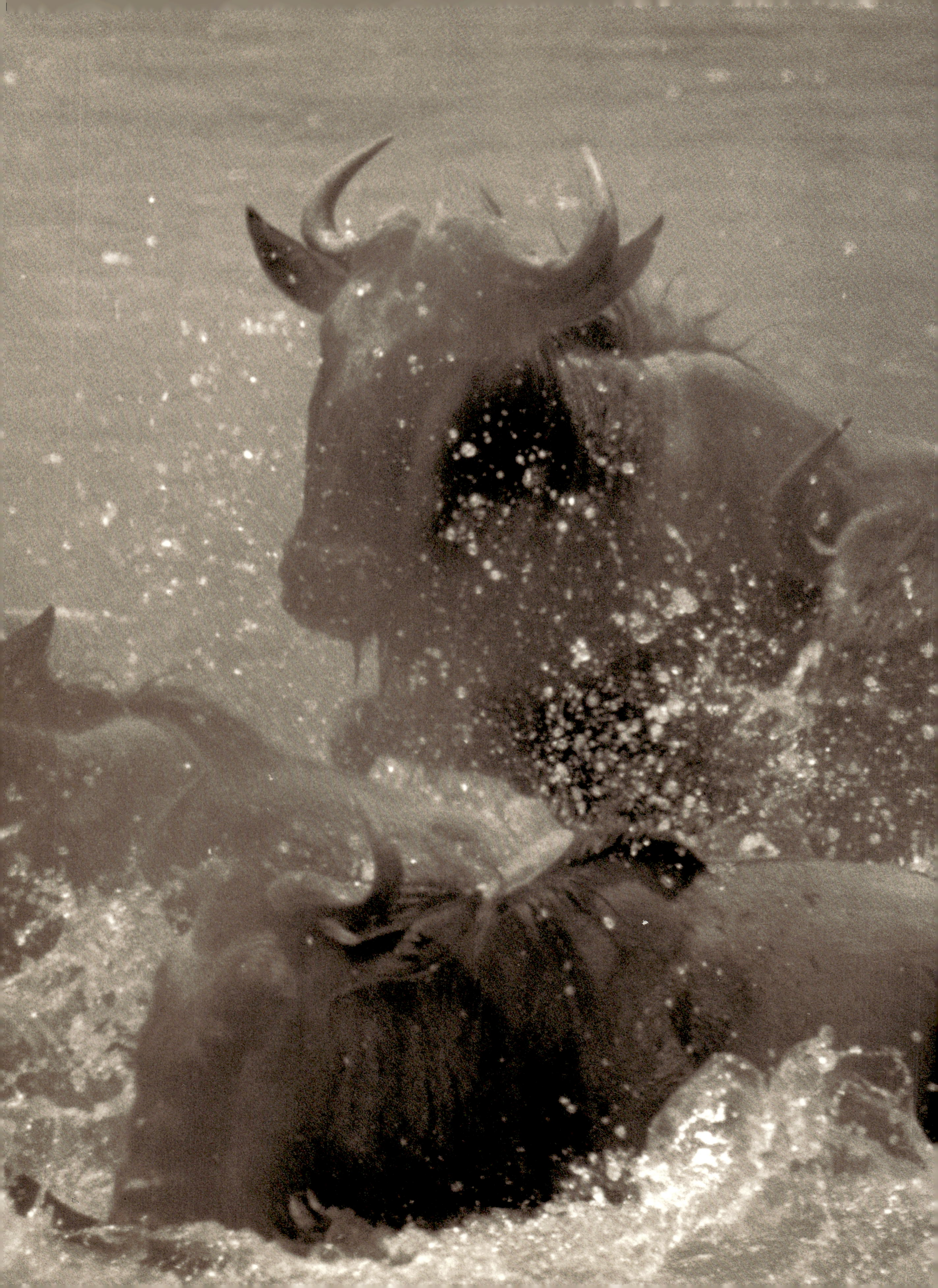

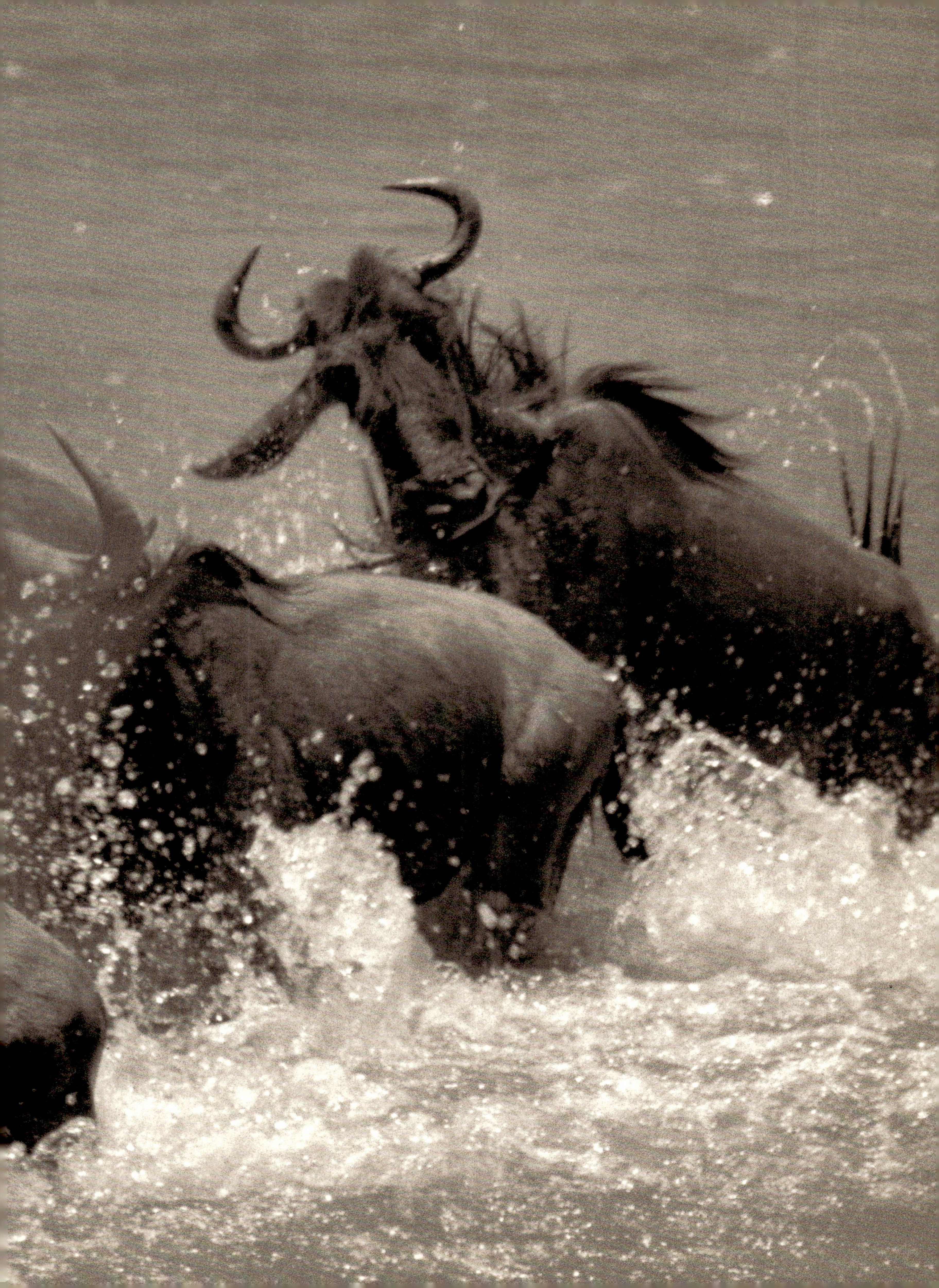

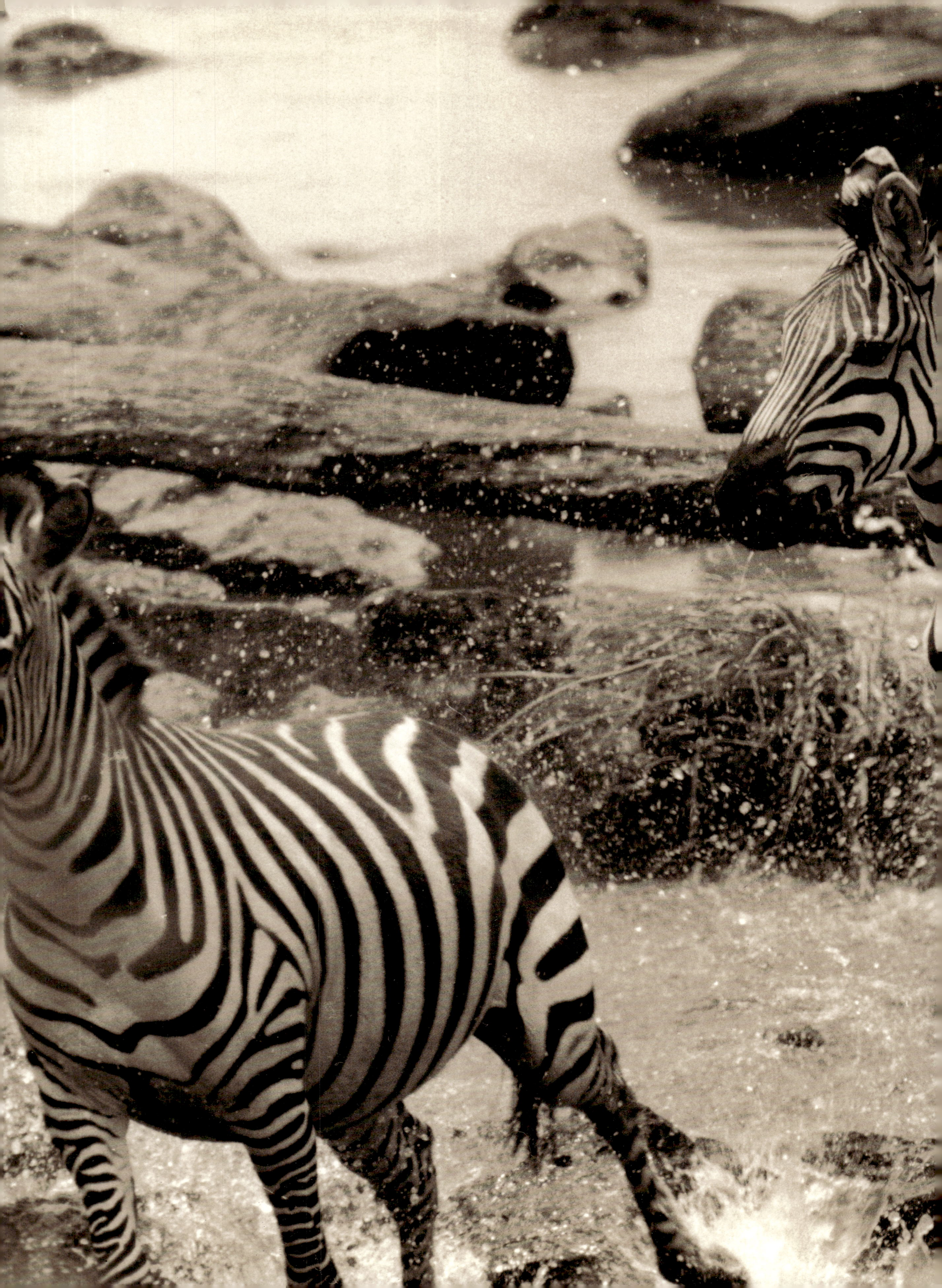

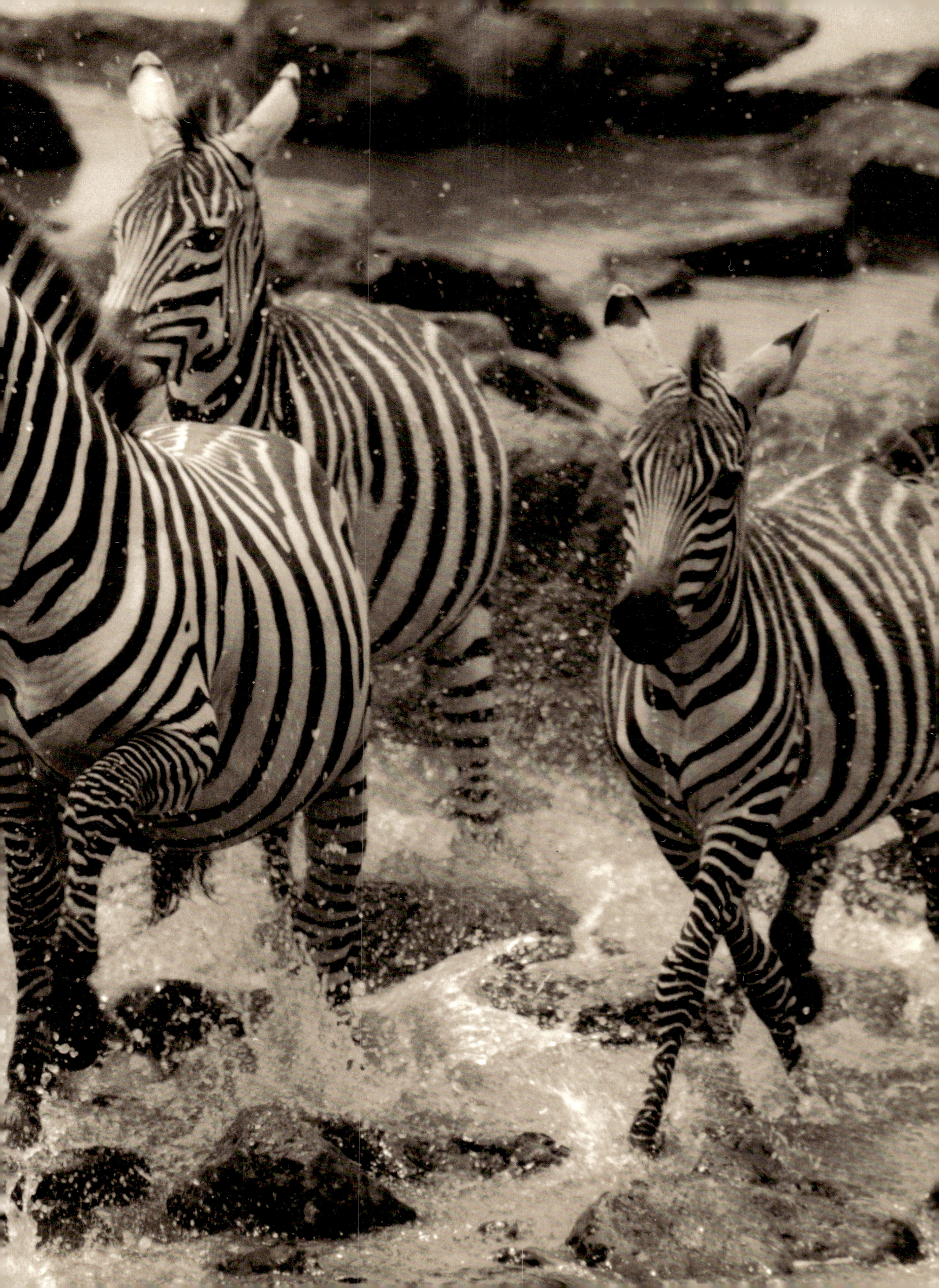

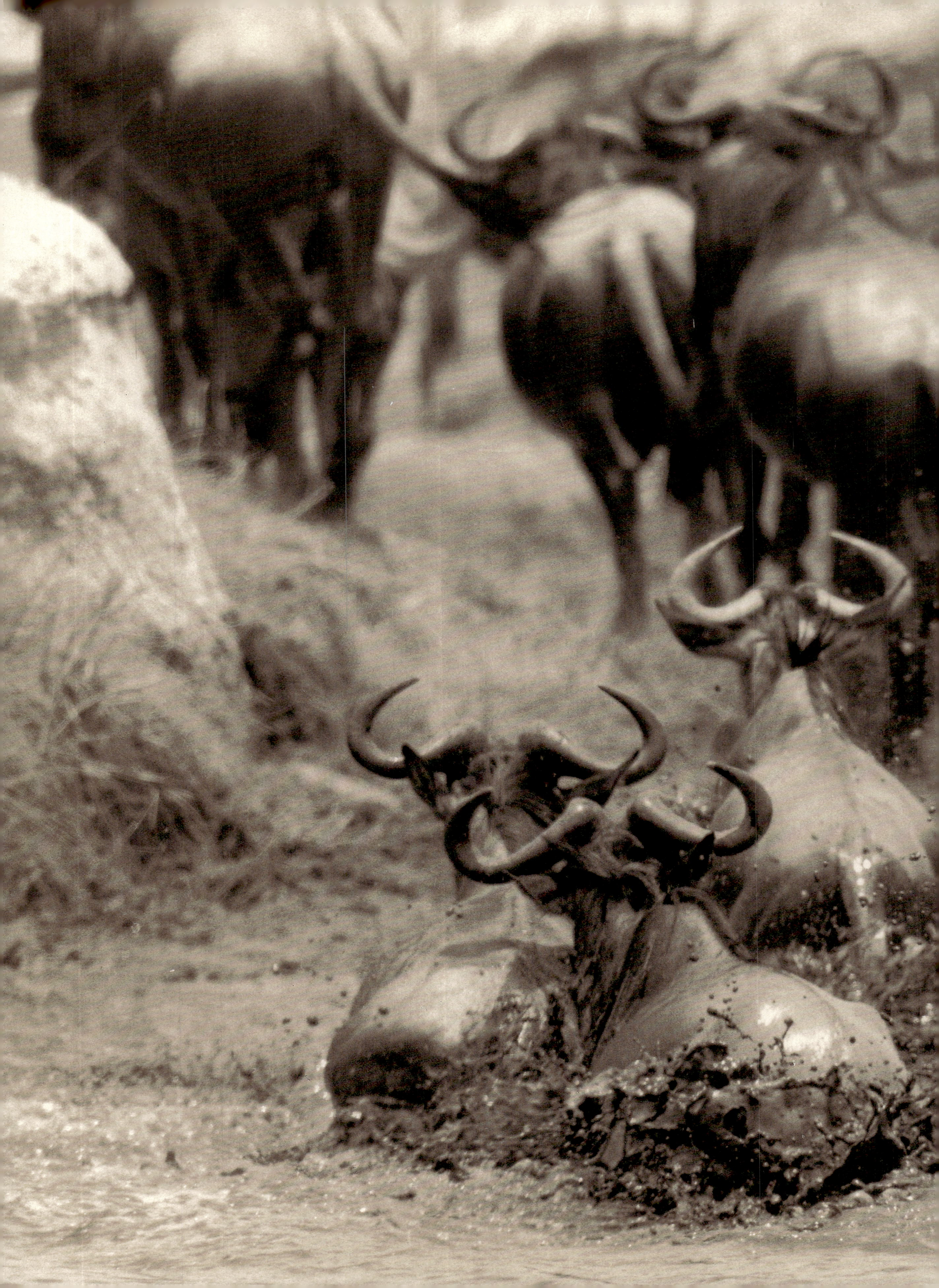

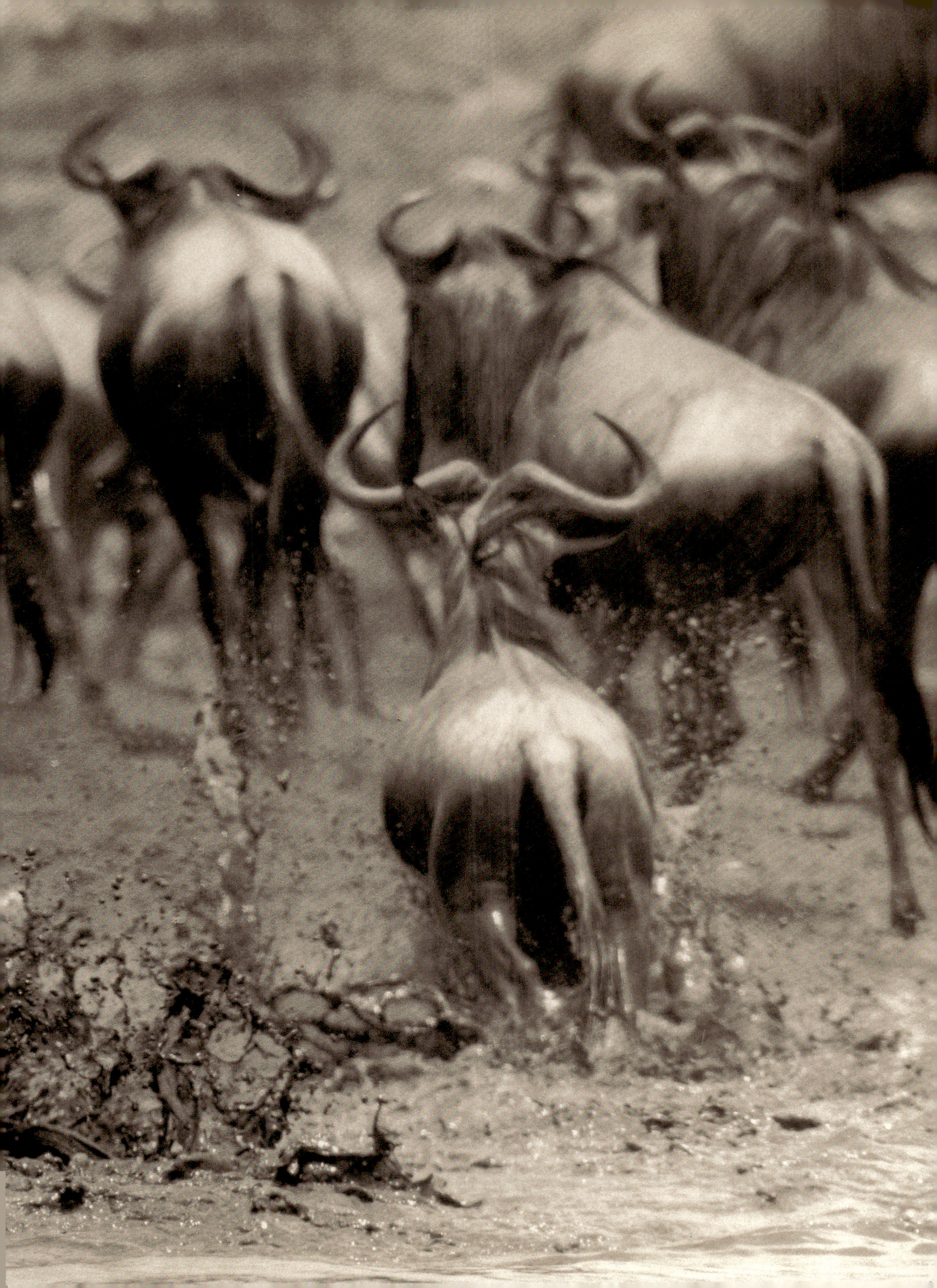

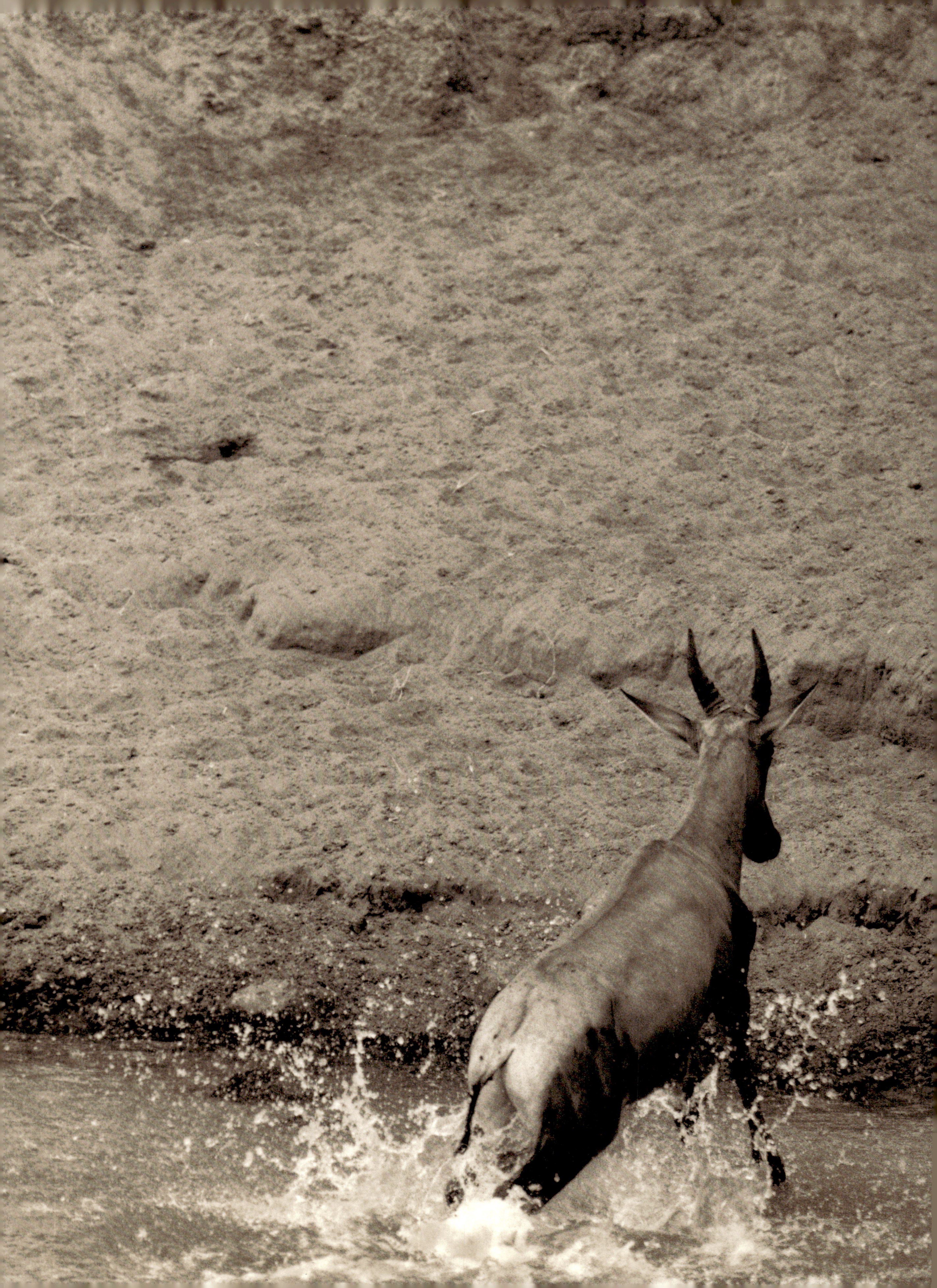

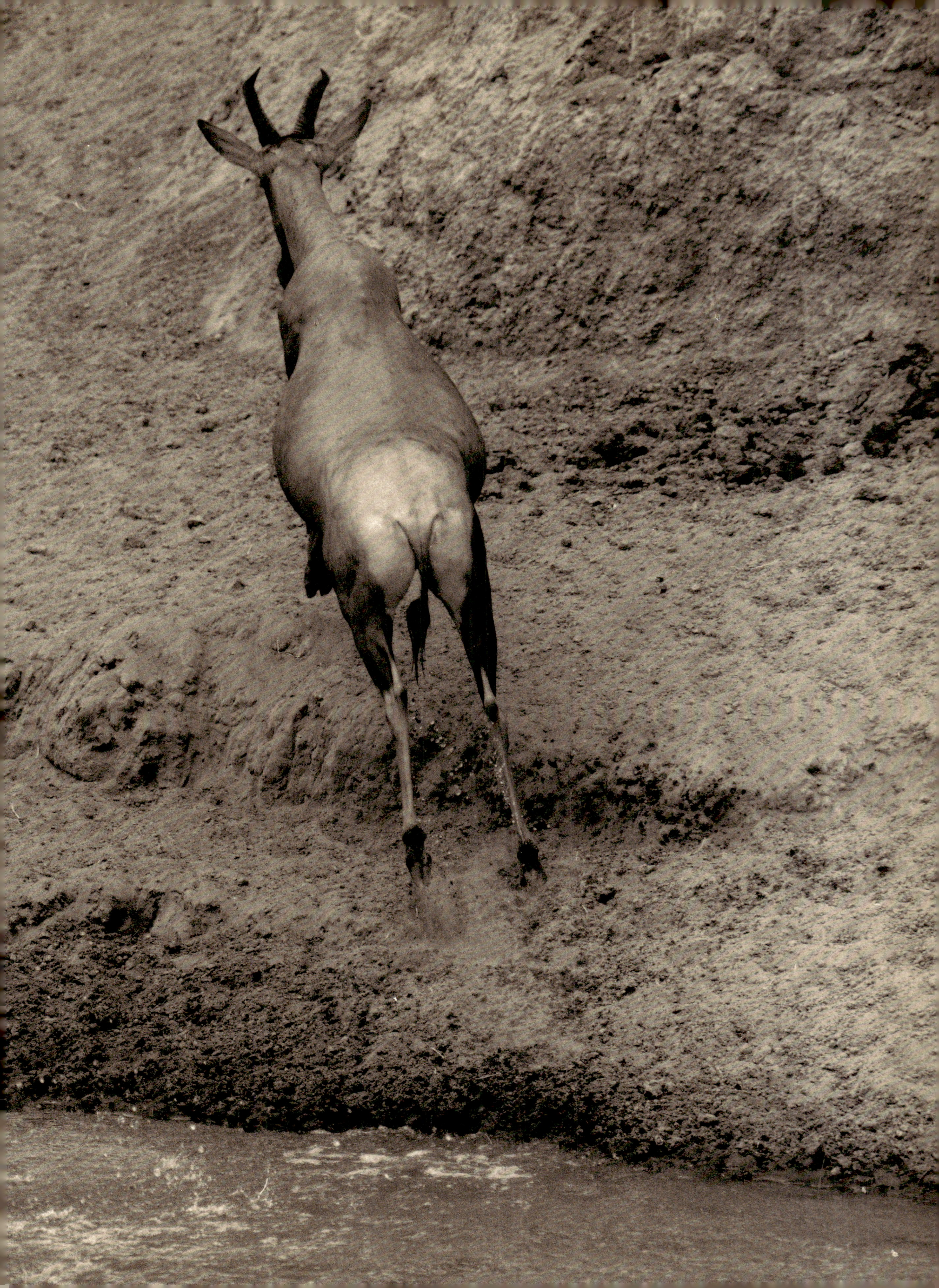

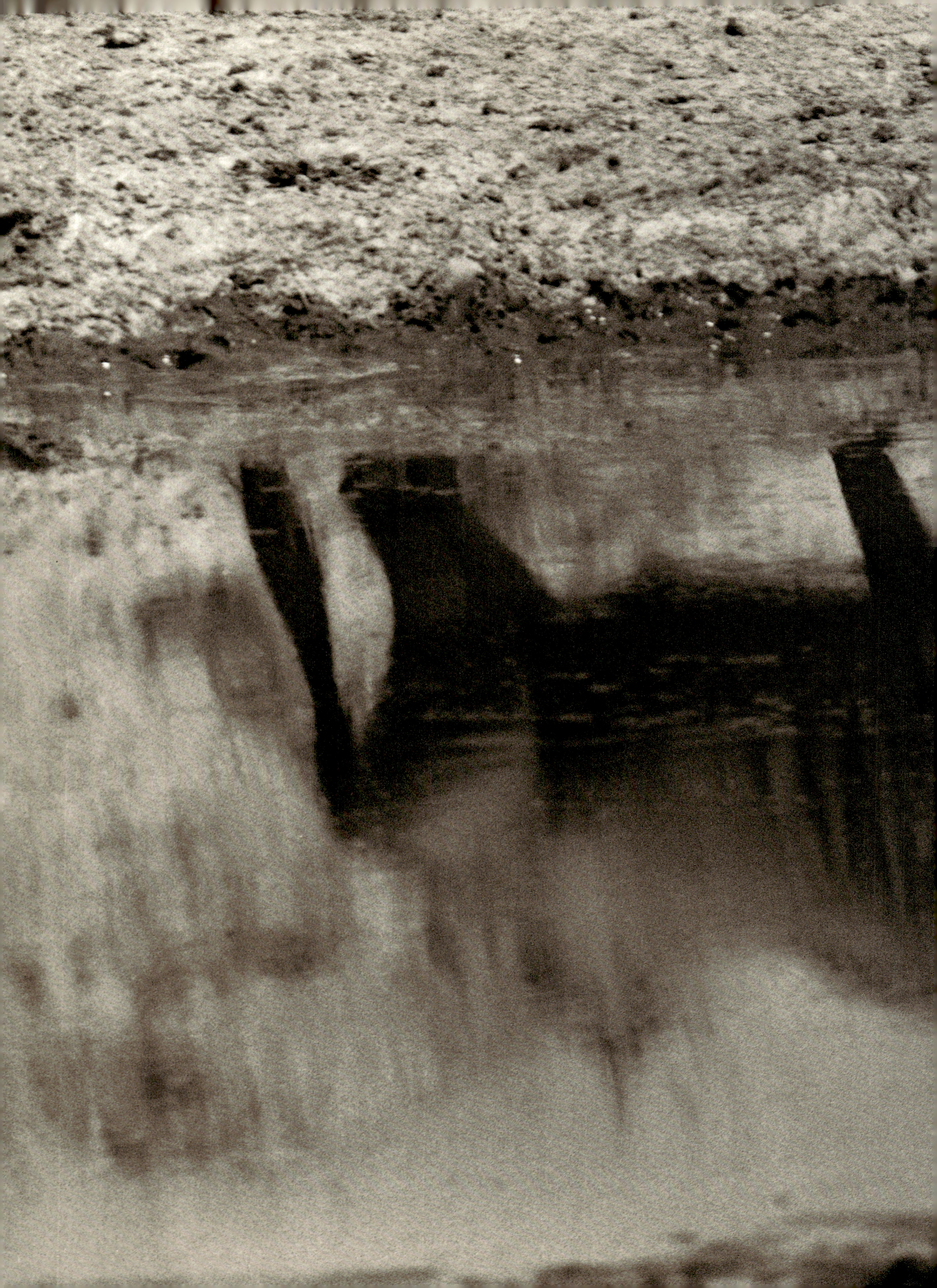

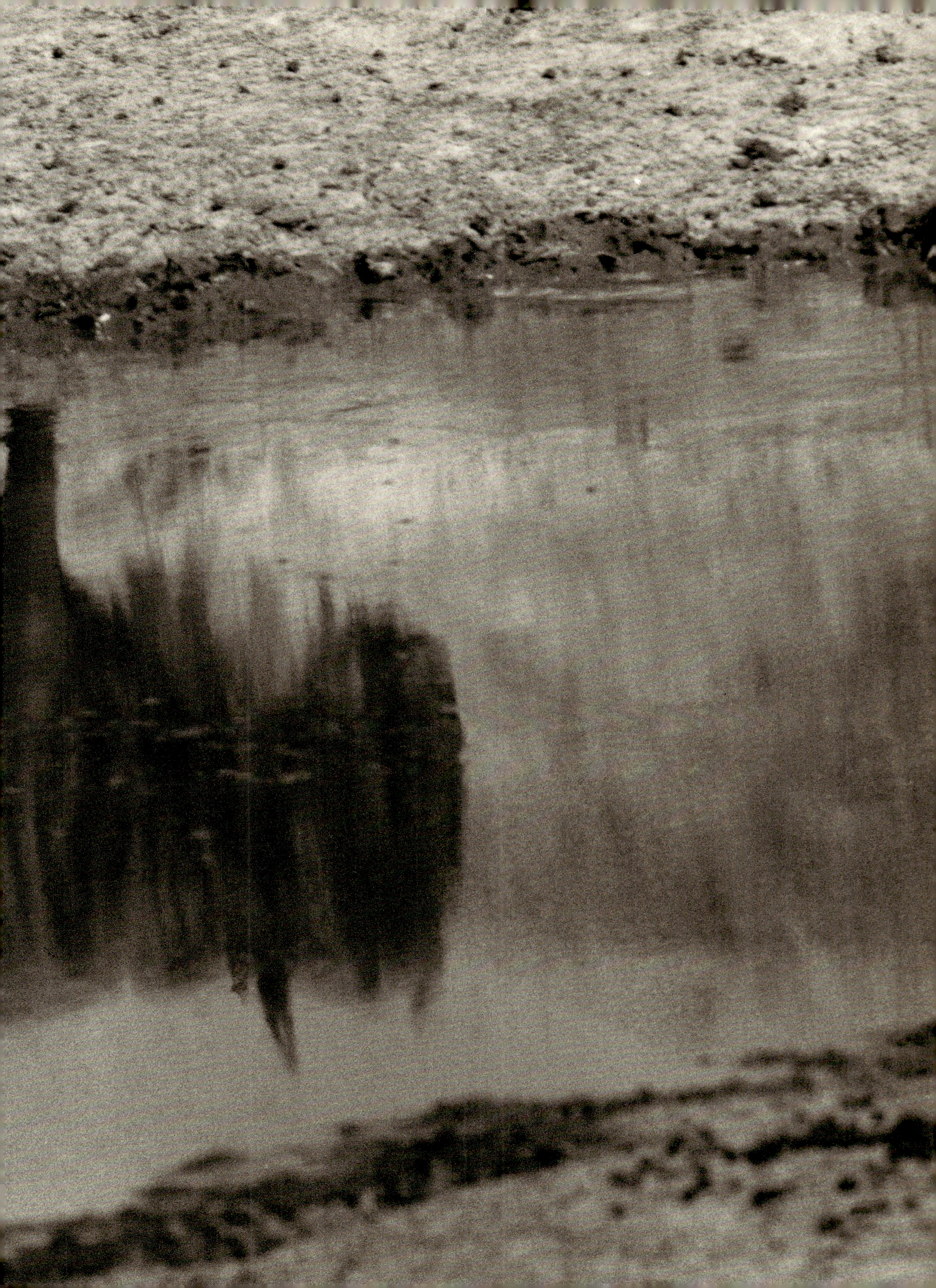

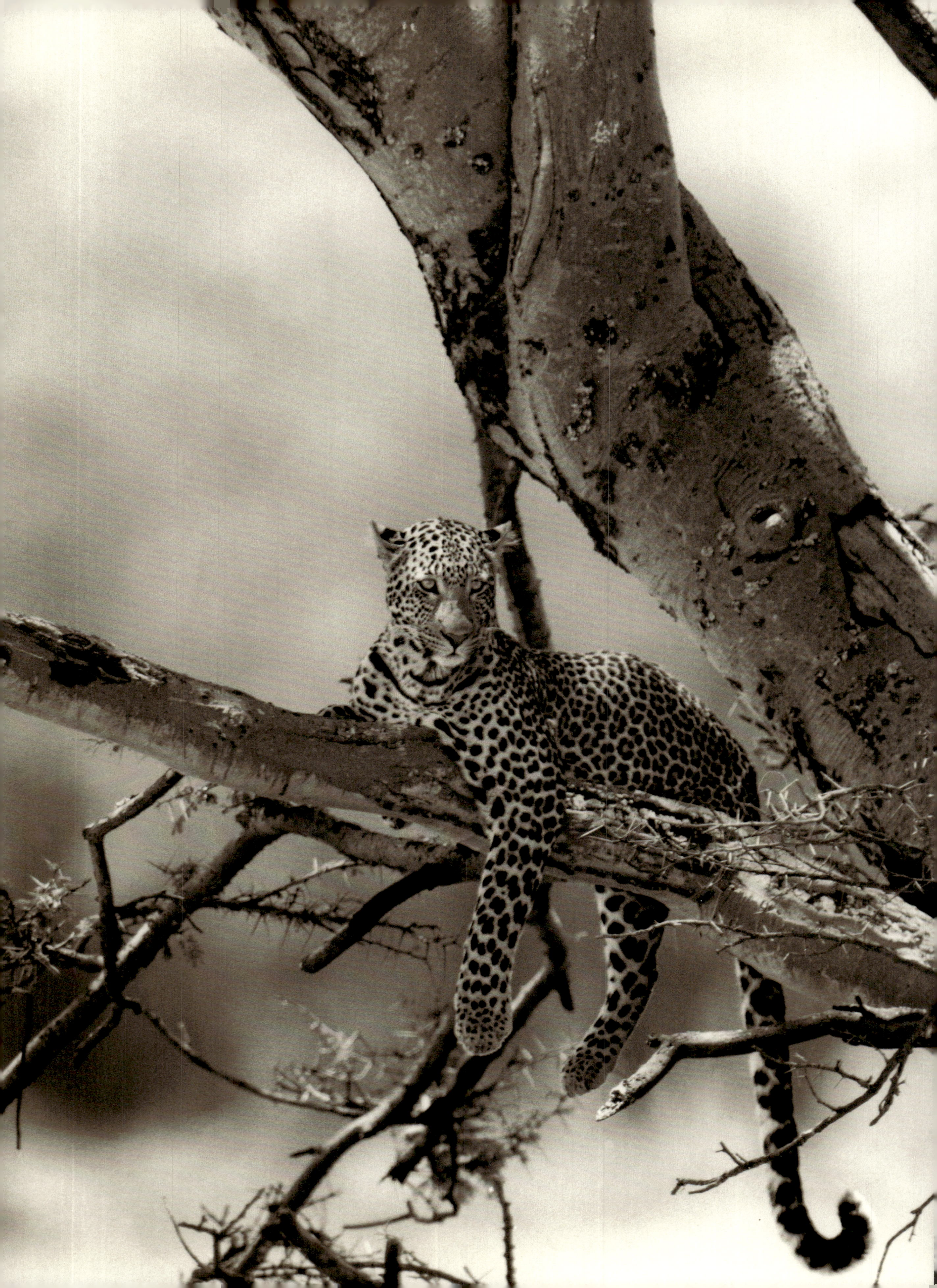

3

EASTERN WOODLANDS

Short Dry Season, December–February

HALF PUSHED, HALF PULLED, THE WILDEBEESTE'S STOMACHS AND GENES WILL GUIDE THEM BACK SOUTH, THROUGH THE SERENGETI'S EASTERN WOODLANDS. THERE AN ARRAY OF PREDATORS AWAITING THEIR PASSAGE CHANGES SUBTLY: MORE LEOPARDS, FEWER CHEETAHS. LEOPARDS MAKE GOOD USE OF TREES AS FURNITURE. A TREE IS A GOOD PLACE TO LIE IN THE SHADE IN THE HEAT OF THE DAY. IT PROVIDES A GOOD LOOK-OUT VANTAGE TO SPOT PASSING MIGRATION STRAGGLERS AND PLAN THE BEST ATTACK ROUTE FOR THE FOLLOWING NIGHT'S AMBUSH. TREES ALSO PROVIDE LARDERS TO STACK AWAY A PARTLY EATEN KILL, KEEPING IT SAFE FROM THE PURLOINING LIONS AND OBNOXIOUS HYENAS.

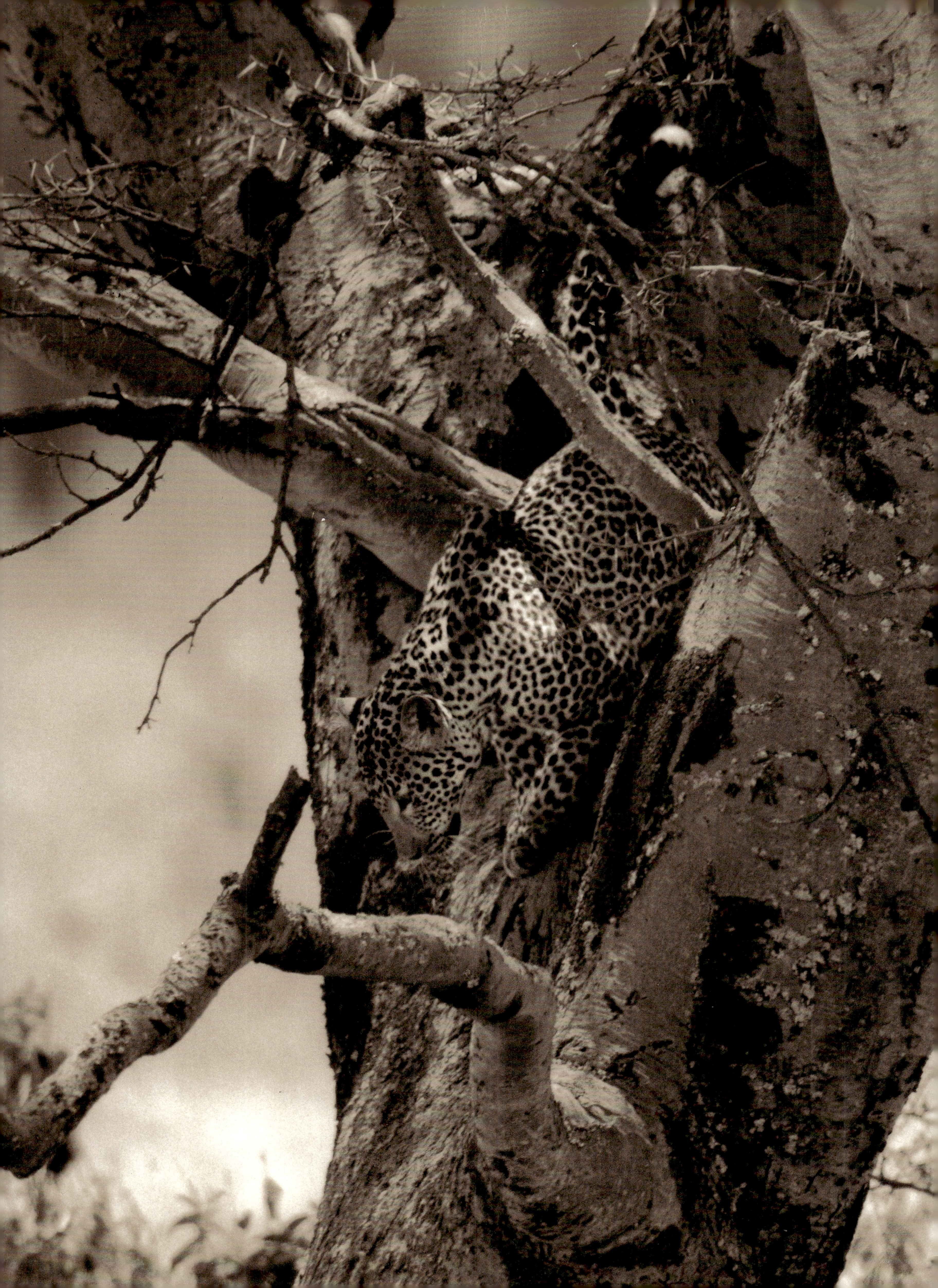

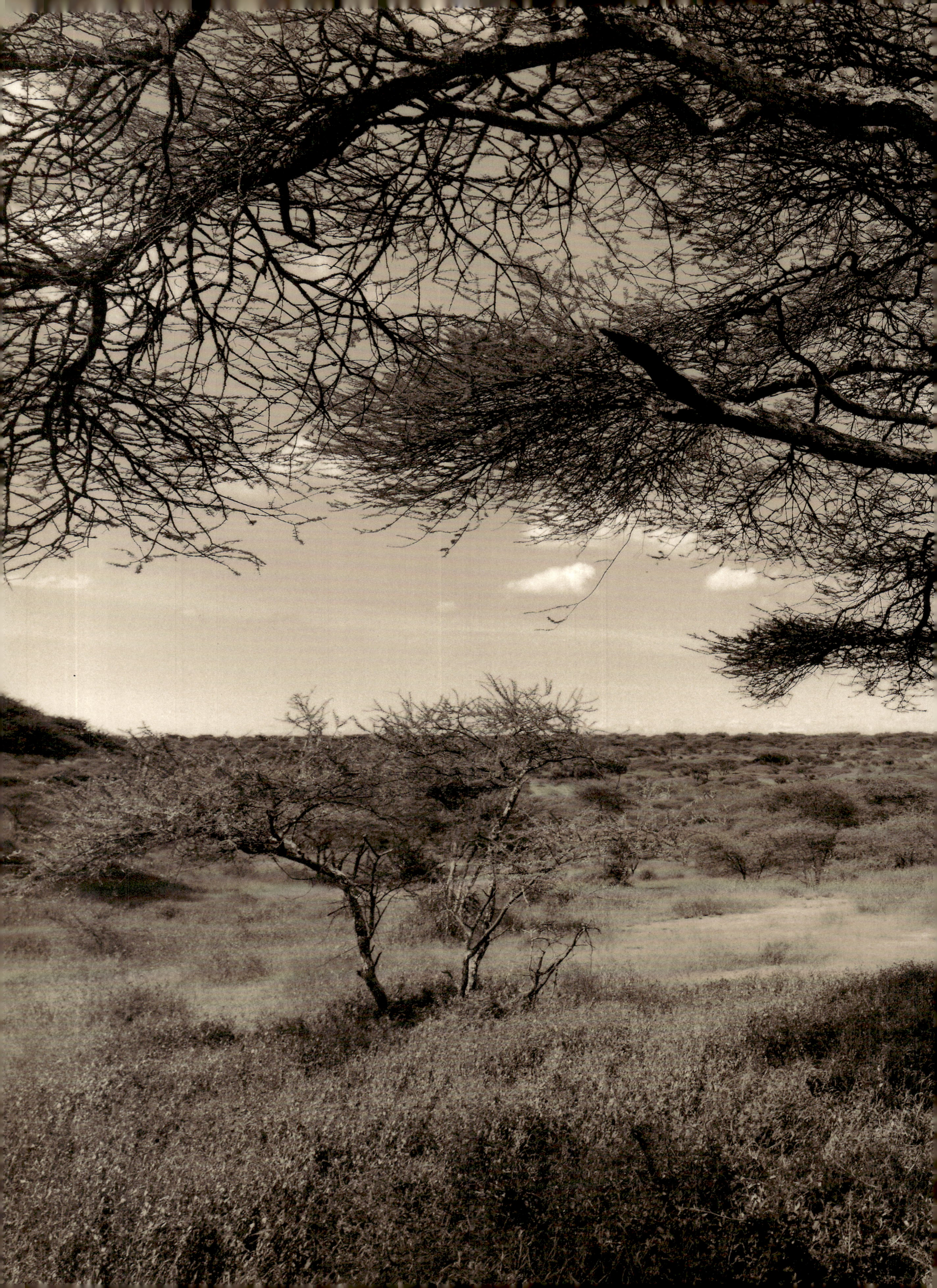

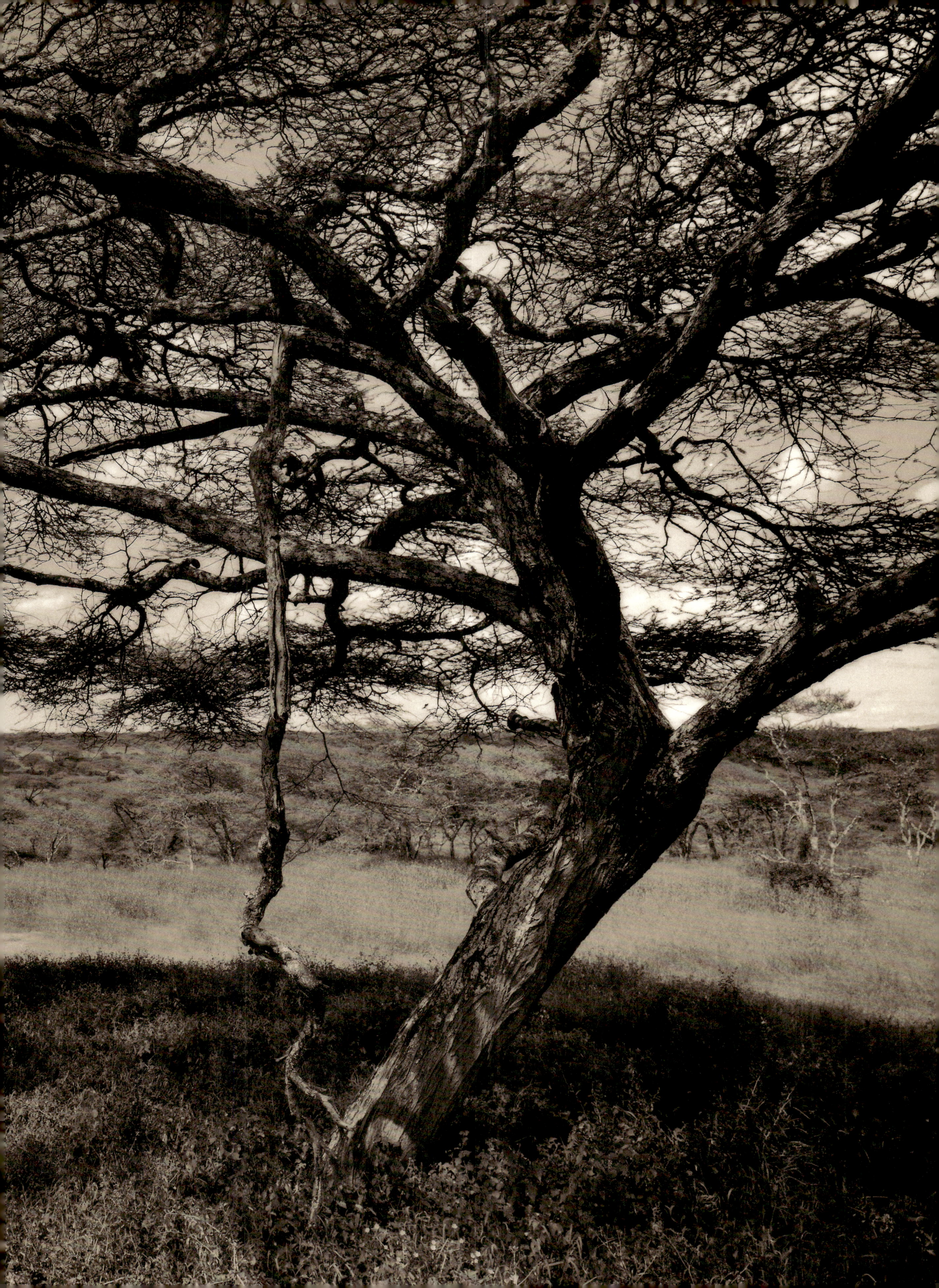

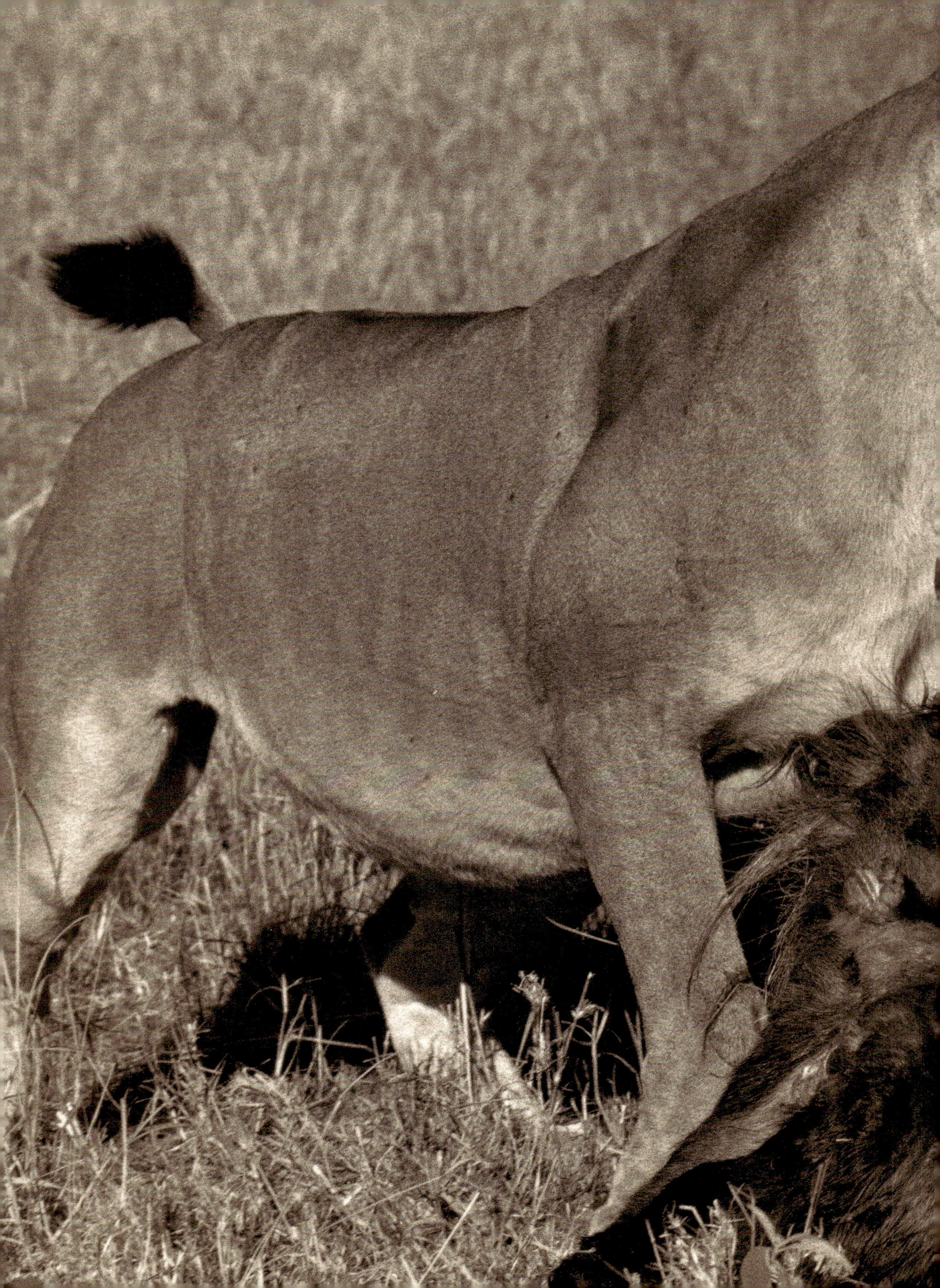

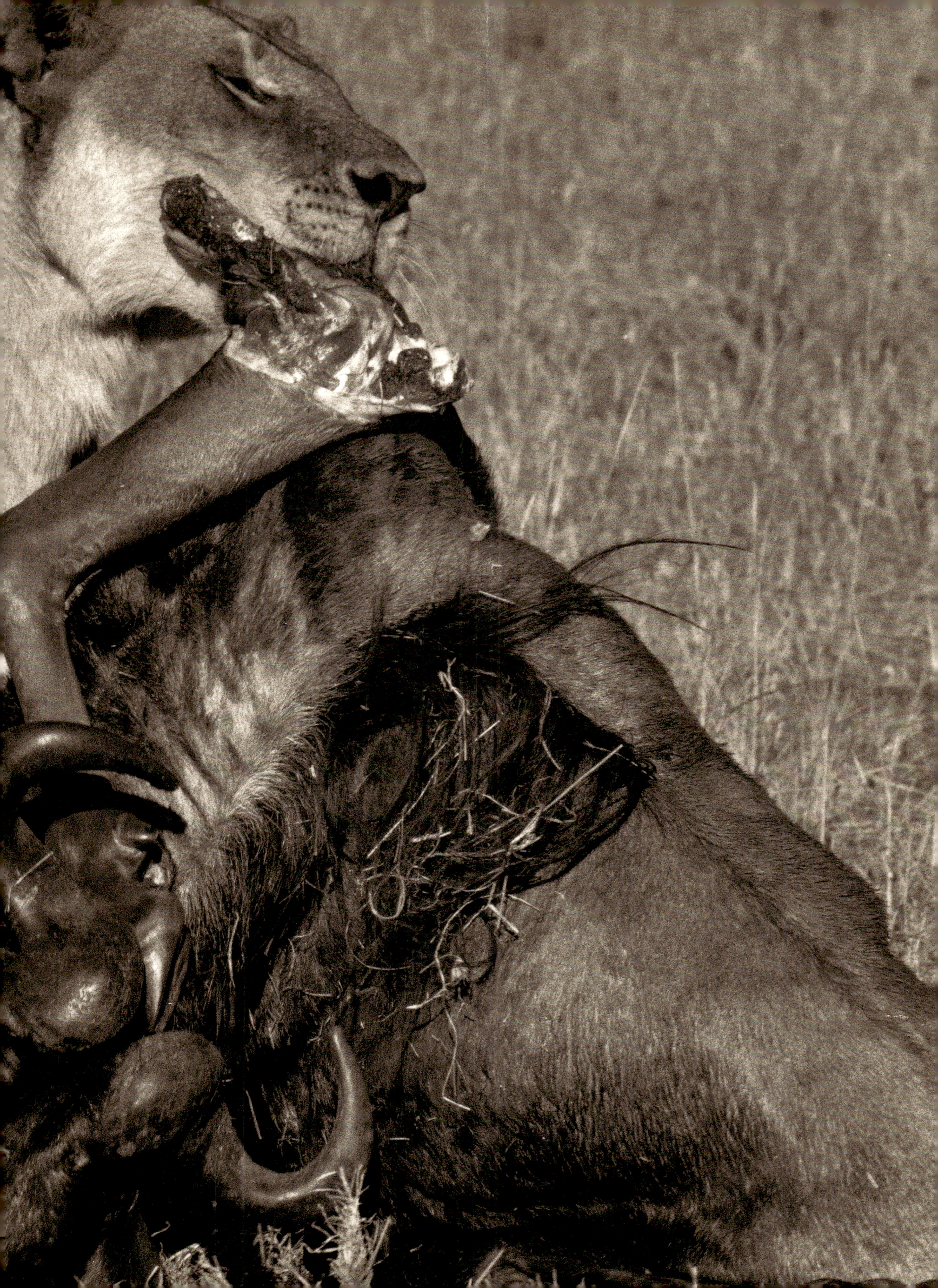

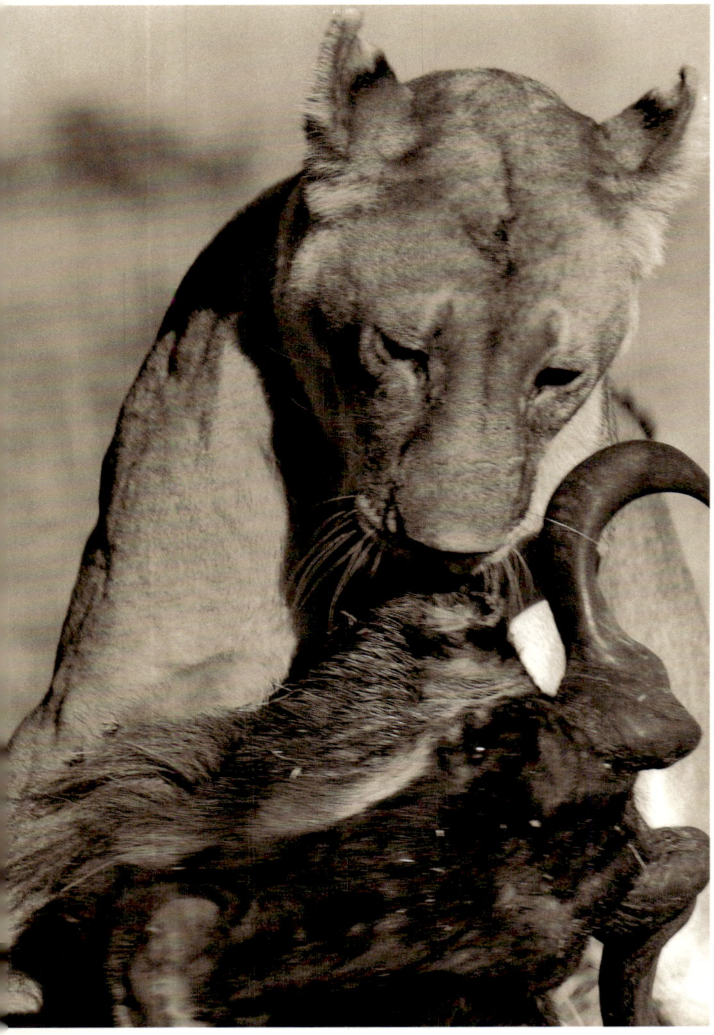

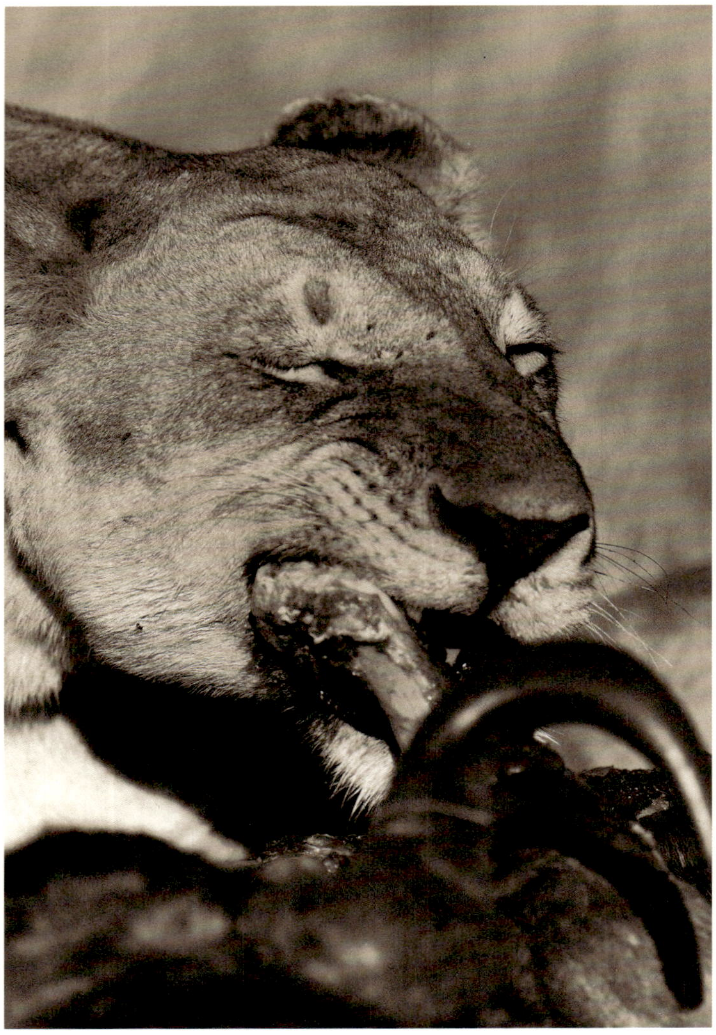

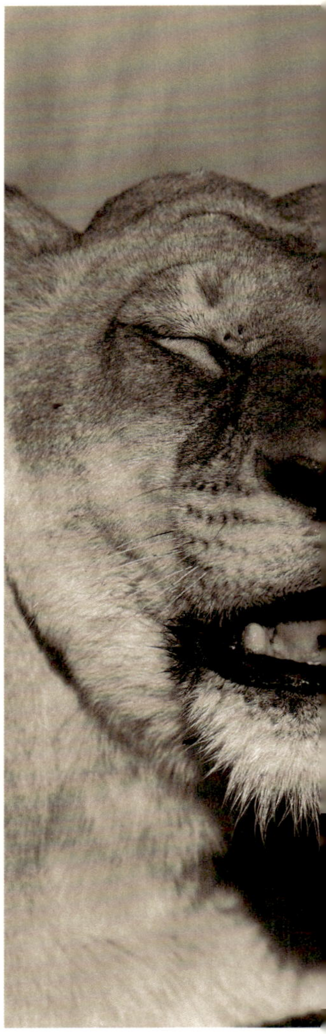

ONE-ON-ONE, WITH A FAIR START, A WILDEBEEST OR A ZEBRA CAN OUTRUN A LION. IF LIONS WERE SOLITARY HUNTERS, THEY WOULD BE HUNGRY MORE OFTEN THAN NOT. THE SOLUTION, AS WITH MANY OF LIFE'S CHALLENGES, LIES IN COOPERATION. THE SHARE FOR EACH MAY BE SMALLER, BUT IT IS BETTER THAN NONE AT ALL. LIONESSES ON PERMANENT TERRITORIES THROUGHOUT THE SERENGETI JOIN FORCES TO STALK, CHASE AND DRIVE POTENTIAL PREY INTO THE CLAWS AND JAWS OF THEIR SISTERS WAITING IN AMBUSH. THE WEAKER AND SLOWER INDIVIDUAL HERBIVORES NATURALLY FALL

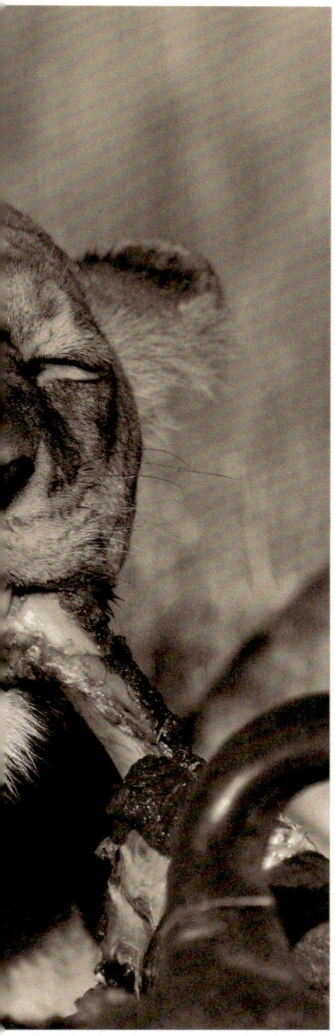

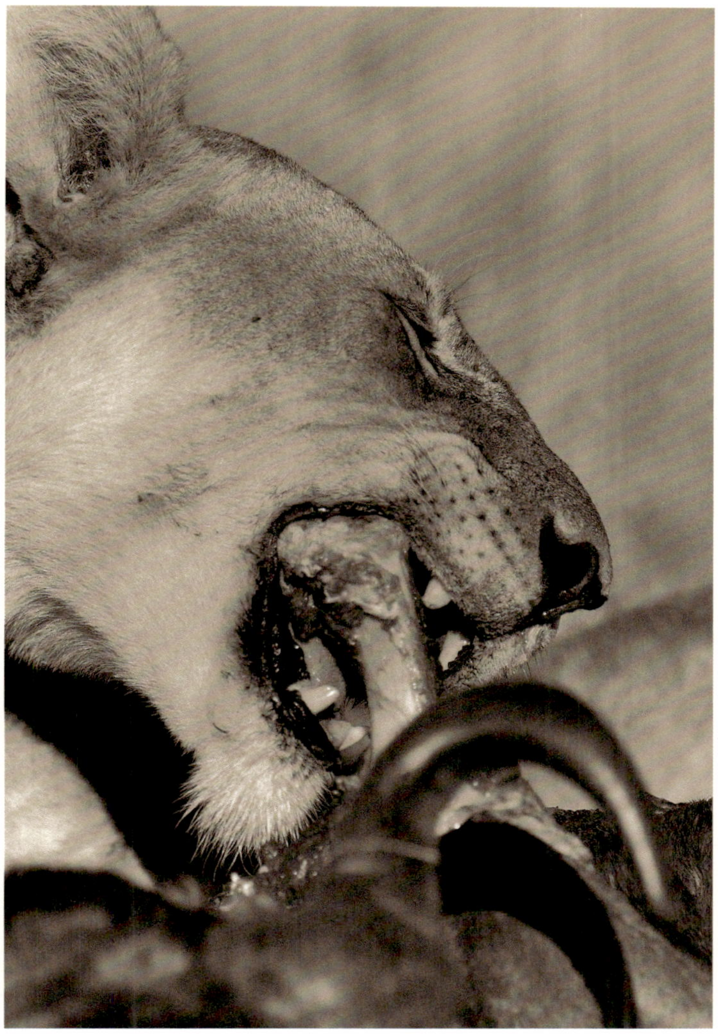

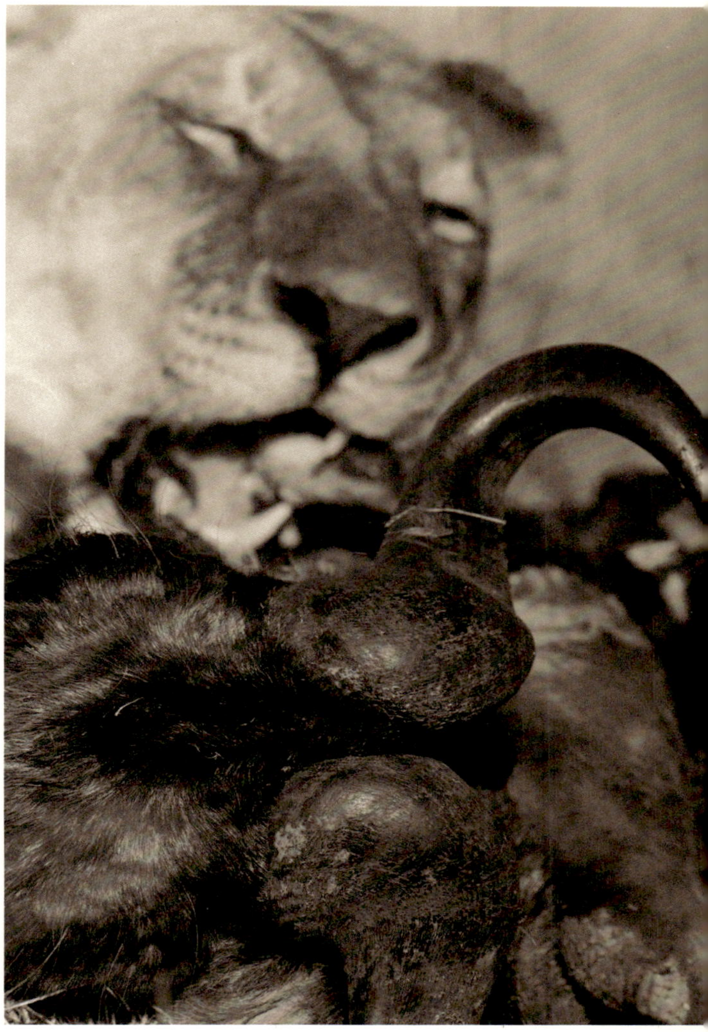

VICTIM FIRST. THE CAT THAT WAS GREETING AND RUBBING HEADS WITH HER PRIDEMATES EARLIER IN THE DAY BECOMES AN INTOLERANT BEAST AT MEAL TIME. AT A KILL, SNARLING THREATS AND WARNING GROWLS ARE FREQUENT, ALL-OUT FIGHTS LESS SO. YOUNGSTERS SNATCH WHAT SCRAPS THEY CAN, DODGING IN AND OUT. NONE OF THE SOLICITOUS FEEDING OF YOUNG EXHIBITED BY, SAY, WILD DOGS. AND, IF THE PRIDE IS LUCKY, A PAIR OF LARGE NOMADIC MANED MALES WILL NOT ARRIVE TO STEAL THE ENTIRE MEAL FROM THE FEMALES AND YOUNG. THE KING OF BEASTS FREQUENTLY BEHAVES LIKE A COMMON THIEF.

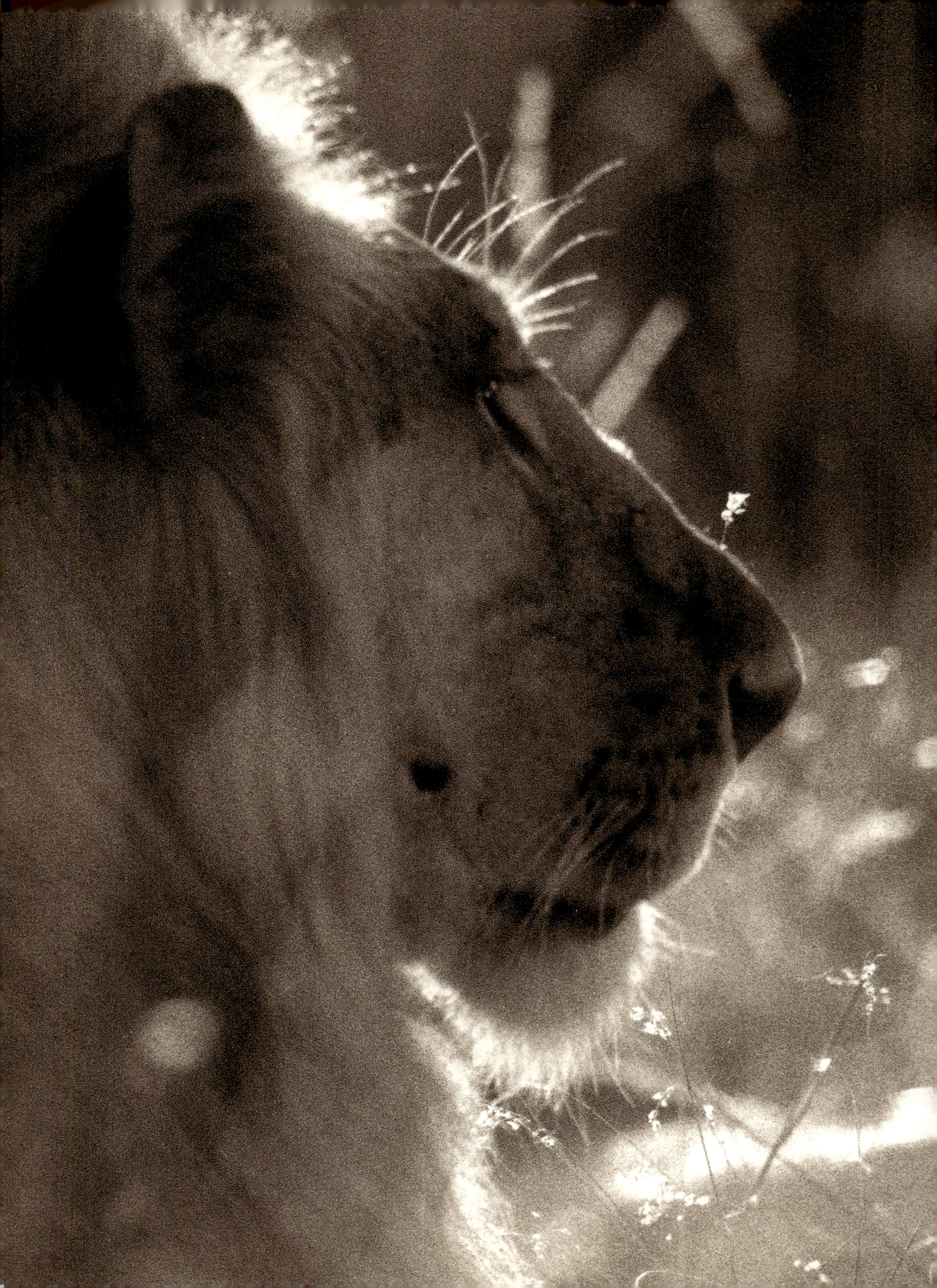

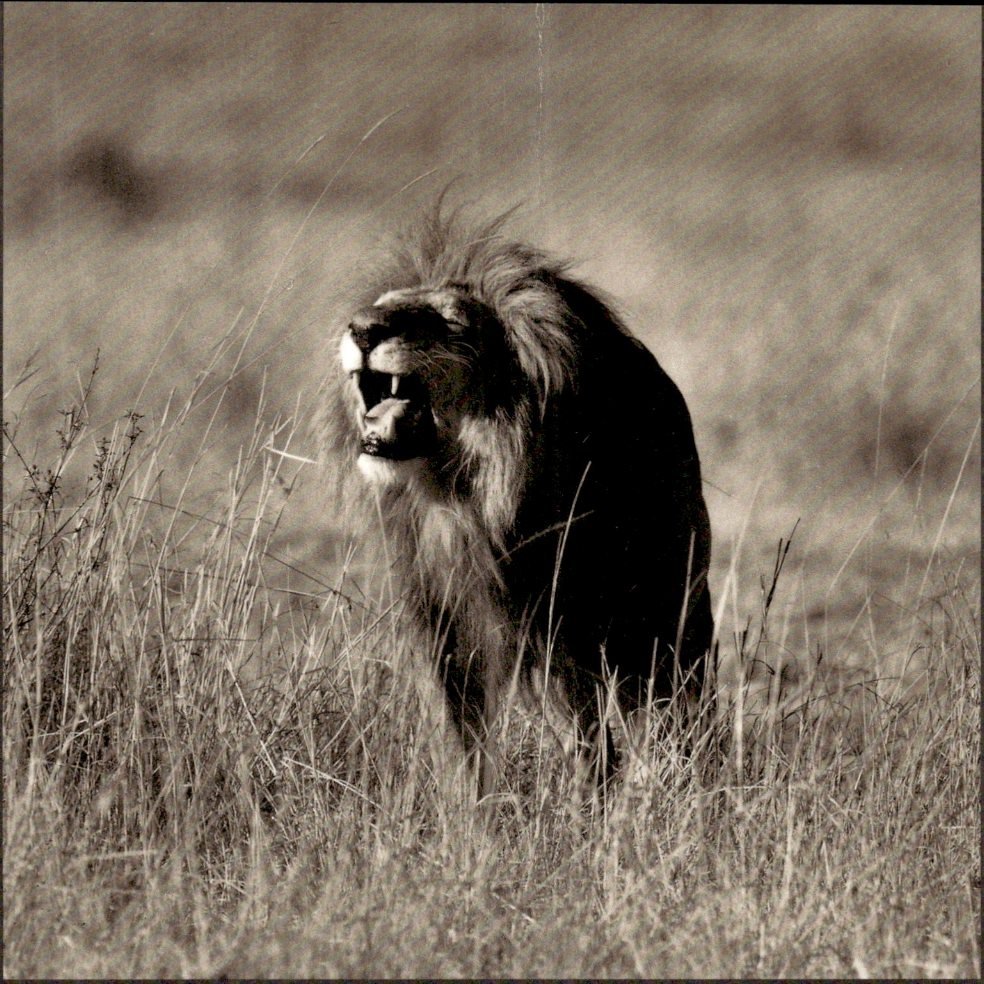

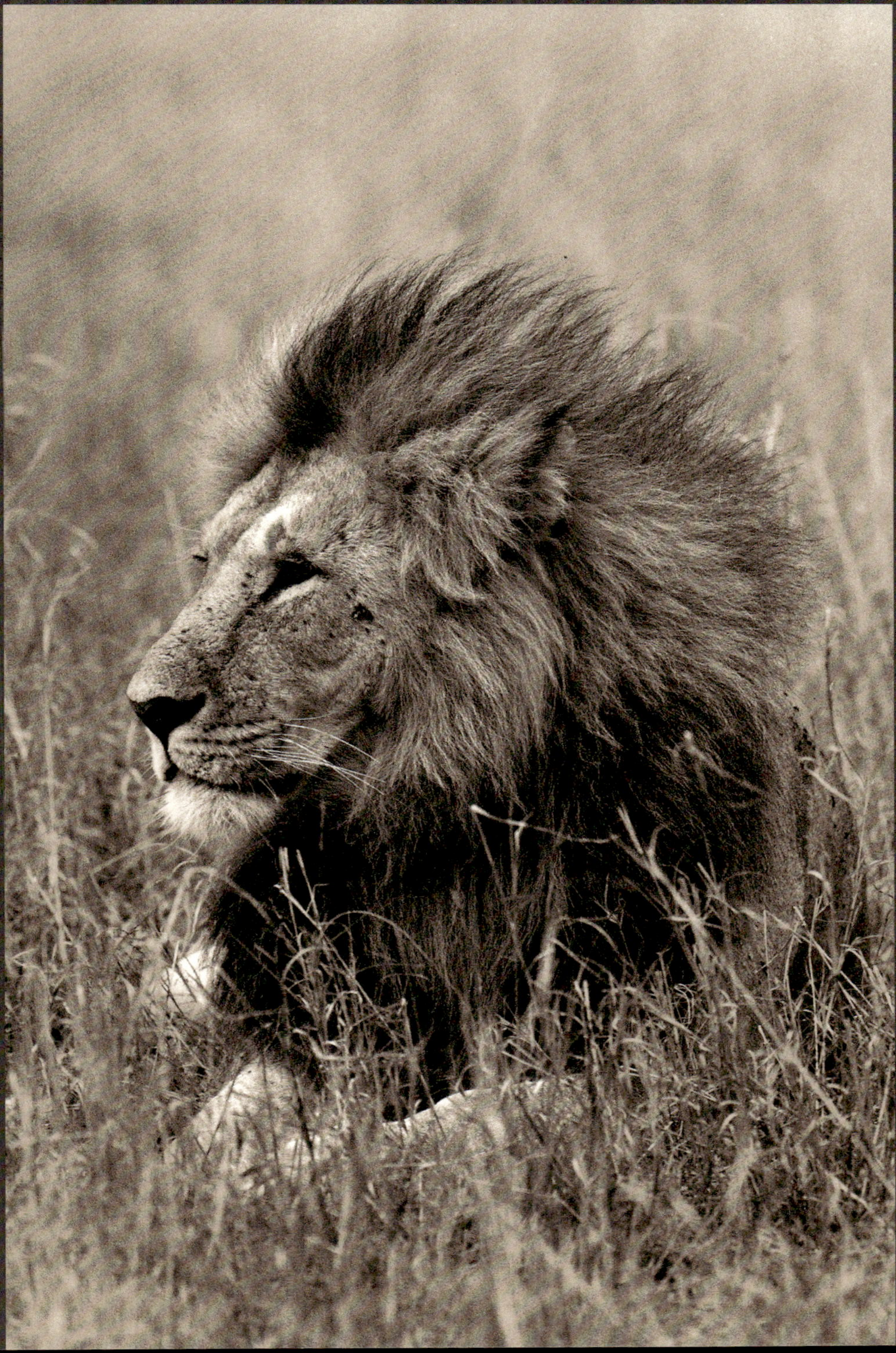

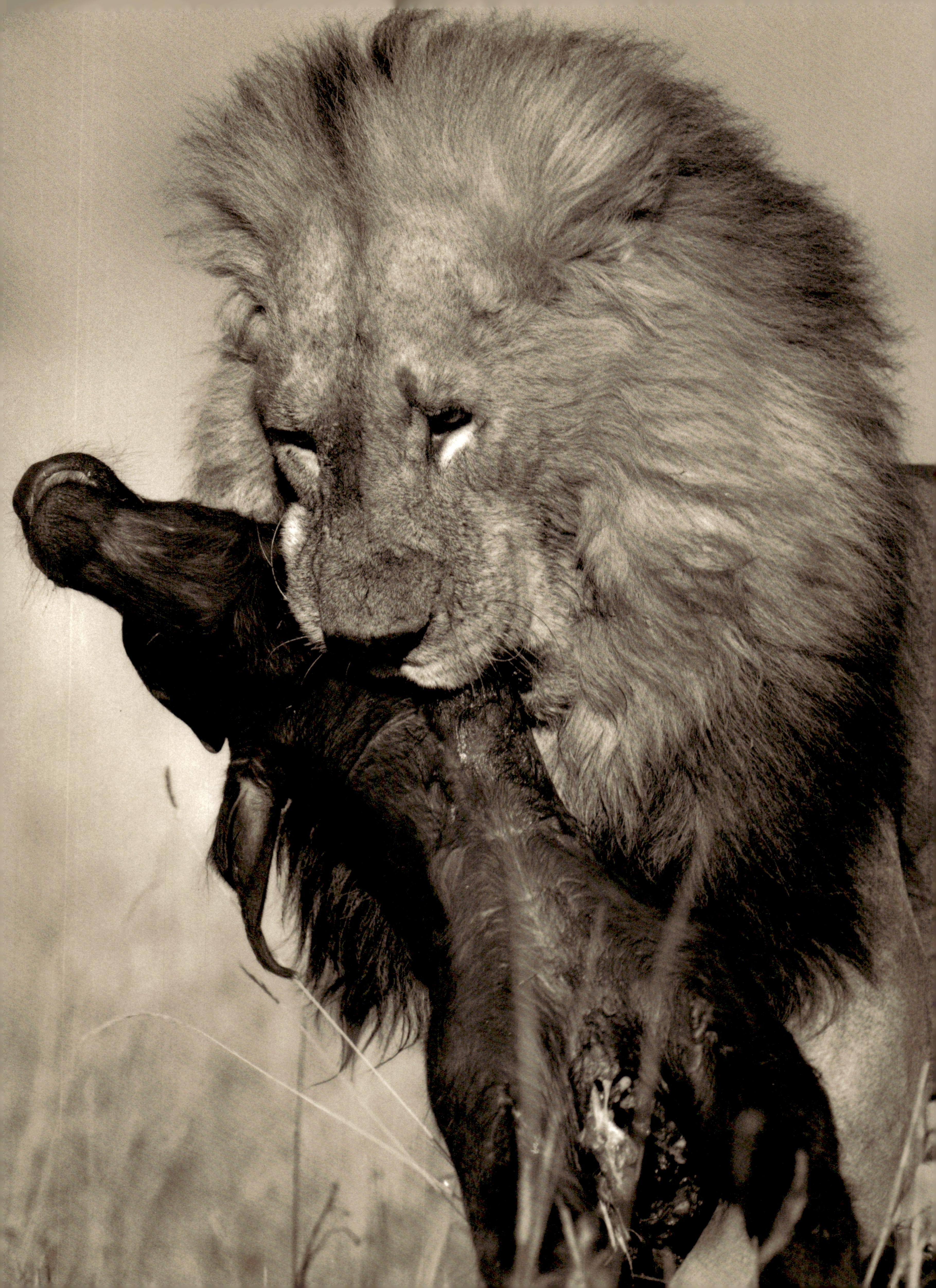

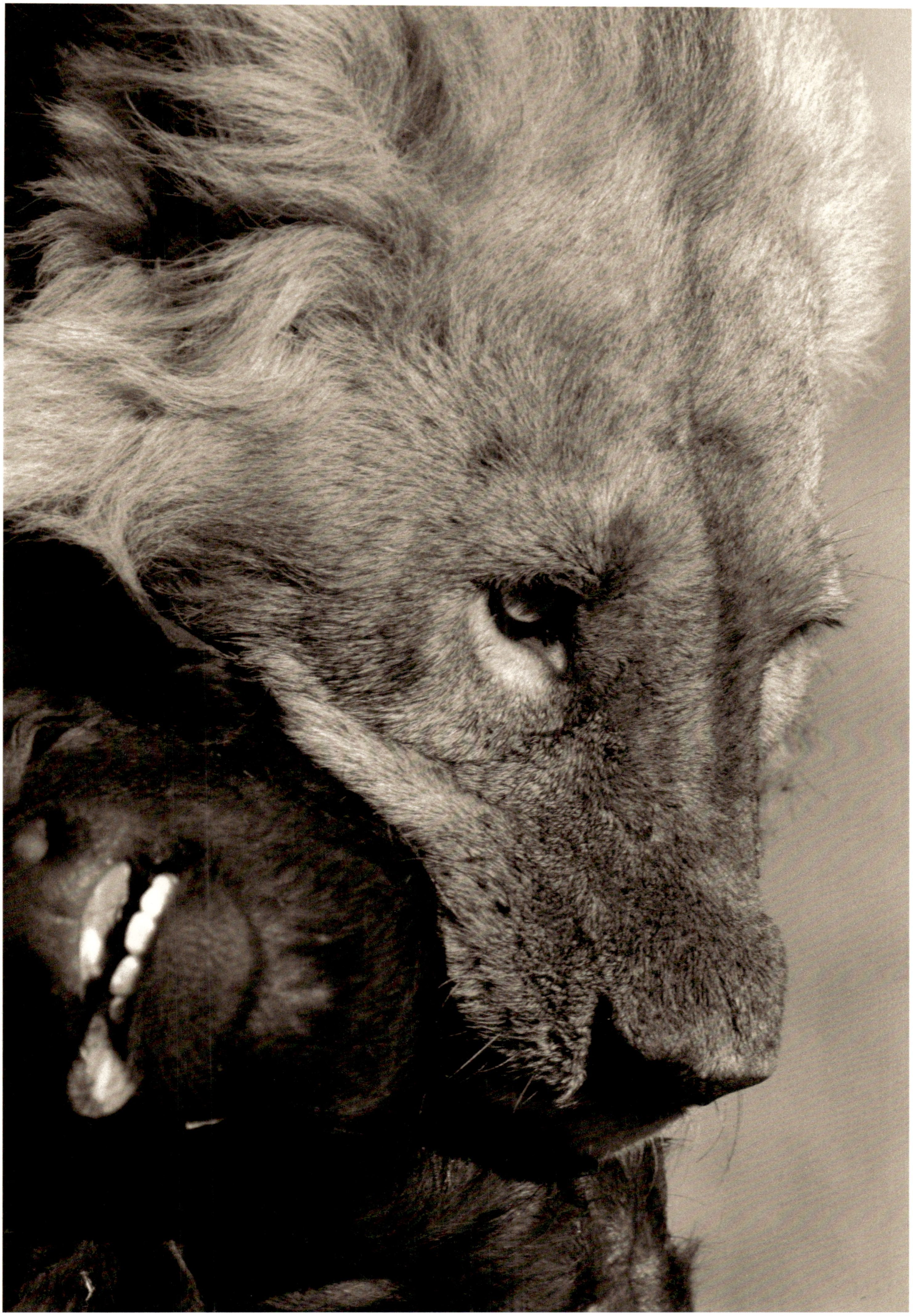

INEVITABLY, WHERE LIFE THRIVES, DEATH ABOUNDS, INEXORABLY COMPLETING THE CYCLE OF LIFE. THE ECOSYSTEM WOULD OTHERWISE BE HOPELESSLY OVERCROWDED AND UNSUSTAINABLE. HAPPILY, YOU CAN ALWAYS RELY ON THE SCAVENGER CLUB INCLUDING THE OPPORTUNISTIC HYENAS TO BE ON HAND, WATCHING AND WAITING TO DO THEIR PART IN THE NECESSARY TASK OF TIDYING UP THE ECOSYSTEM AND HASTENING THE RETURN OF THE MIGRATIONS' CONSTITUENT PARTS BACK TO THE SOIL READY TO BE USED AGAIN.

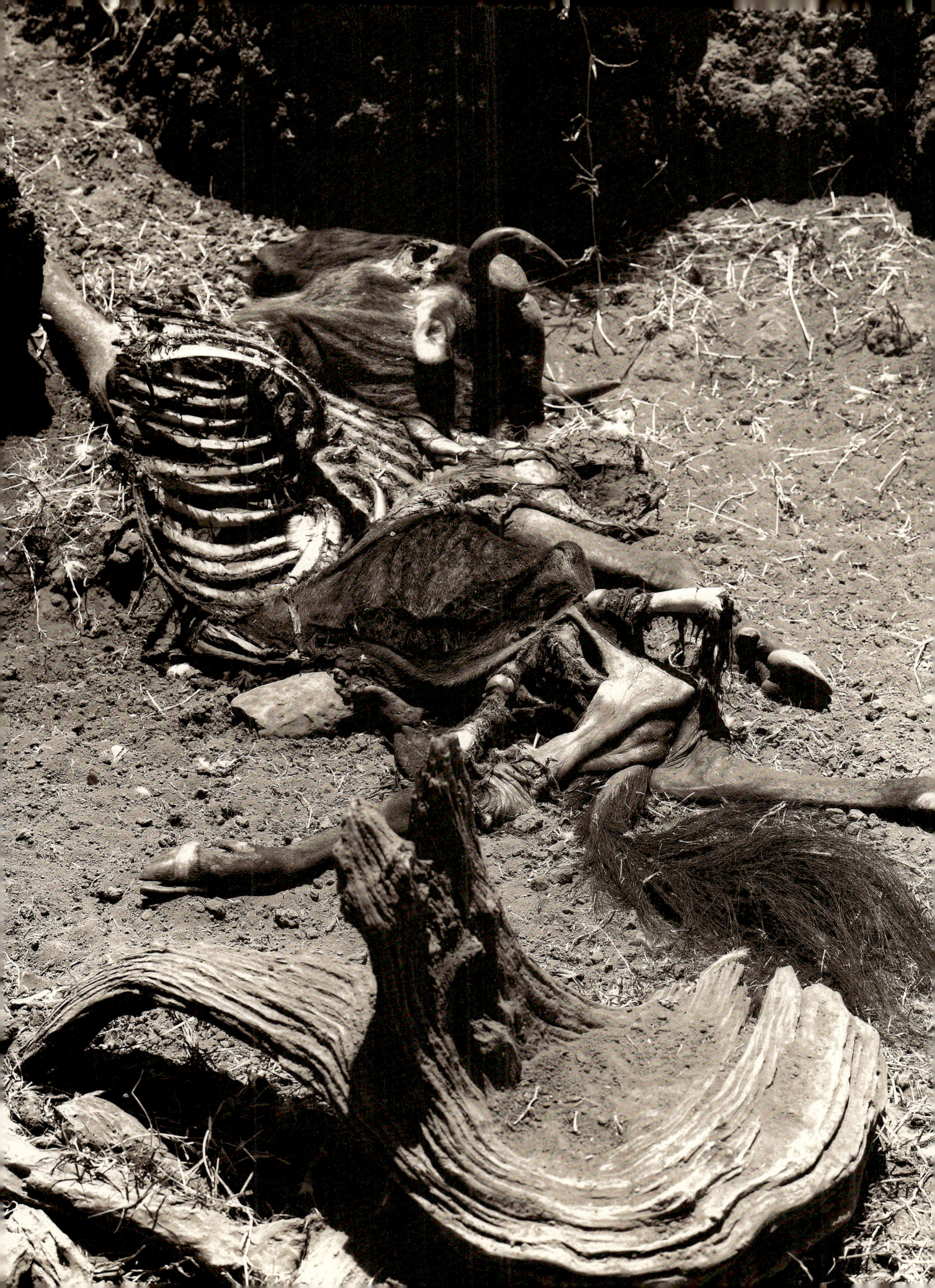

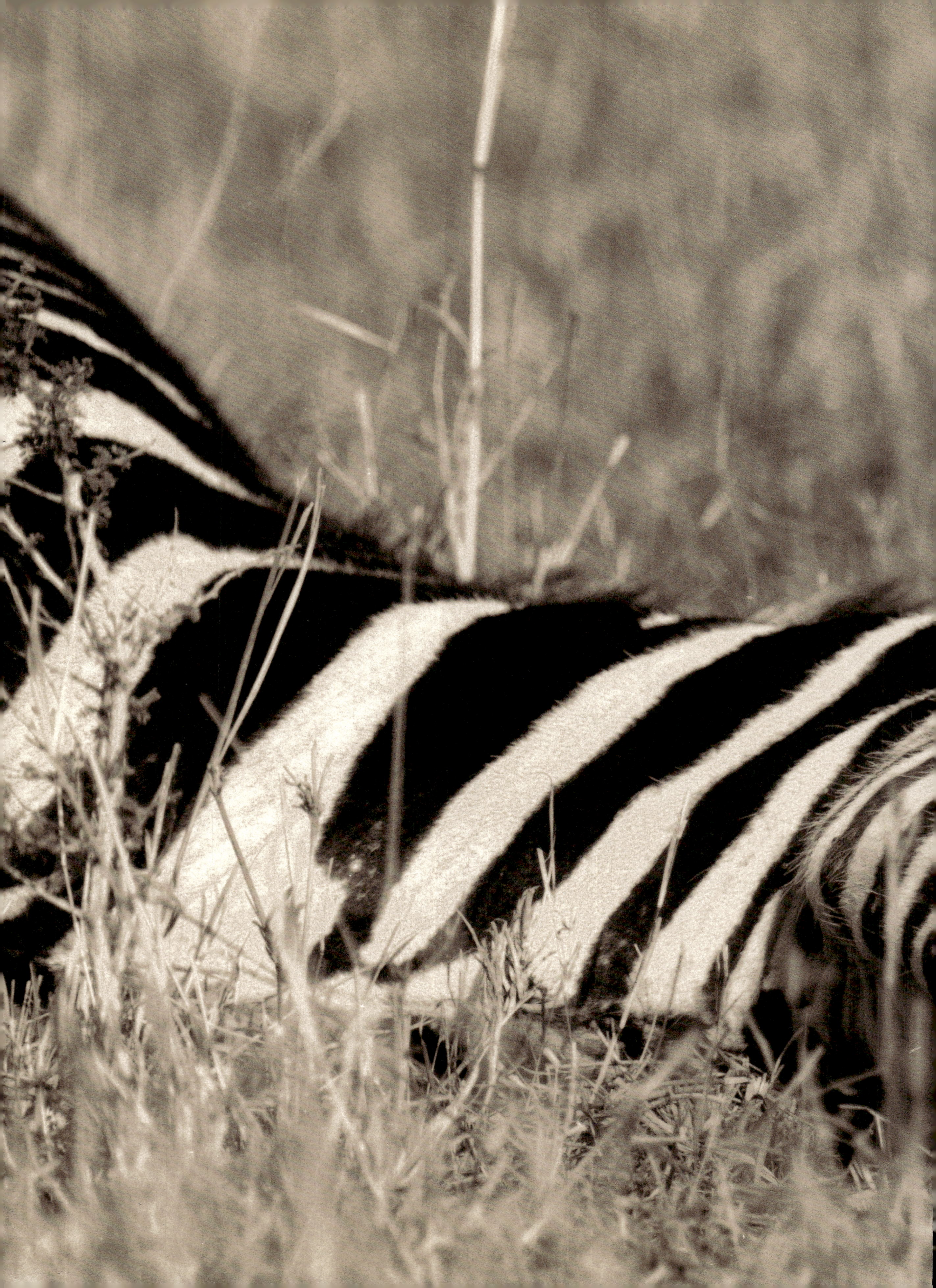

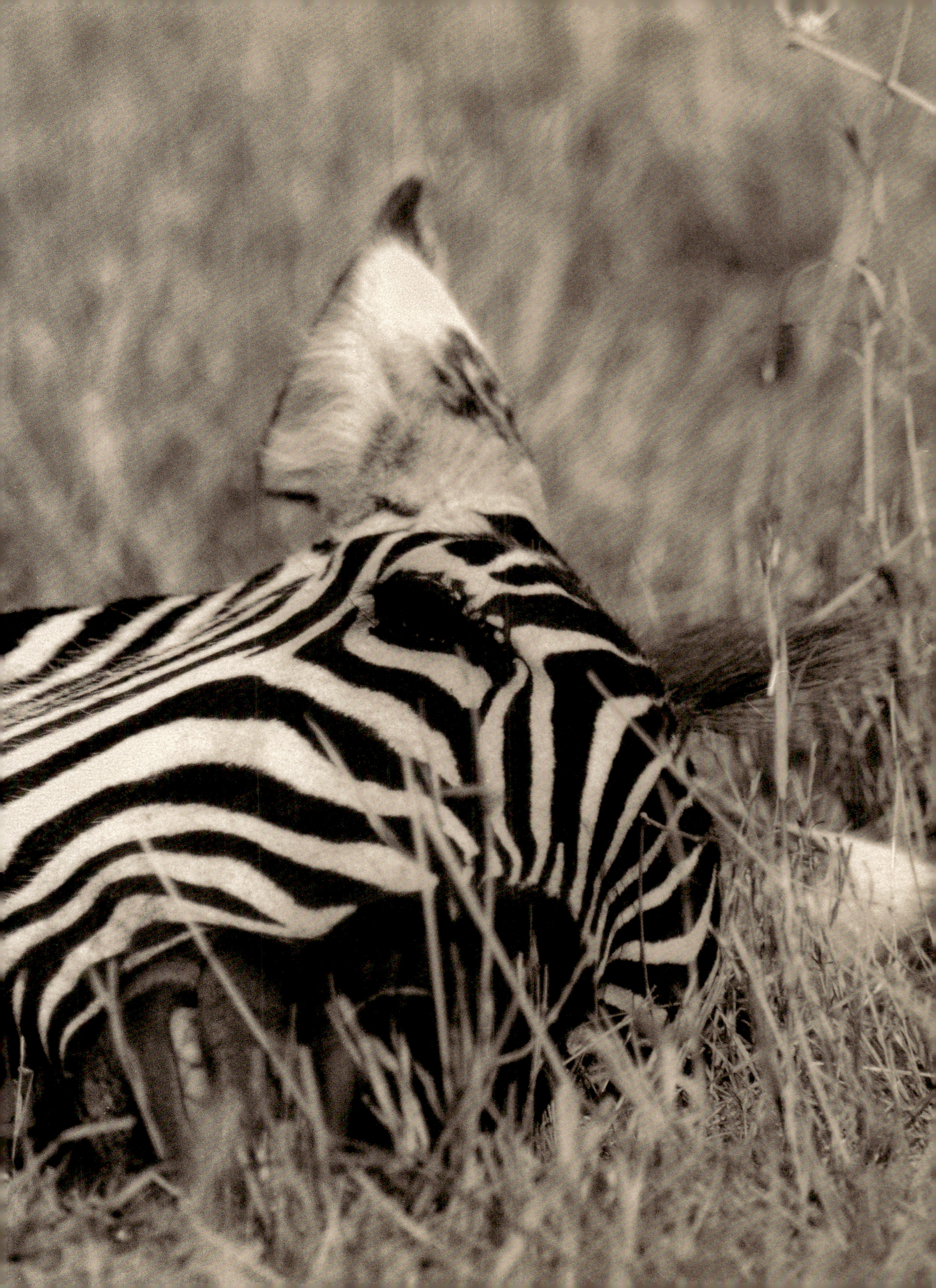

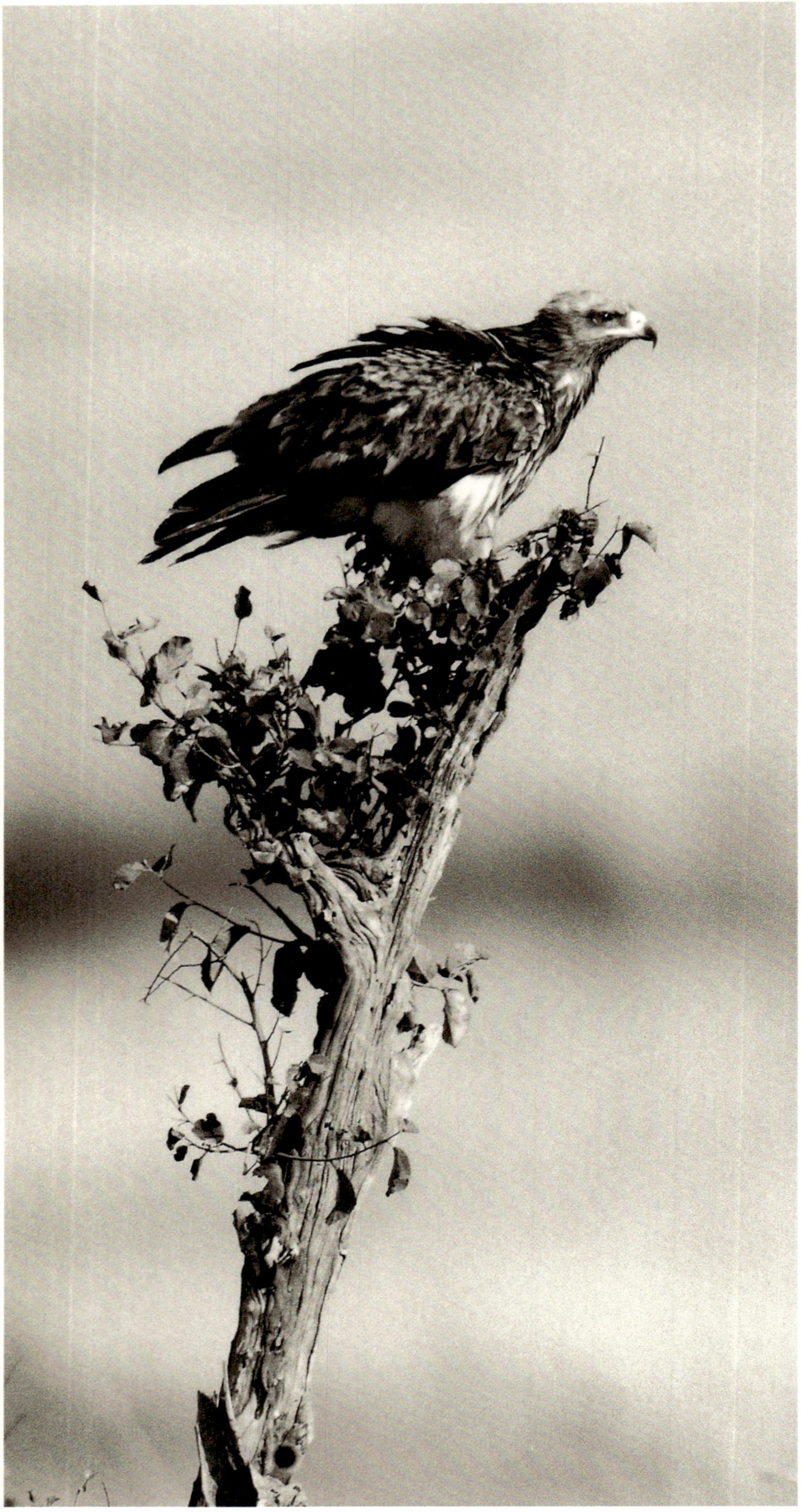

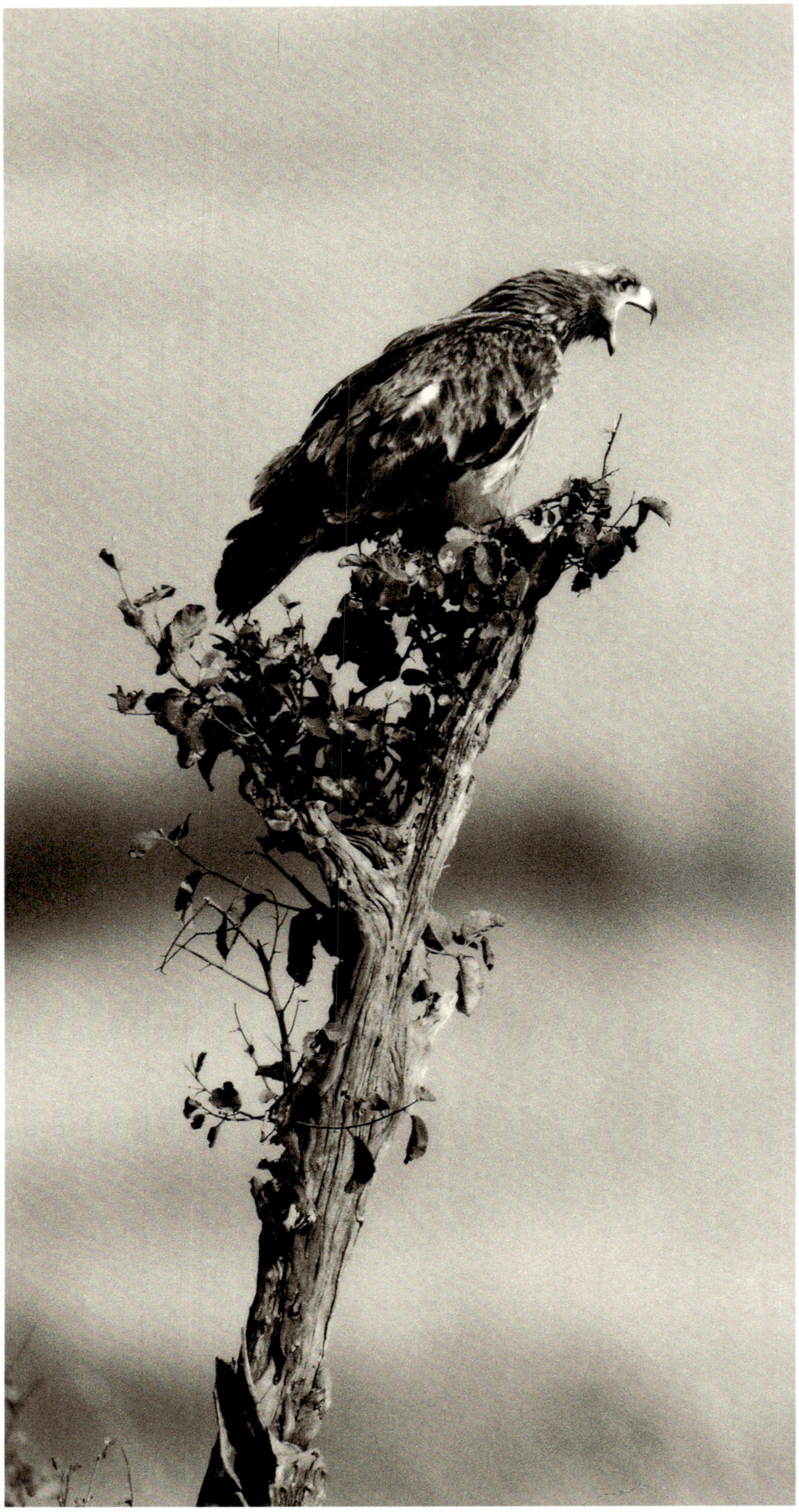

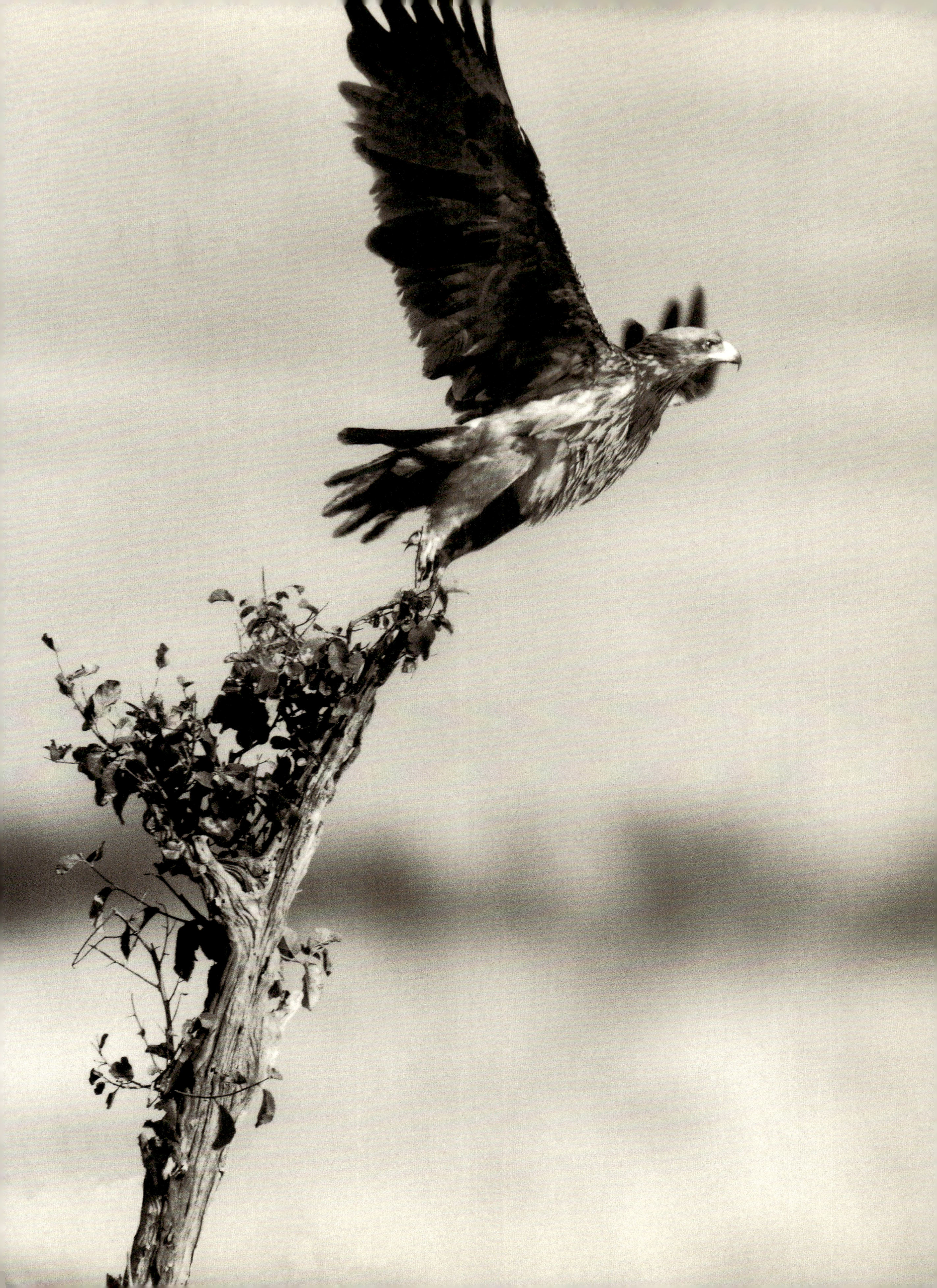

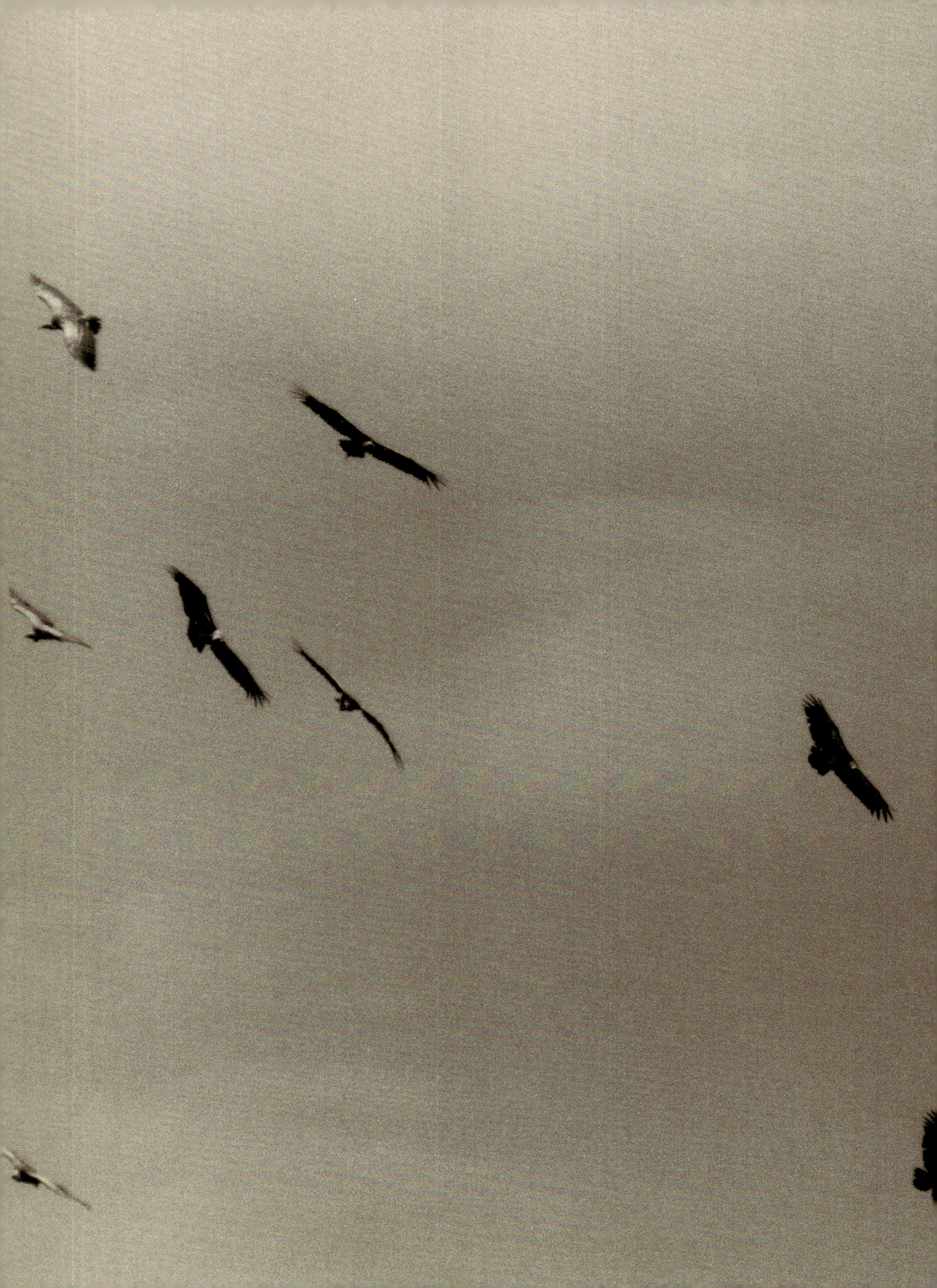

AT FIRST SIGHT, THE SQUAWKING SCRUM OF VULTURES AT A CARCASS MAY SEEM LIKE A TOTALLY CHAOTIC FREE-FOR-ALL. BUT A KIND OF ORDER PREVAILS BETWEEN SEVERAL SPECIES, BASED ON BODY SIZE, BEAK SHAPE AND PUNCTUALITY. BIG LAPPET-FACED AND WHITE-HEADED VULTURES SOARING OVERHEAD, WIDE-RANGING AND KEEN-EYED, CAN SPOT A KILL FROM AFAR AND ARRIVE FIRST TO RIP OPEN WHAT'S LEFT OF THE CARCASS. THEY ARE FOLLOWED BY MEDIUM-SIZED WHITE-BACKED (OVERLEAF) AND RUPPELL'S (OPPOSITE AND FOLLOWING PAGES) GRIFFON VULTURES WHO DASH IN AND OUT, GULPING DOWN WHAT THEY CAN SNATCH FROM THE FRAY. FINALLY THE SMALL HOODED AND EGYPTIAN VULTURES SCRAP WHATEVER'S LEFT OF THE BONES. EVEN SOME EAGLES, LIKE THE OPPORTUNISTIC TAWNY (PP.198–199), WILL DROP IN TO SEE WHAT'S GOING DOWN.

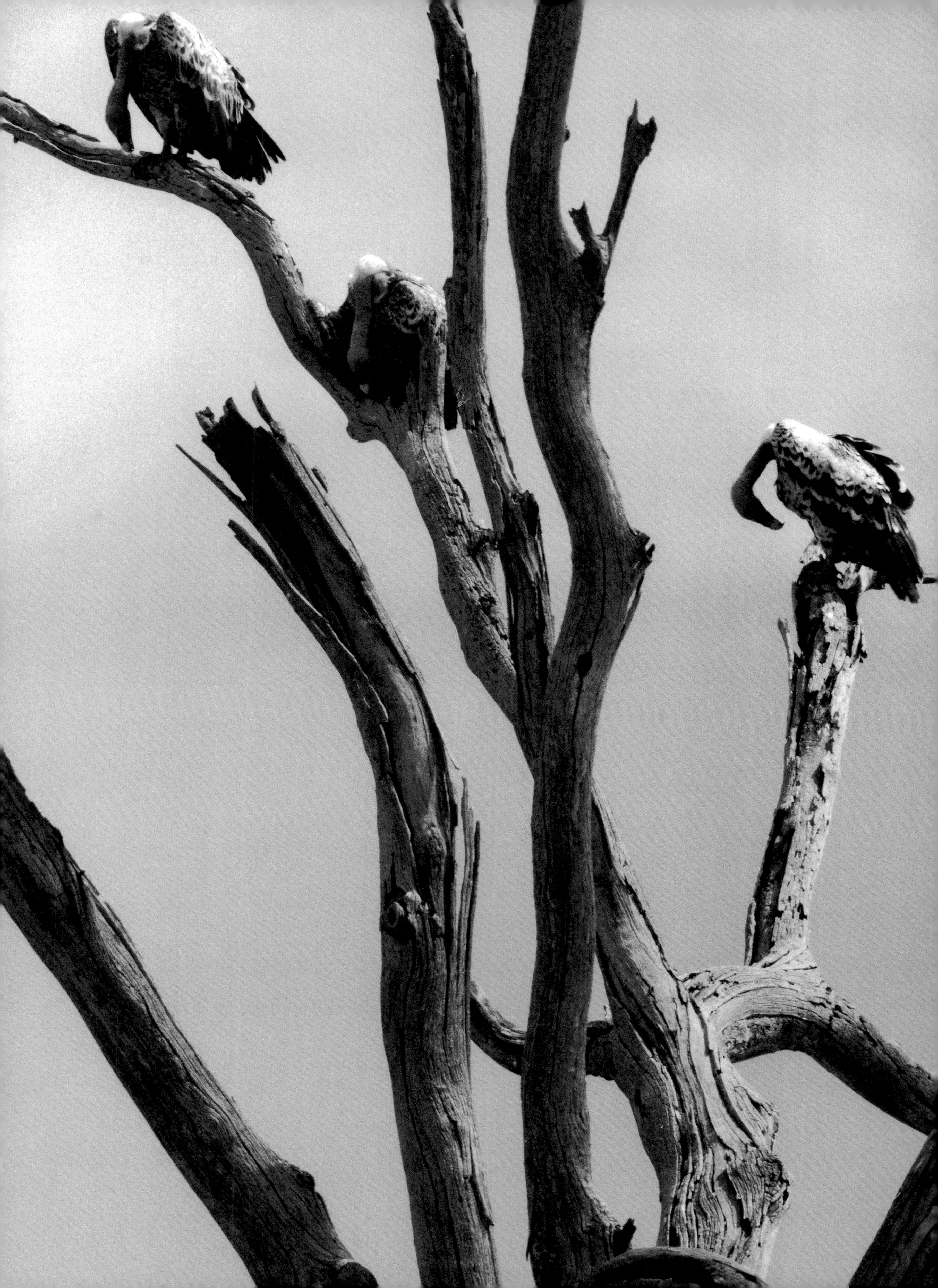

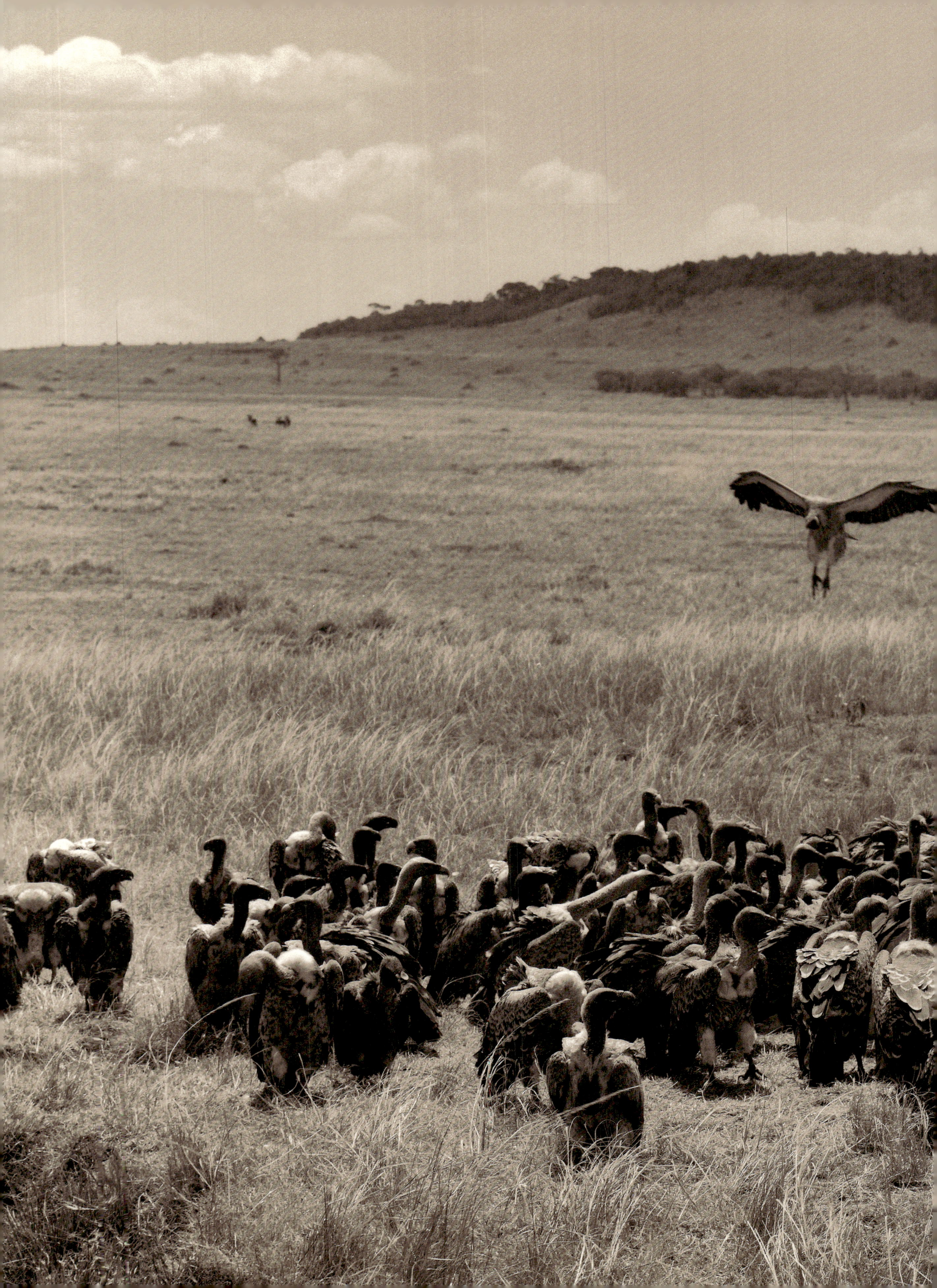

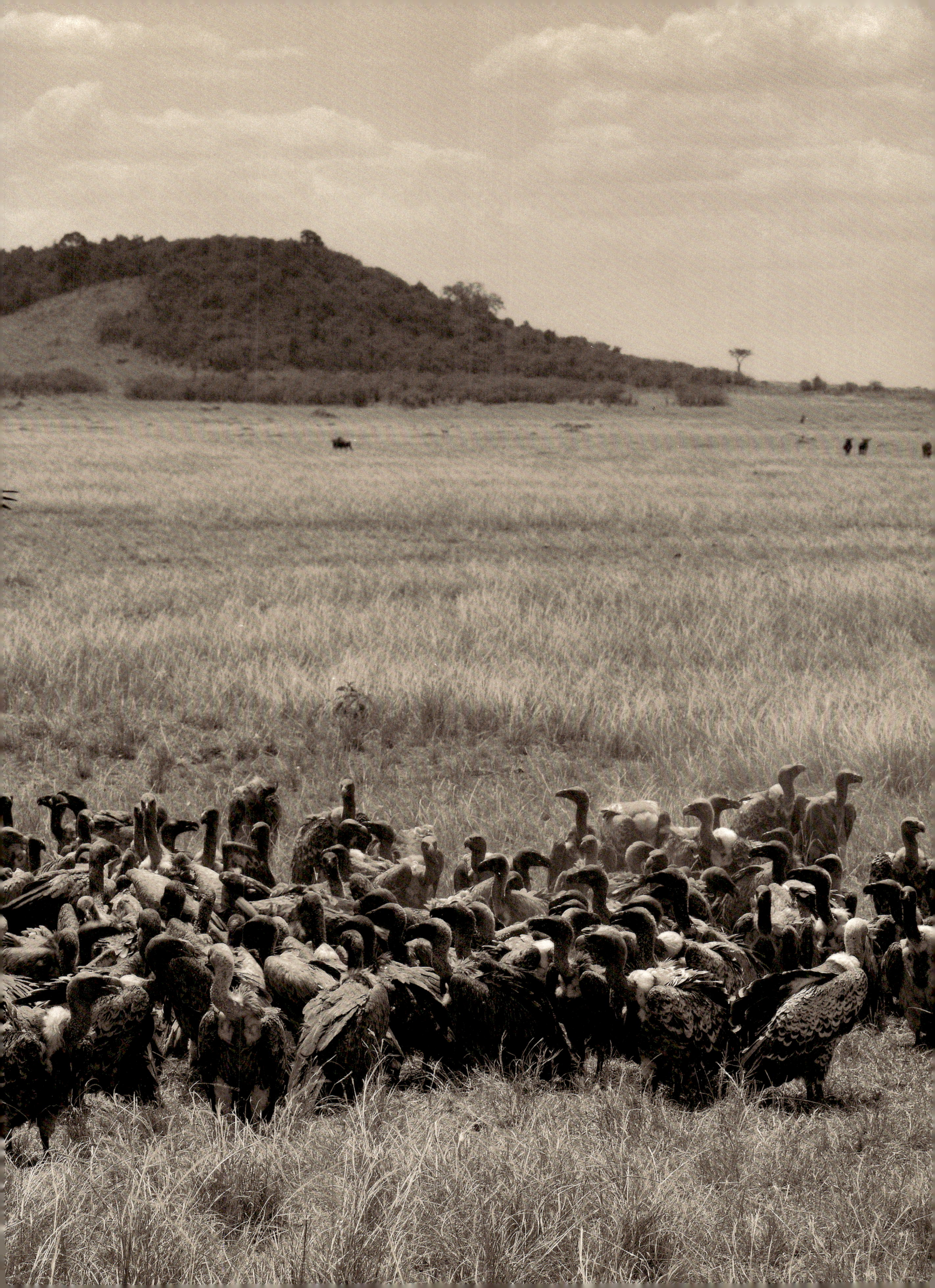

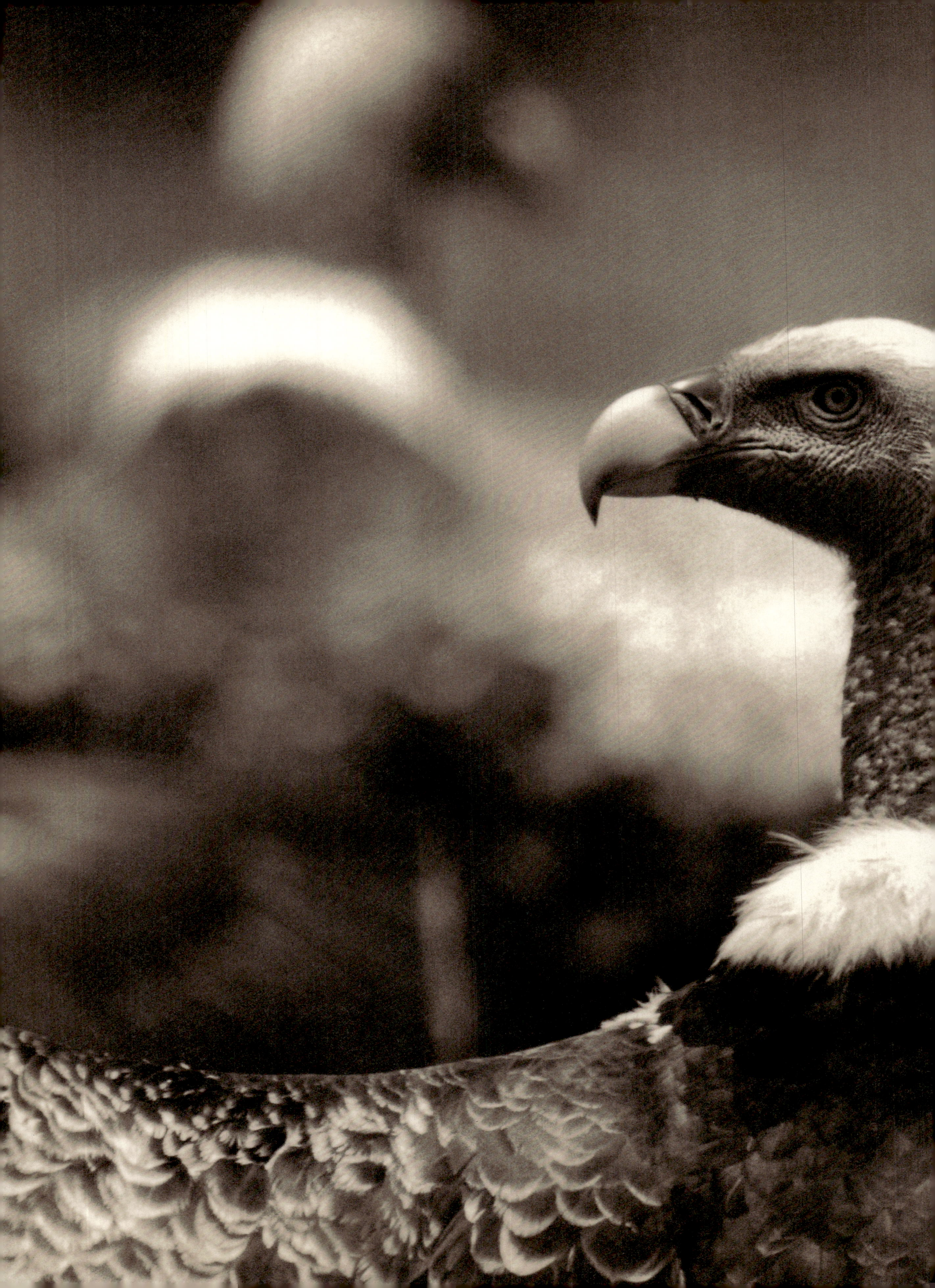

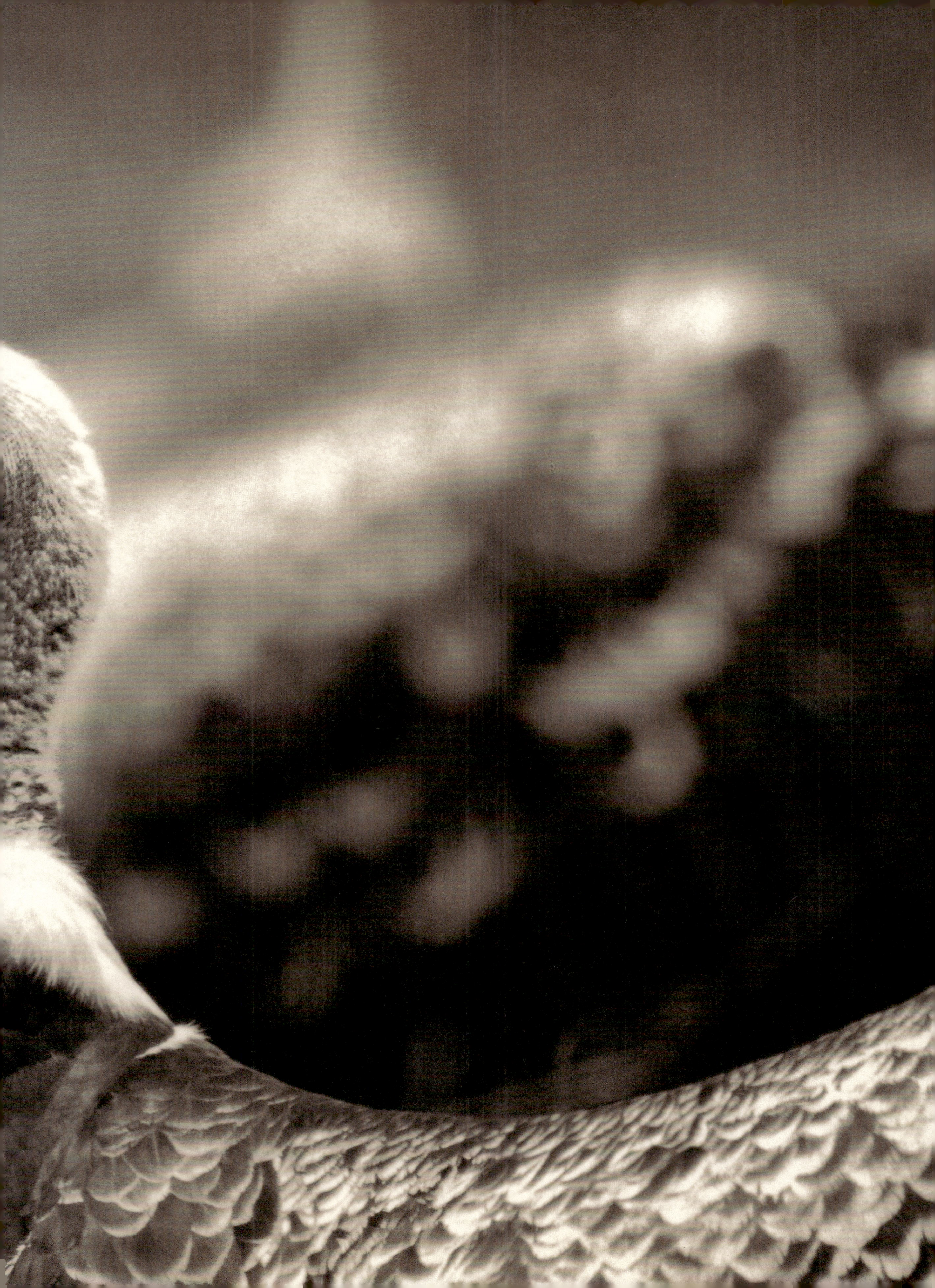

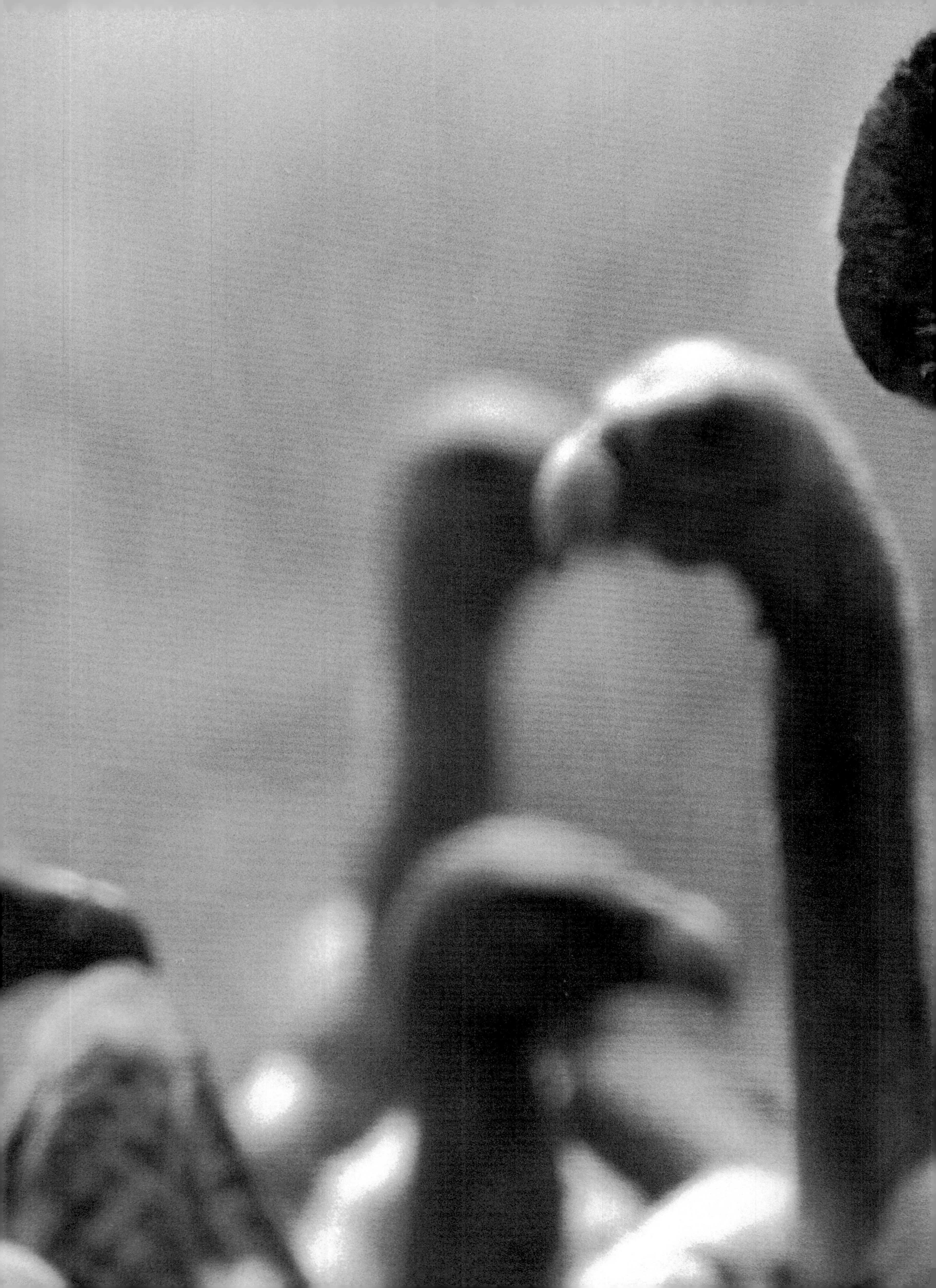

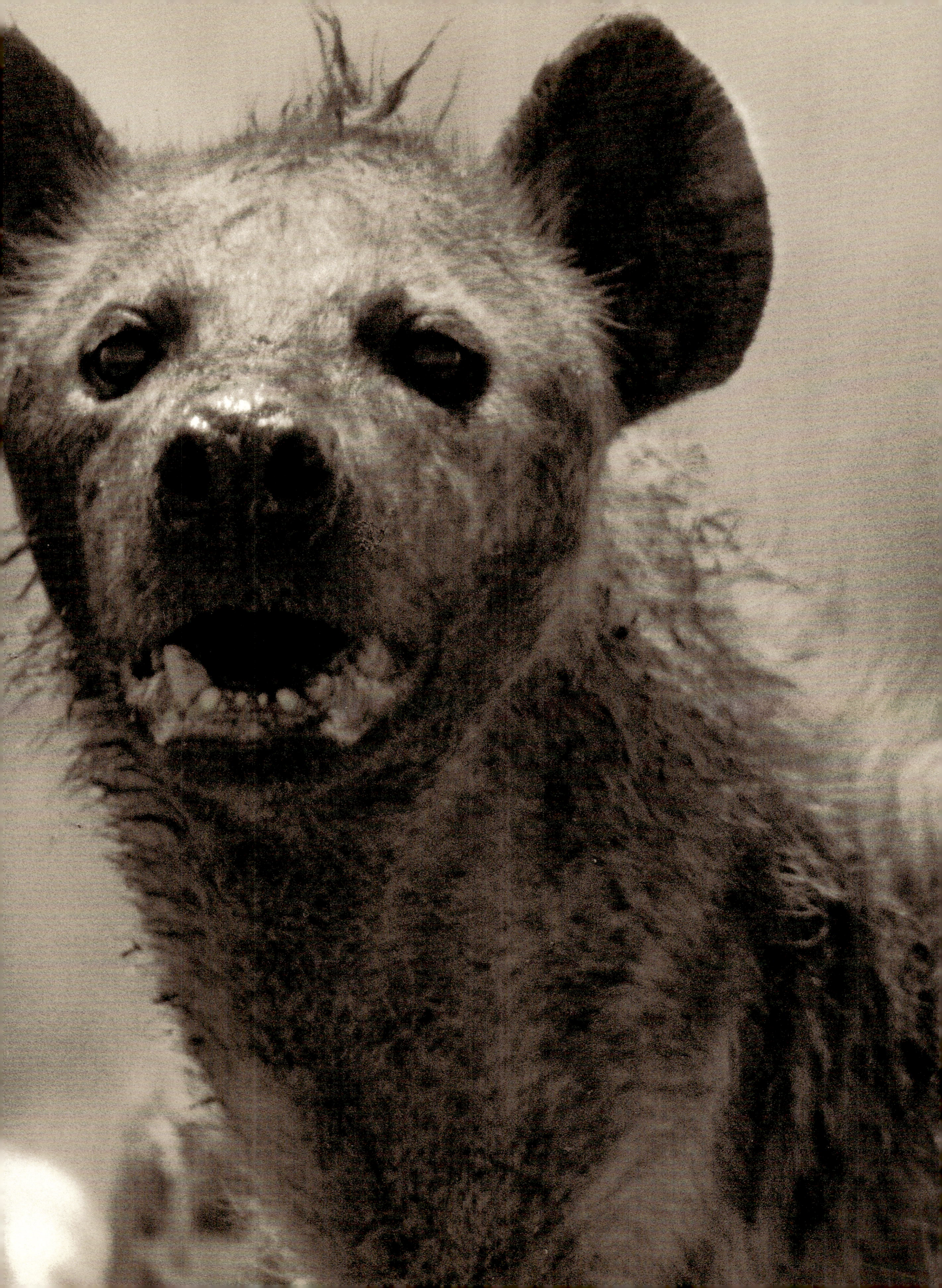

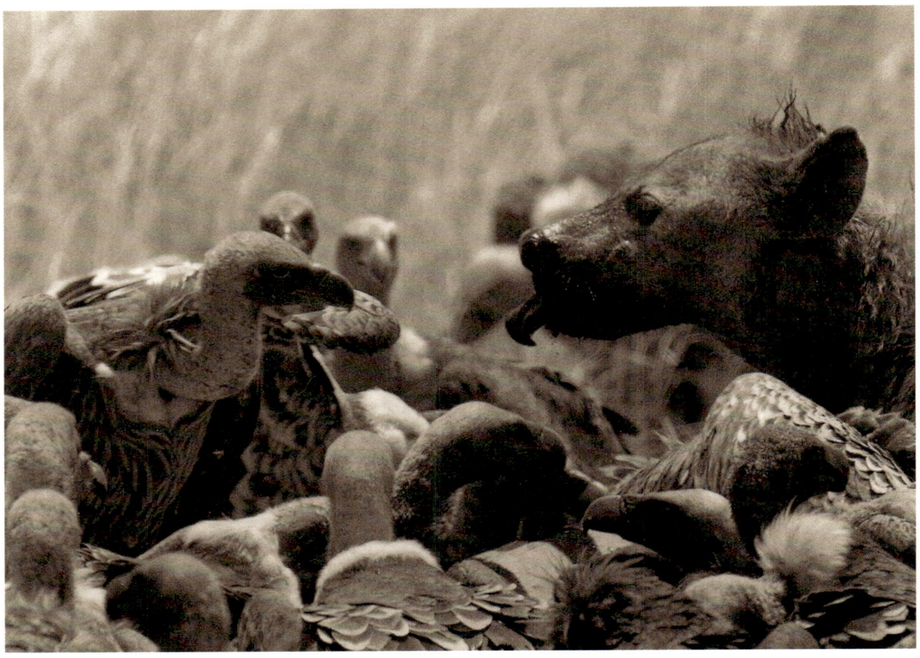

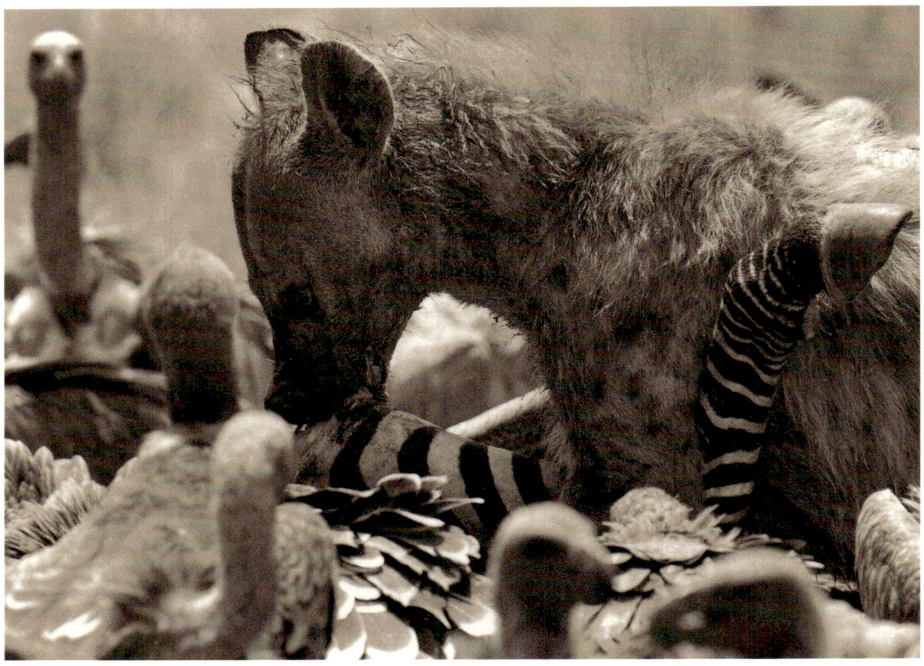

FORGET CONVENTIONAL WISDOM: HYENAS ARE LESS LIKE YOUR COMMON SCAVENGER AND MORE LIKE OUR EARLY ANCESTORS. THEY LIVE IN MIXED CLANS OF ADULTS OF BOTH SEXES AND YOUNG, MUTUALLY DEFEND THEIR TERRITORIES AGAINST INTERLOPERS, ENGAGE IN HIGHLY ORGANISED COOPERATIVE HUNTING FORAYS, AND ONLY SCAVENGE IF ABSOLUTELY NECESSARY. COME TO THINK OF IT, MUCH LIKE US TODAY.

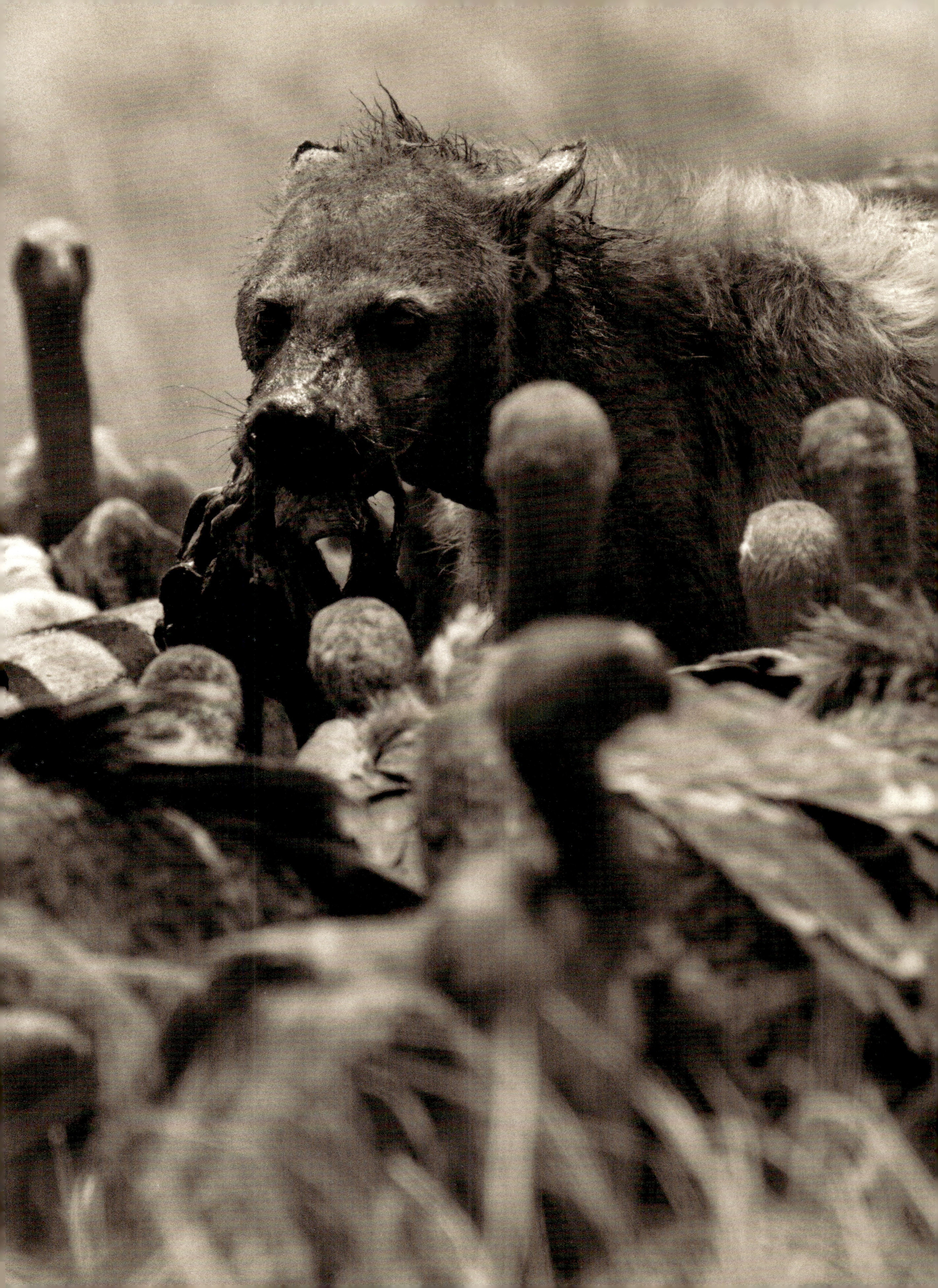

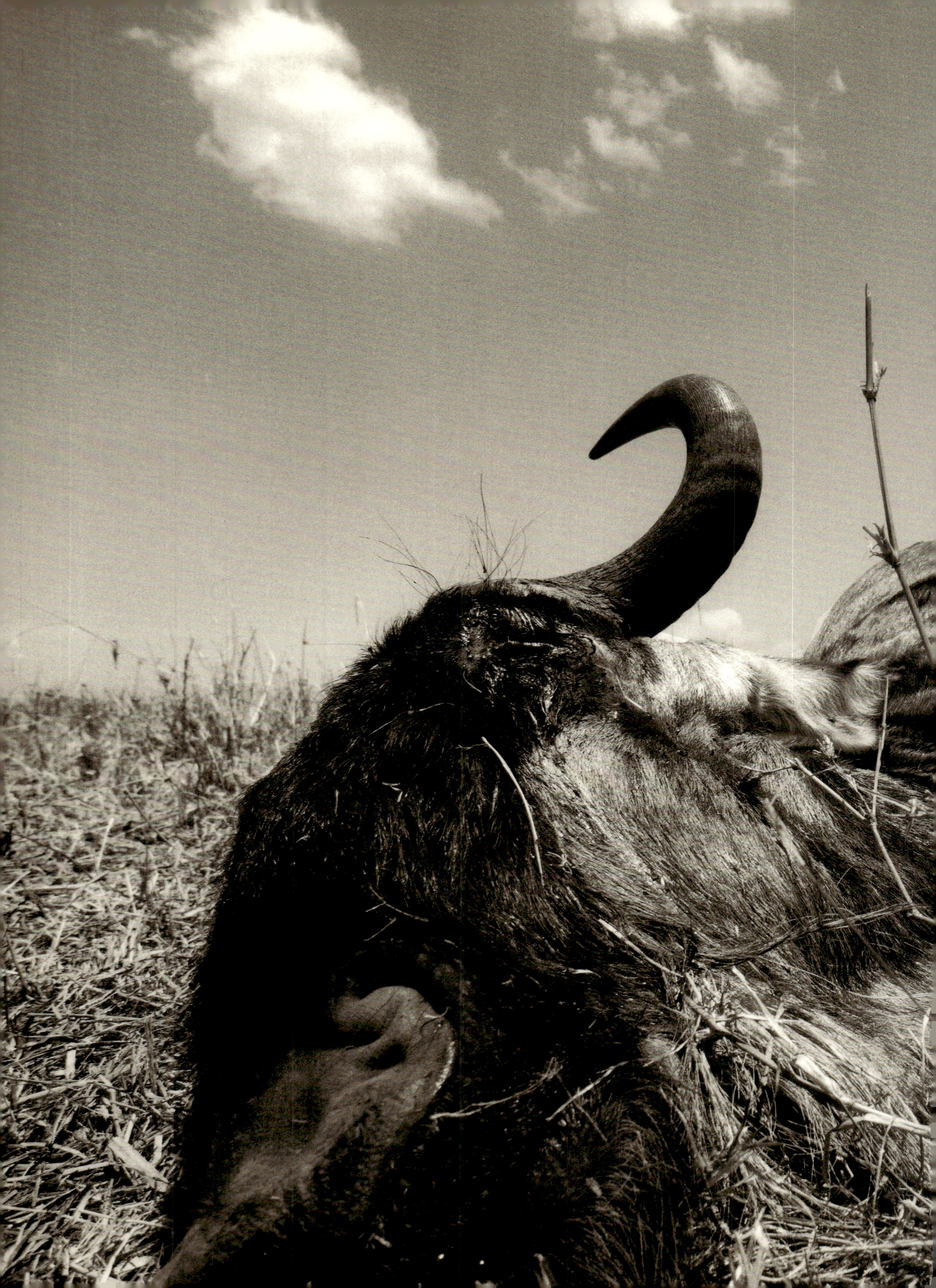

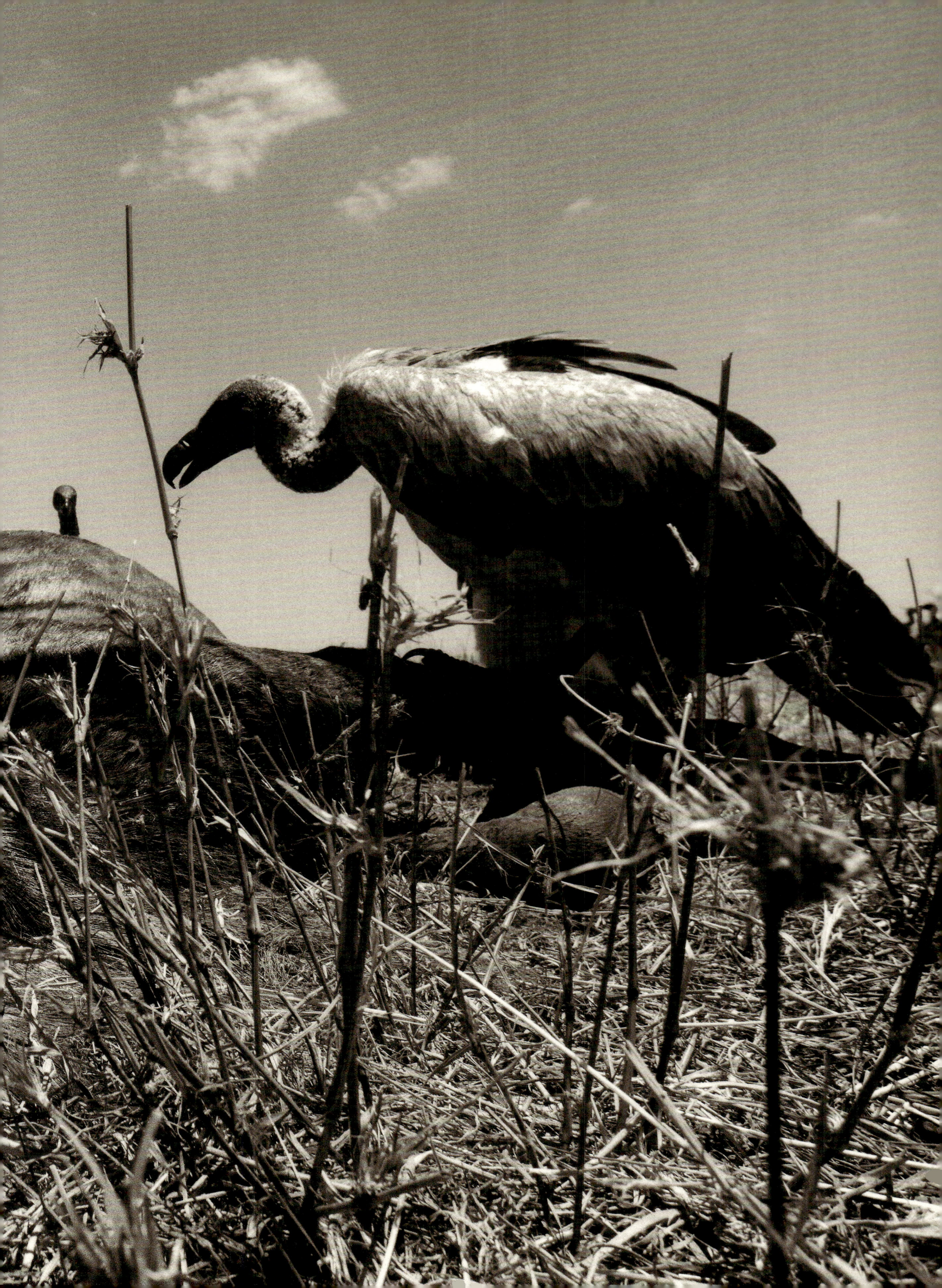

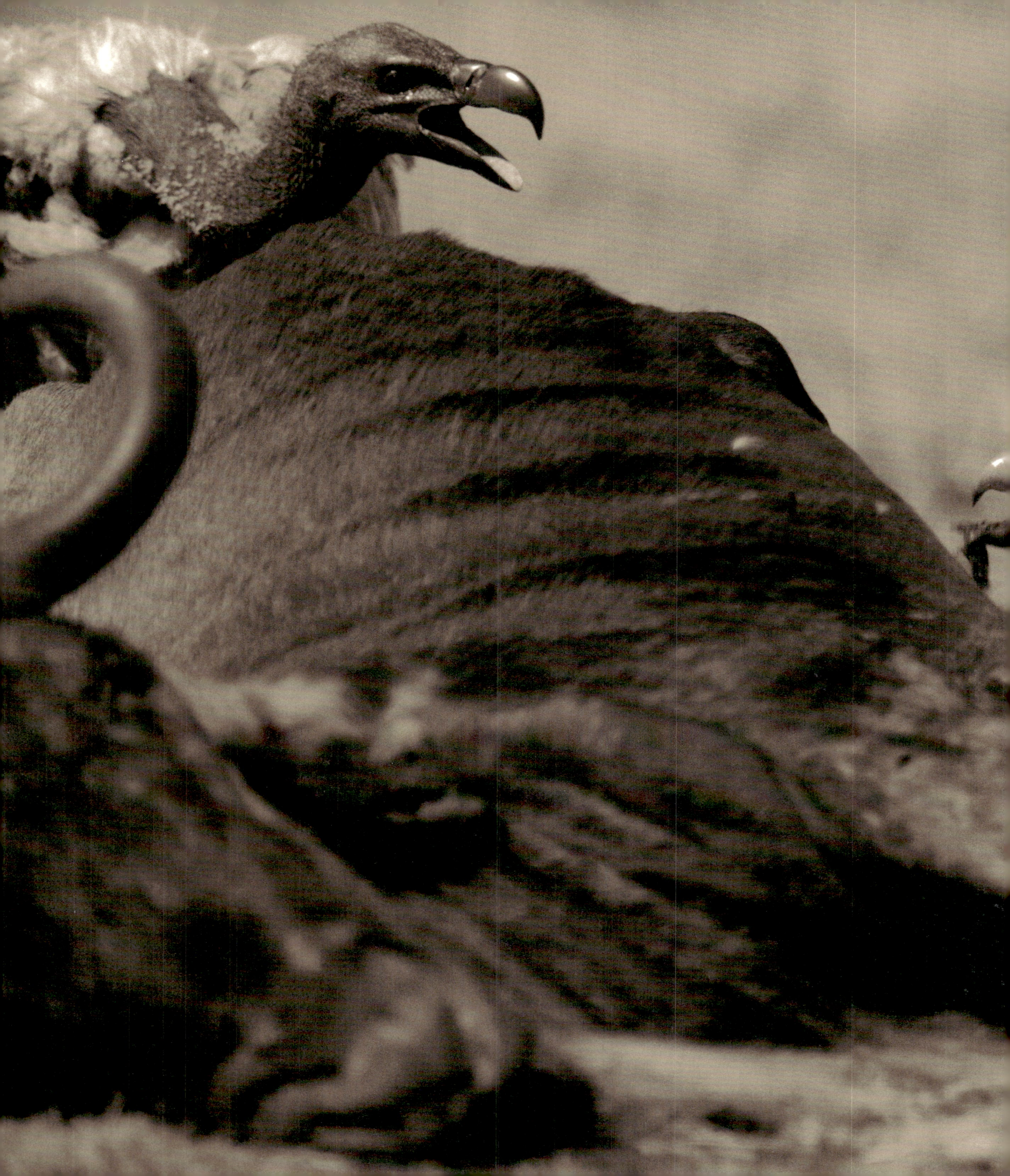

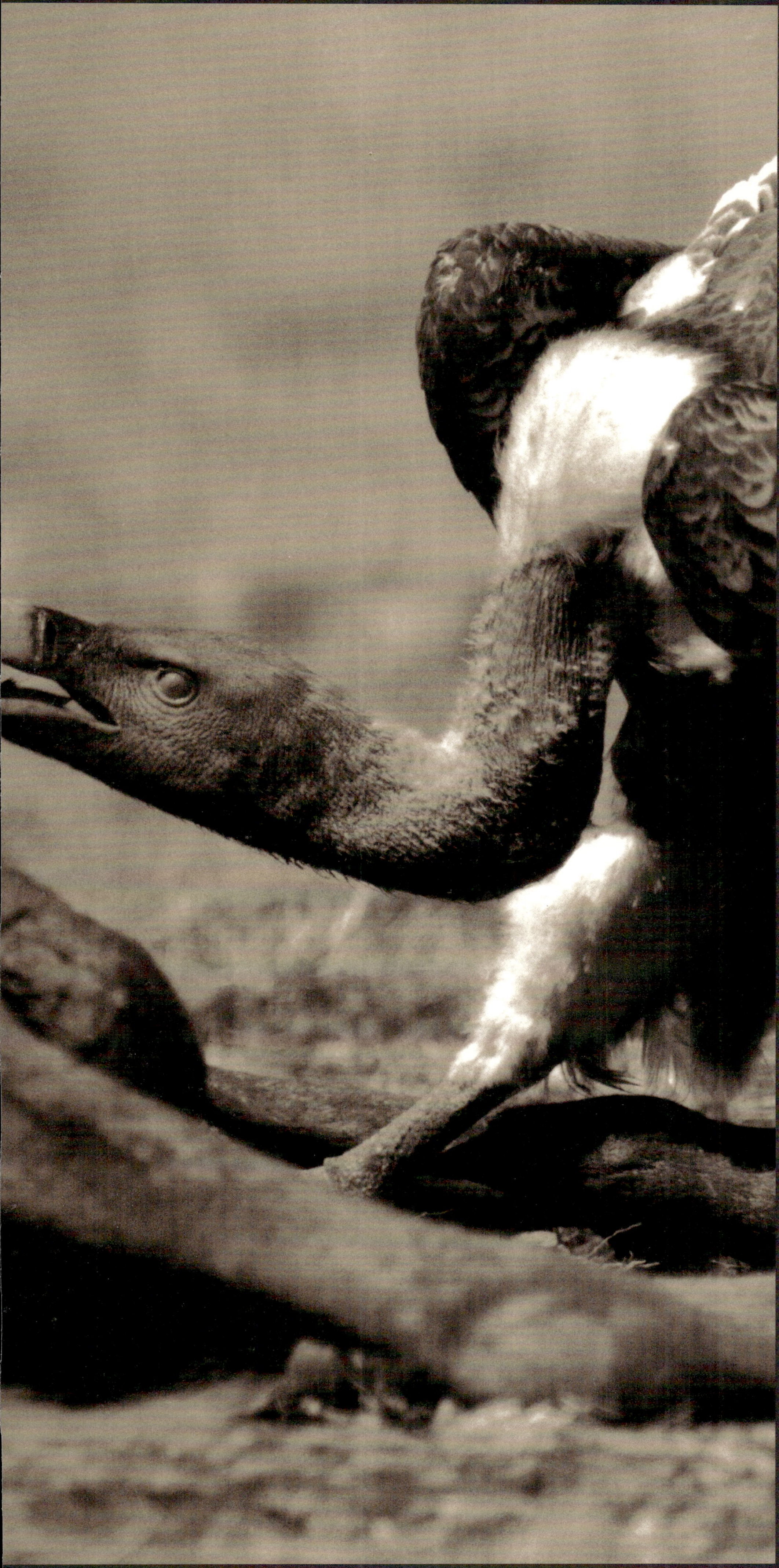
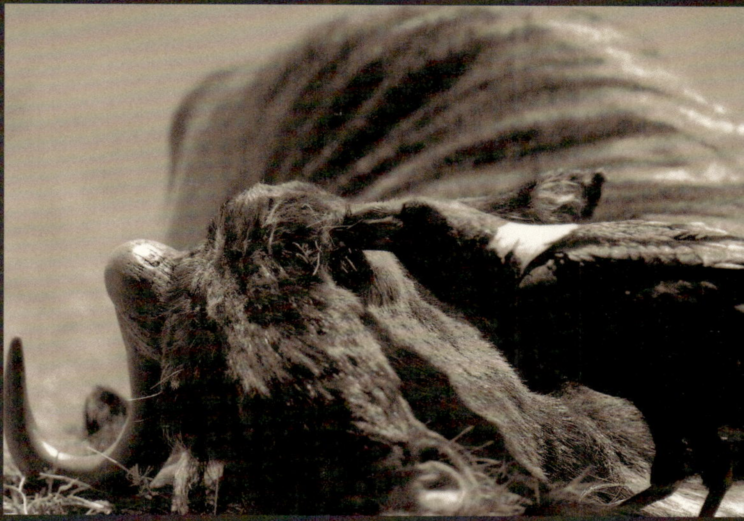
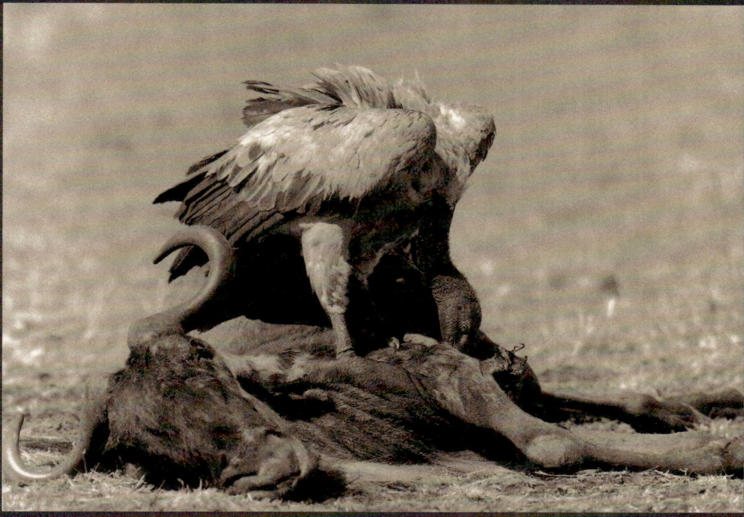
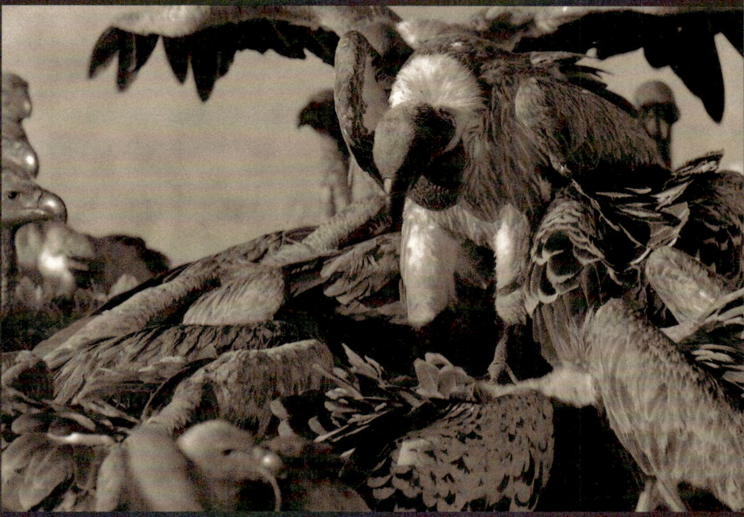
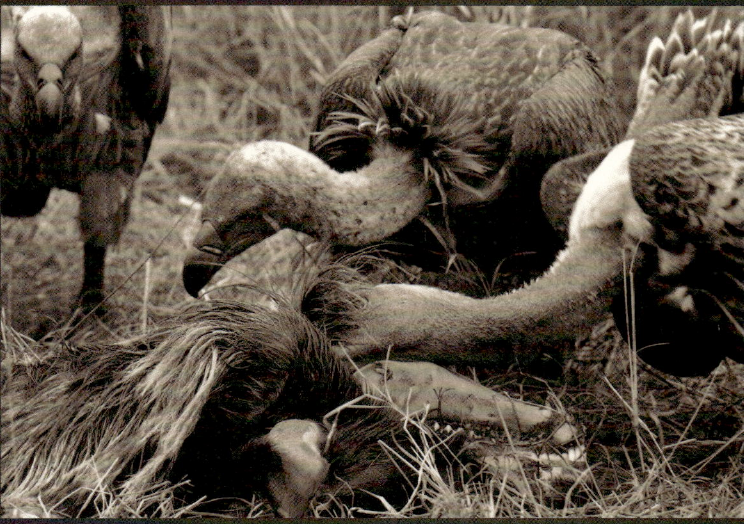

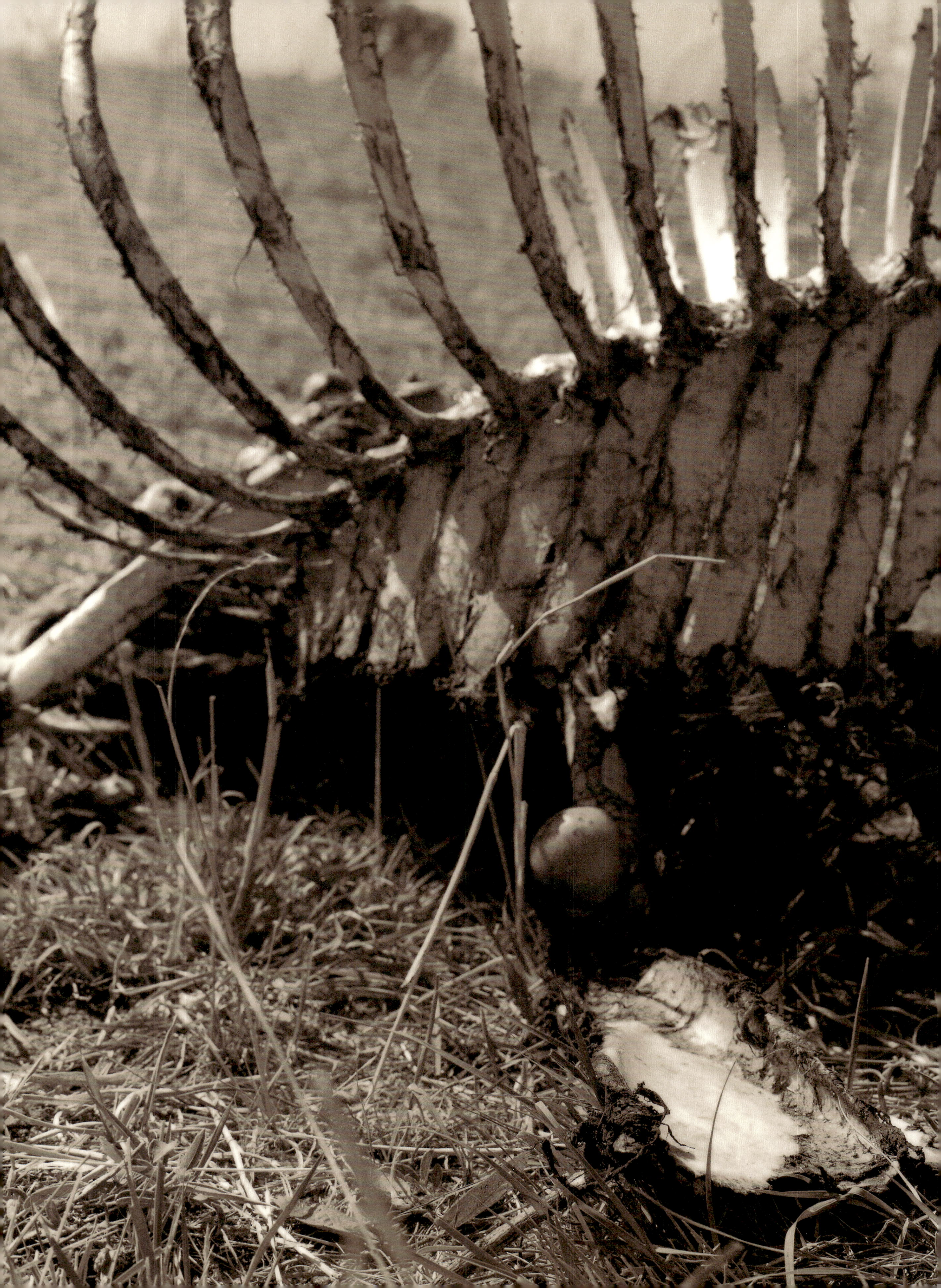

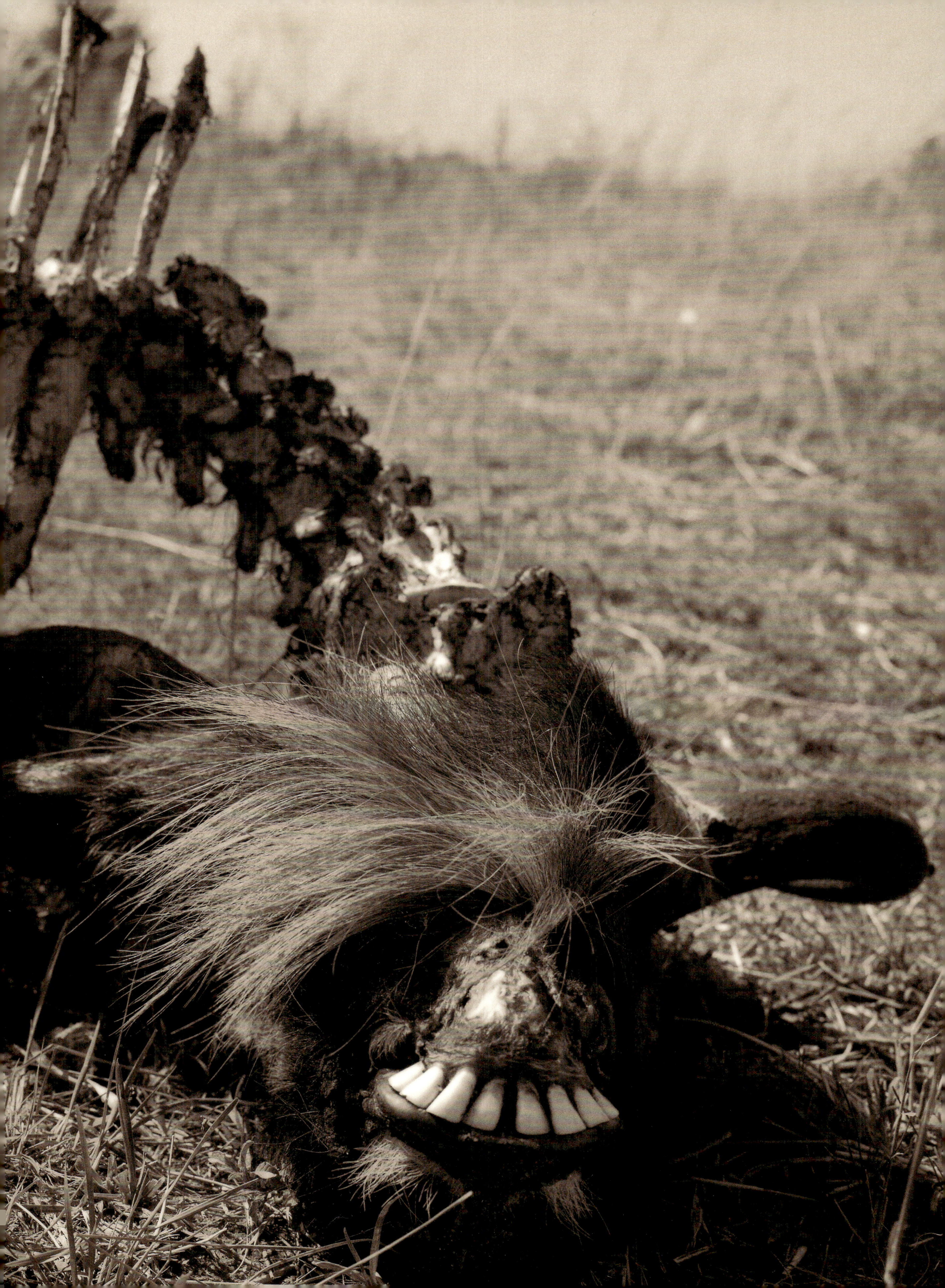

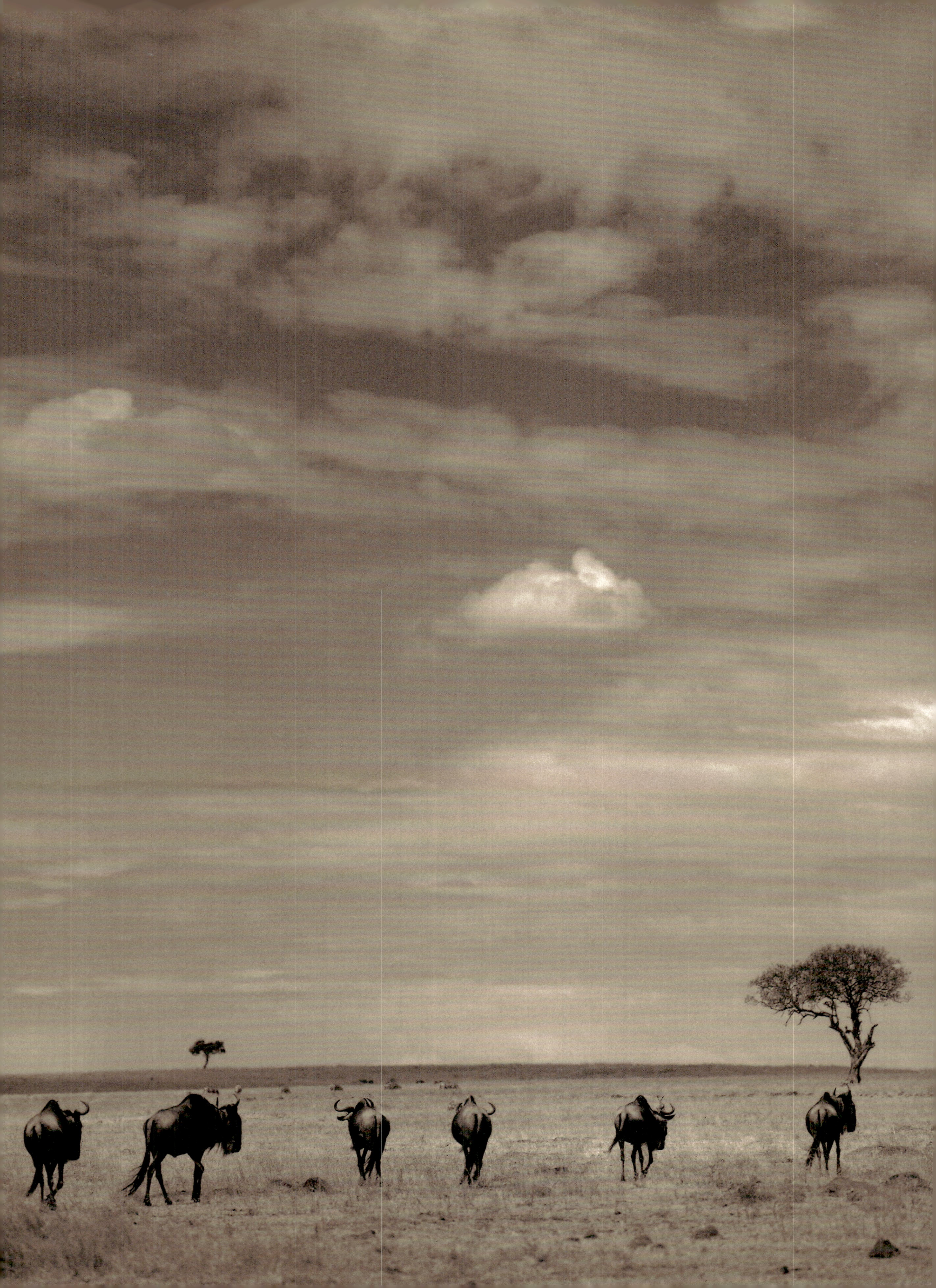

4

SOUTHERN SERENGETI SHORTGRASS PLAINS

Long Rains, March–April

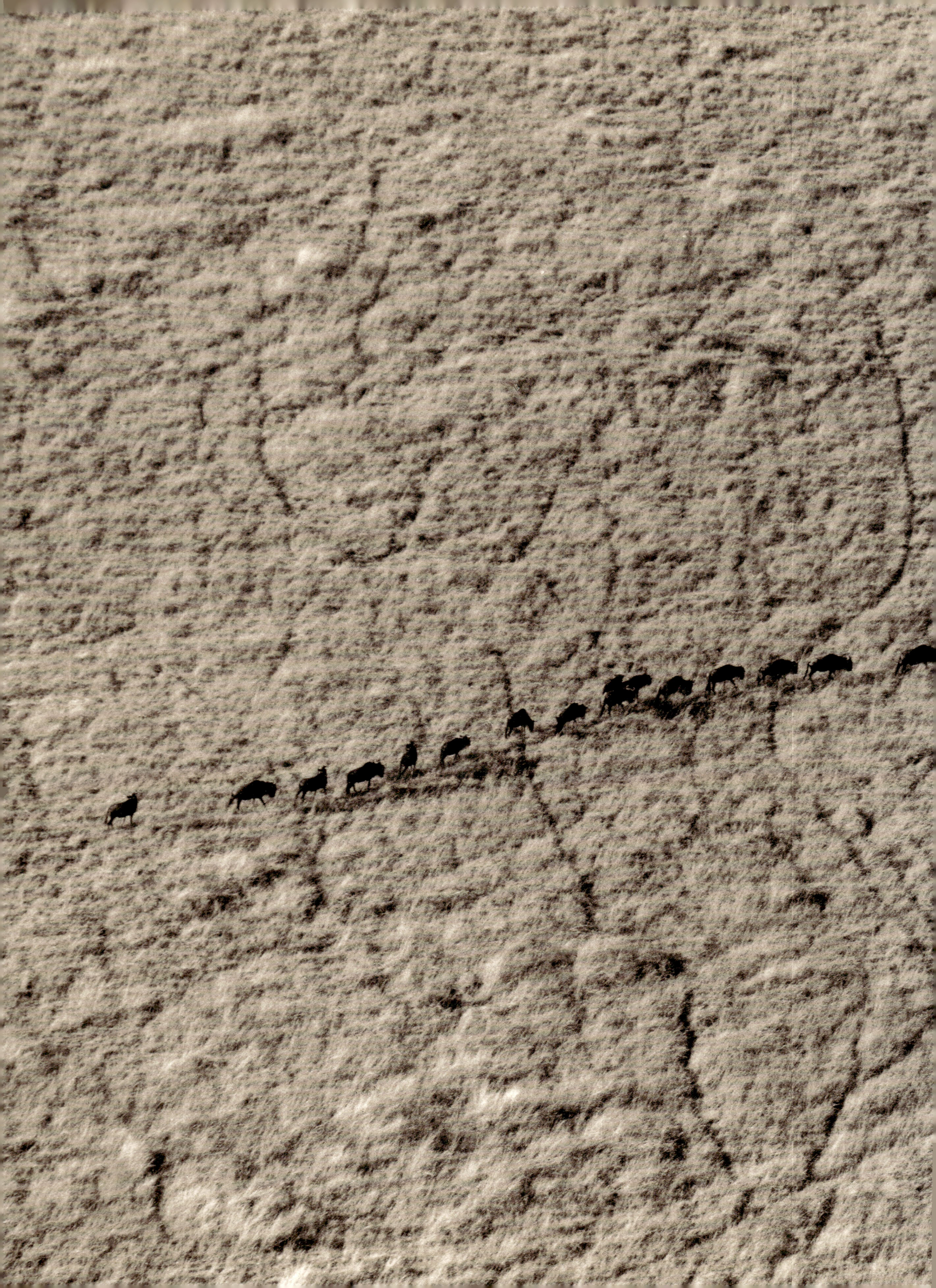

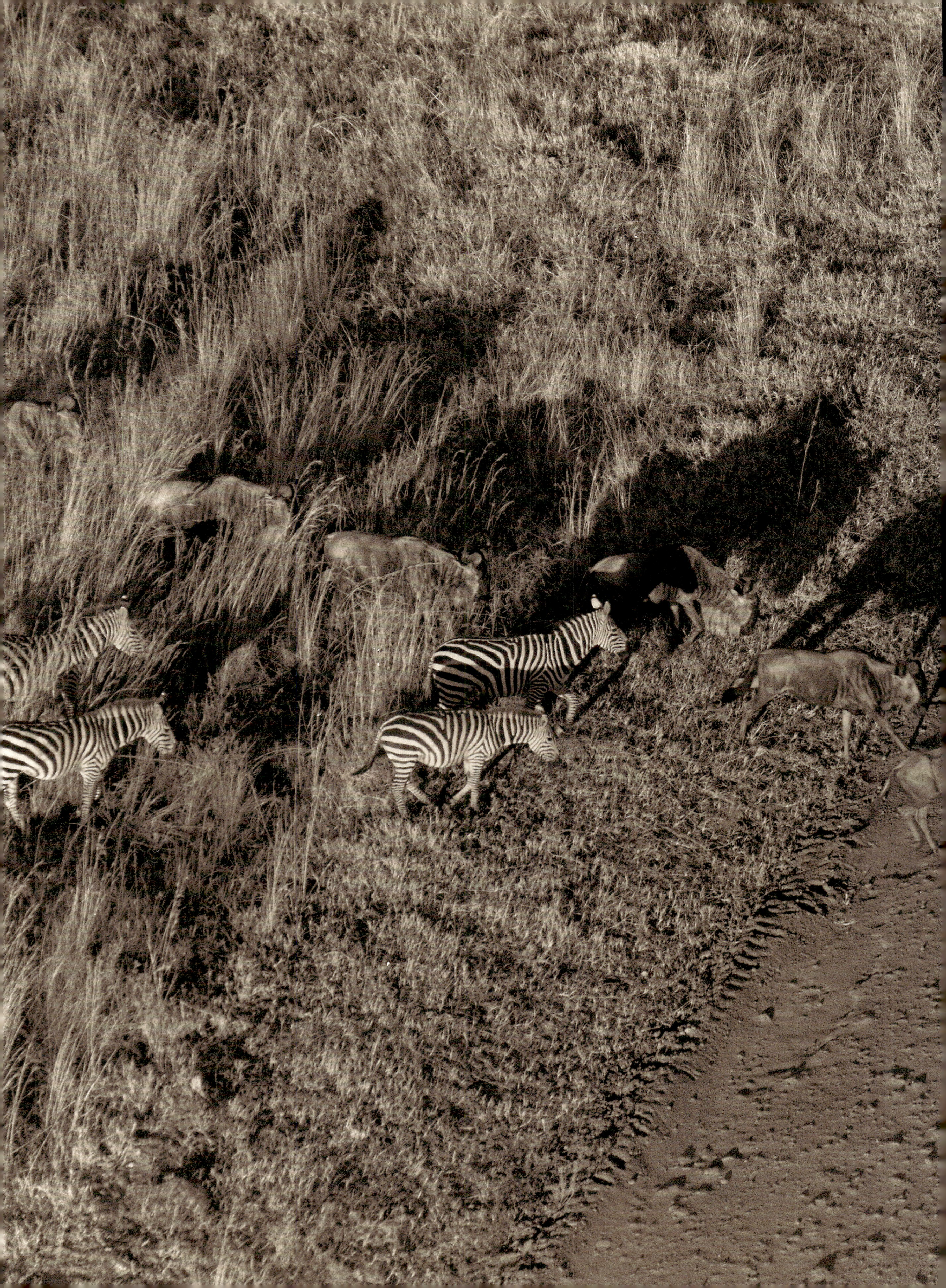

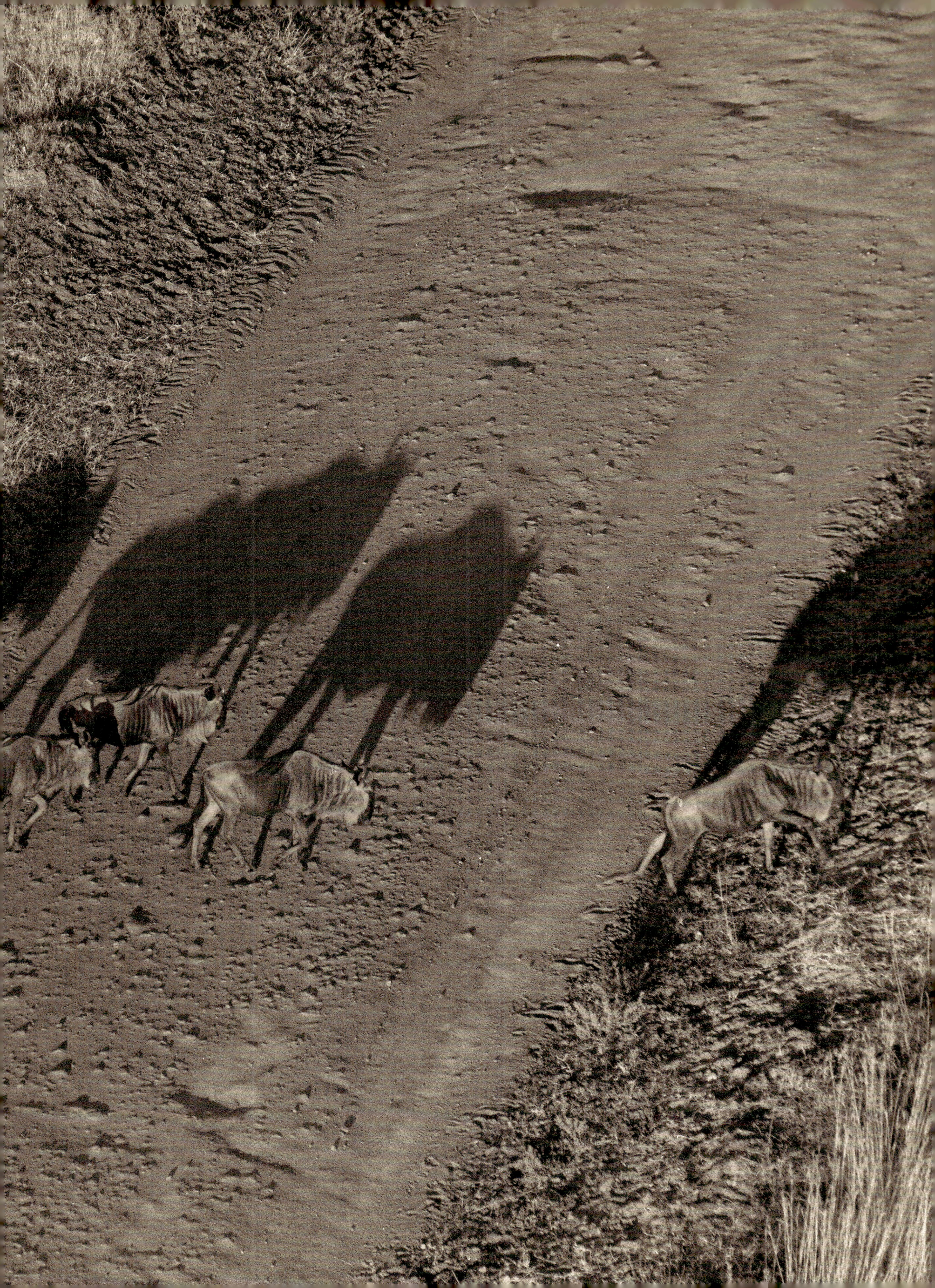

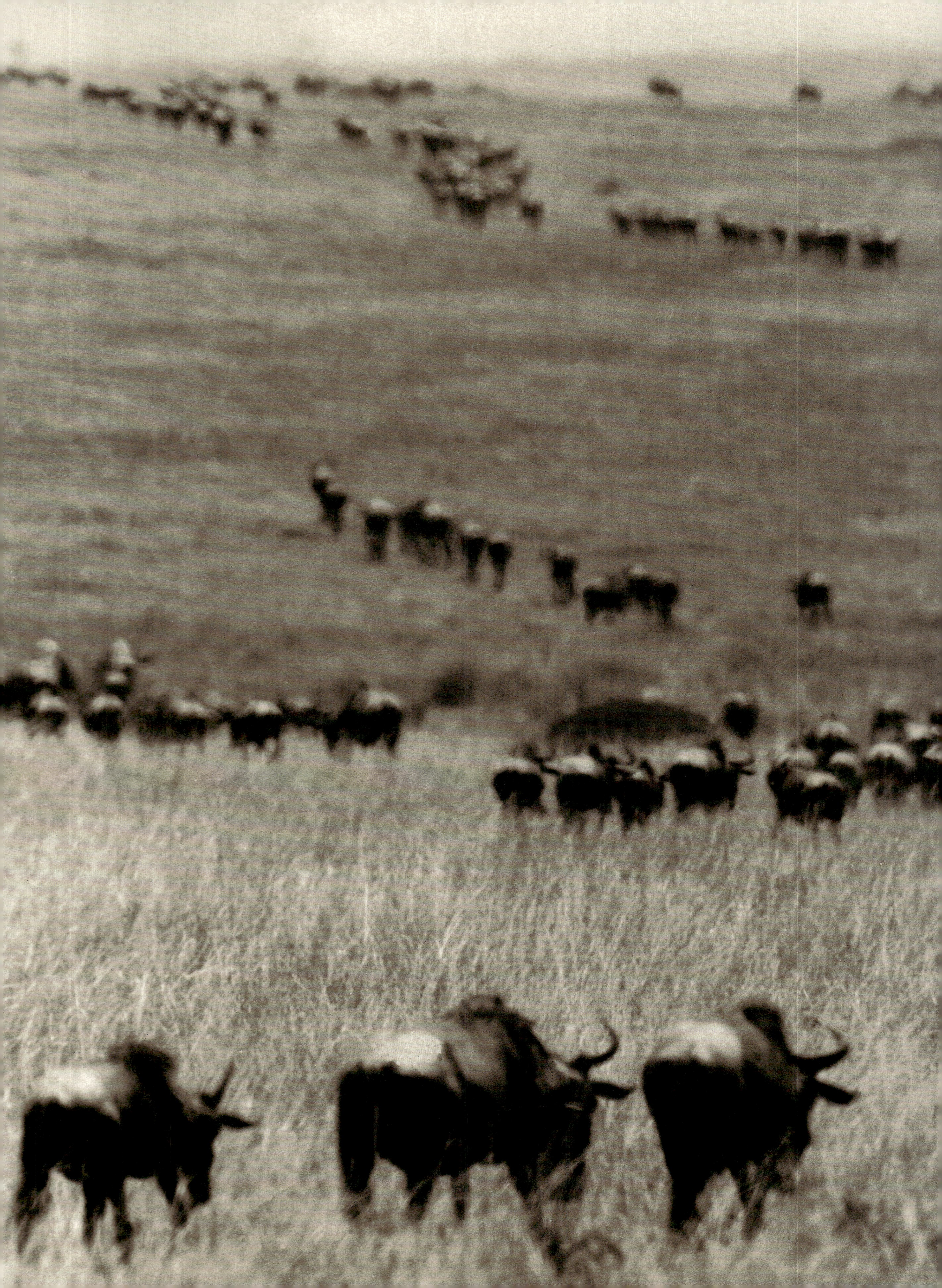

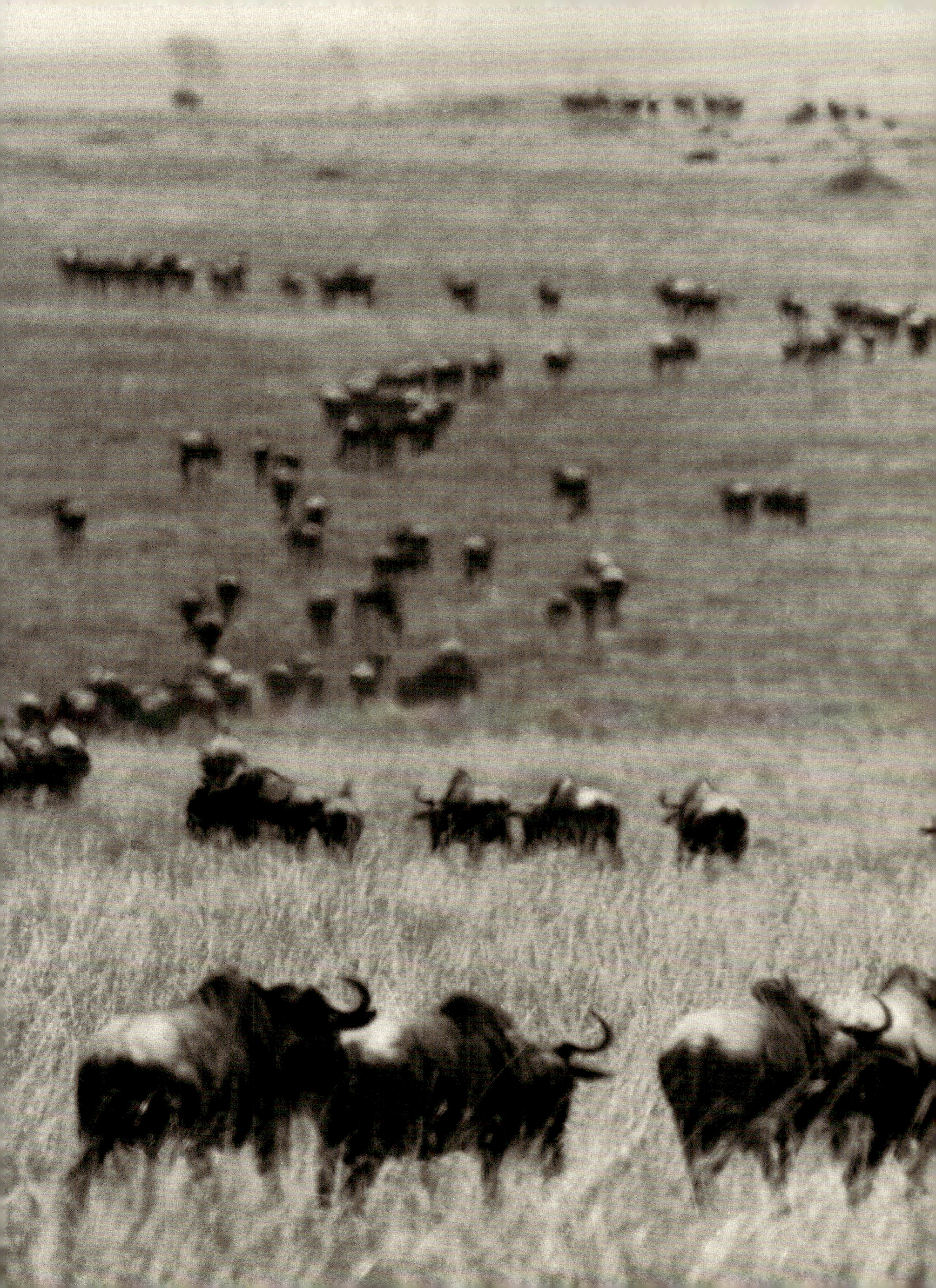

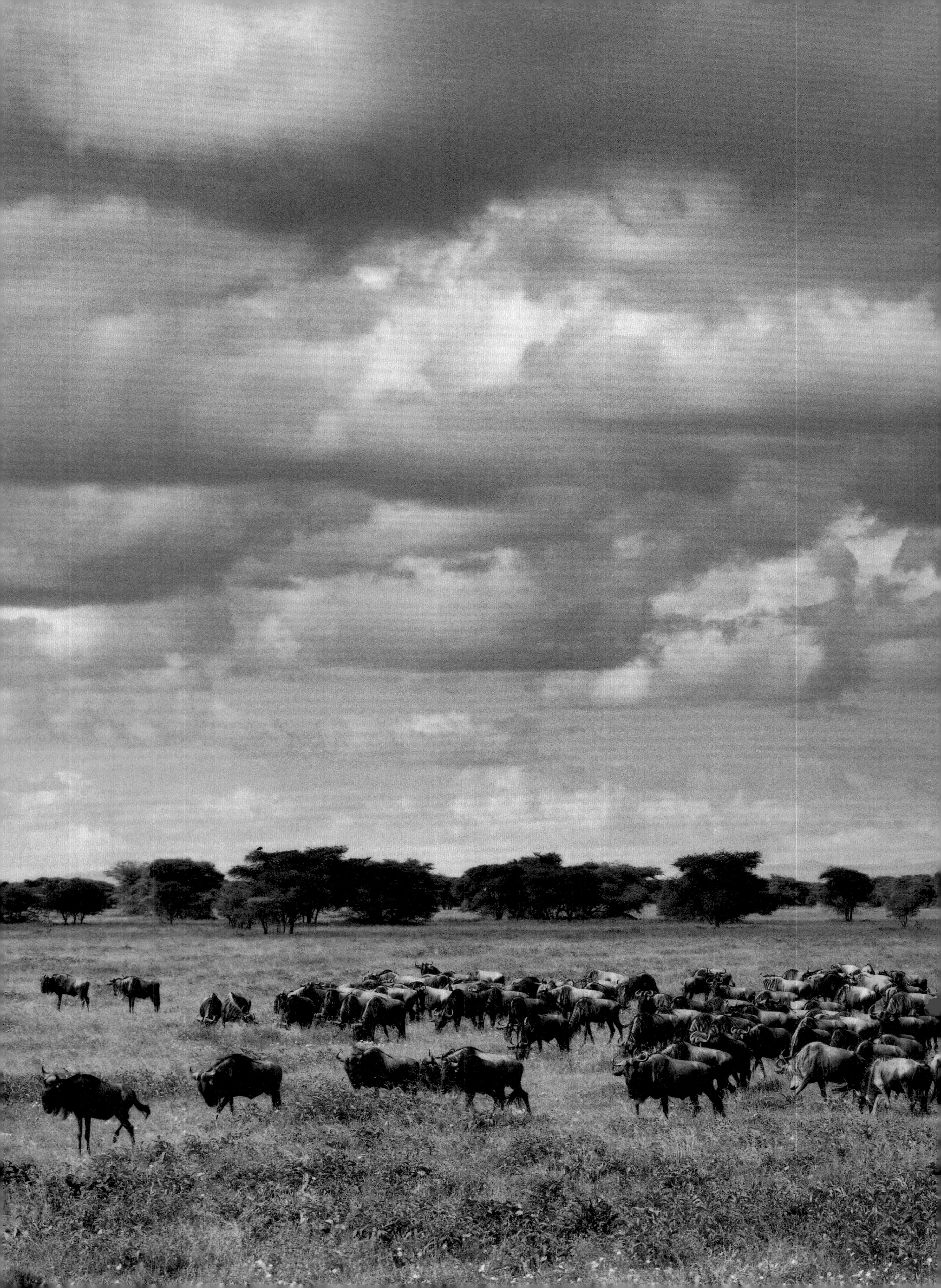

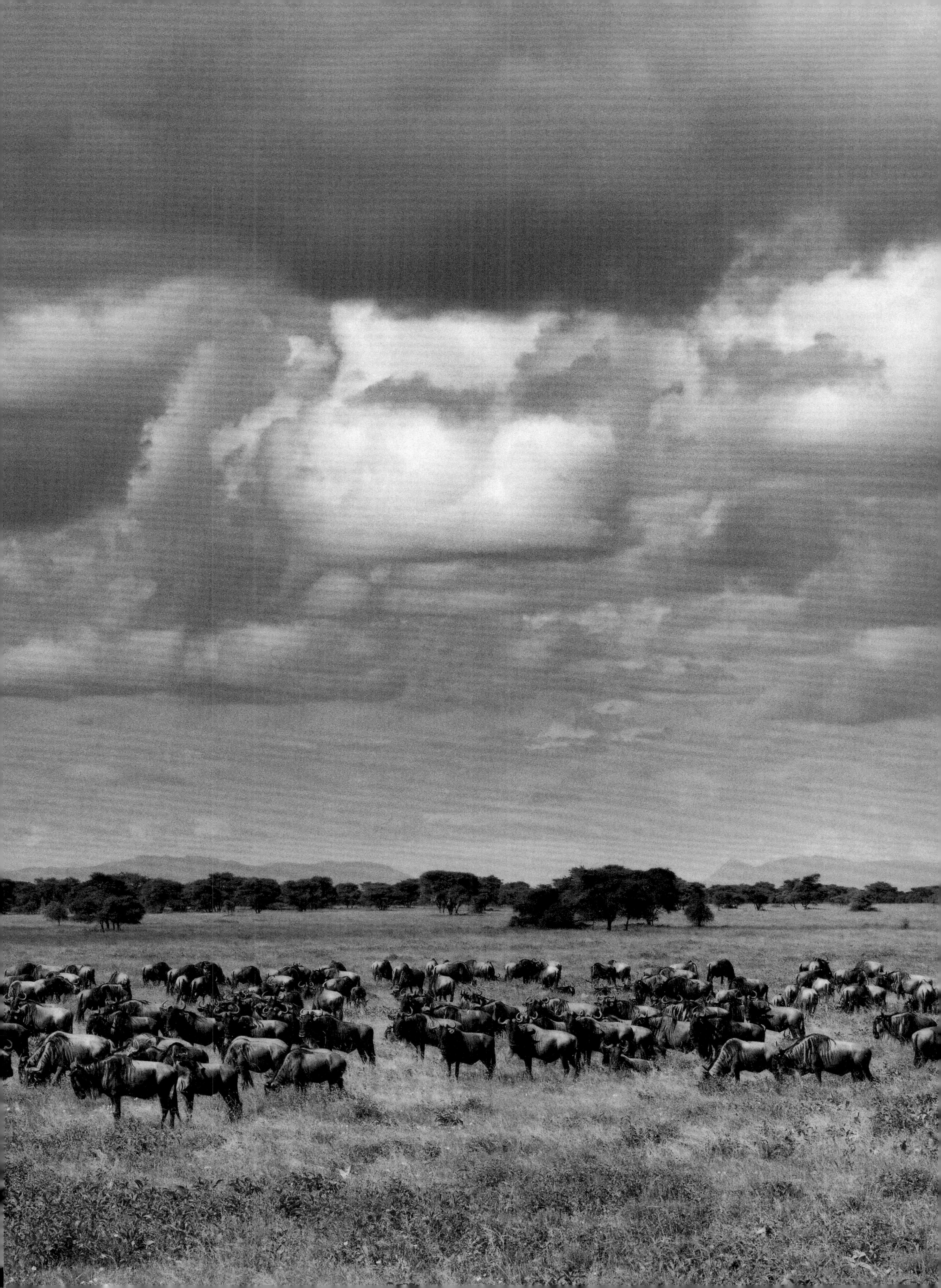

THE APOGEE OF THE GREAT MIGRATION IS THE ANNUAL BABY BOOM ON THE SHORTGRASS PLAINS. THERE IS A VAST LAWN OF NUTRITIOUS GRASS FOR LACTATING MOTHERS AND THERE IS A GLUT OF TASTY CALVES FOR LURKING PREDATORS. THE PREDATORS CAN ONLY EAT SO MANY AND THERE ARE SUCH A GREAT NUMBER OF THEM, THAT FEWER ARE EATEN THAN WOULD OR COULD BE WERE THE BIRTHING SEASON TO STRETCH OVER A LONGER PERIOD. IT IS THE CRAFTIEST PLOY OF THE WILDEBEEST BY FAR.

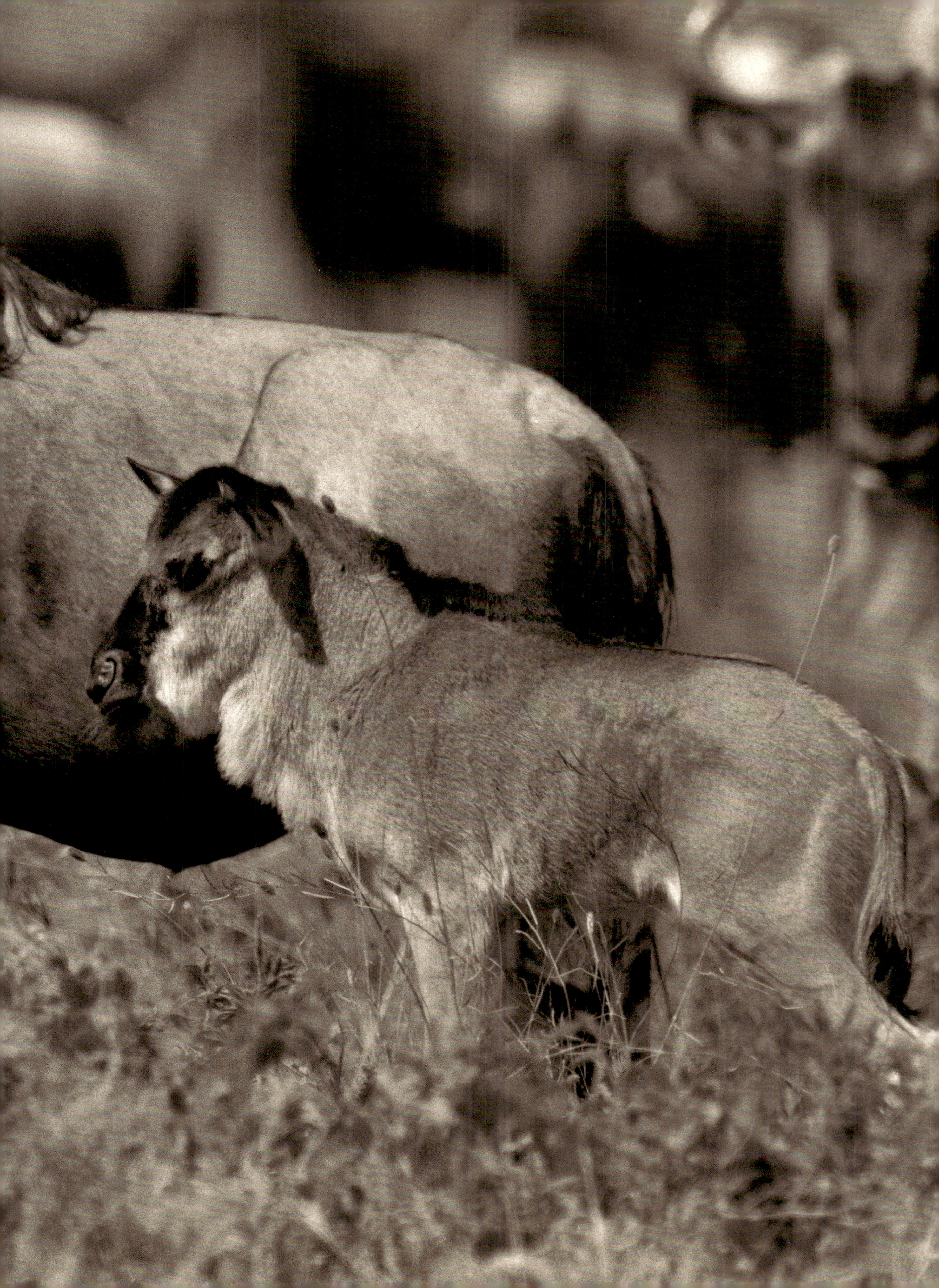

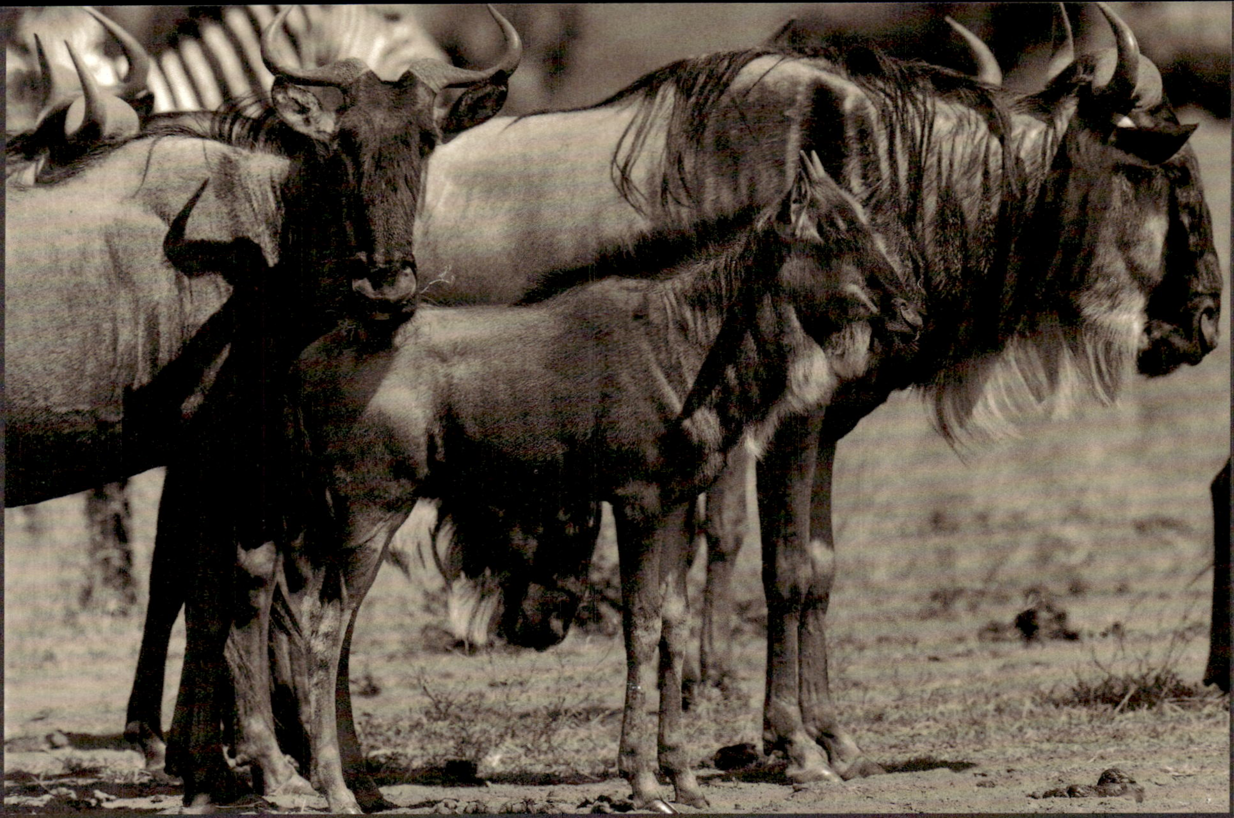

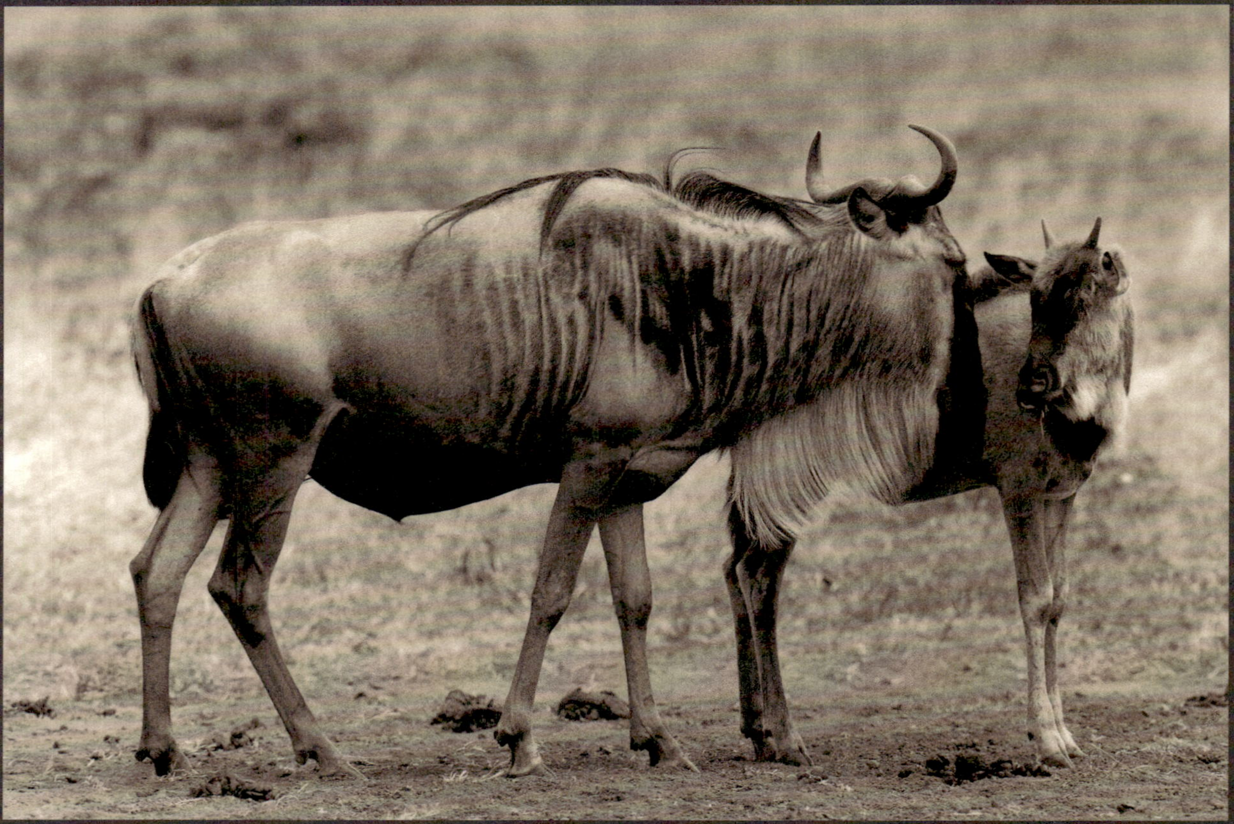

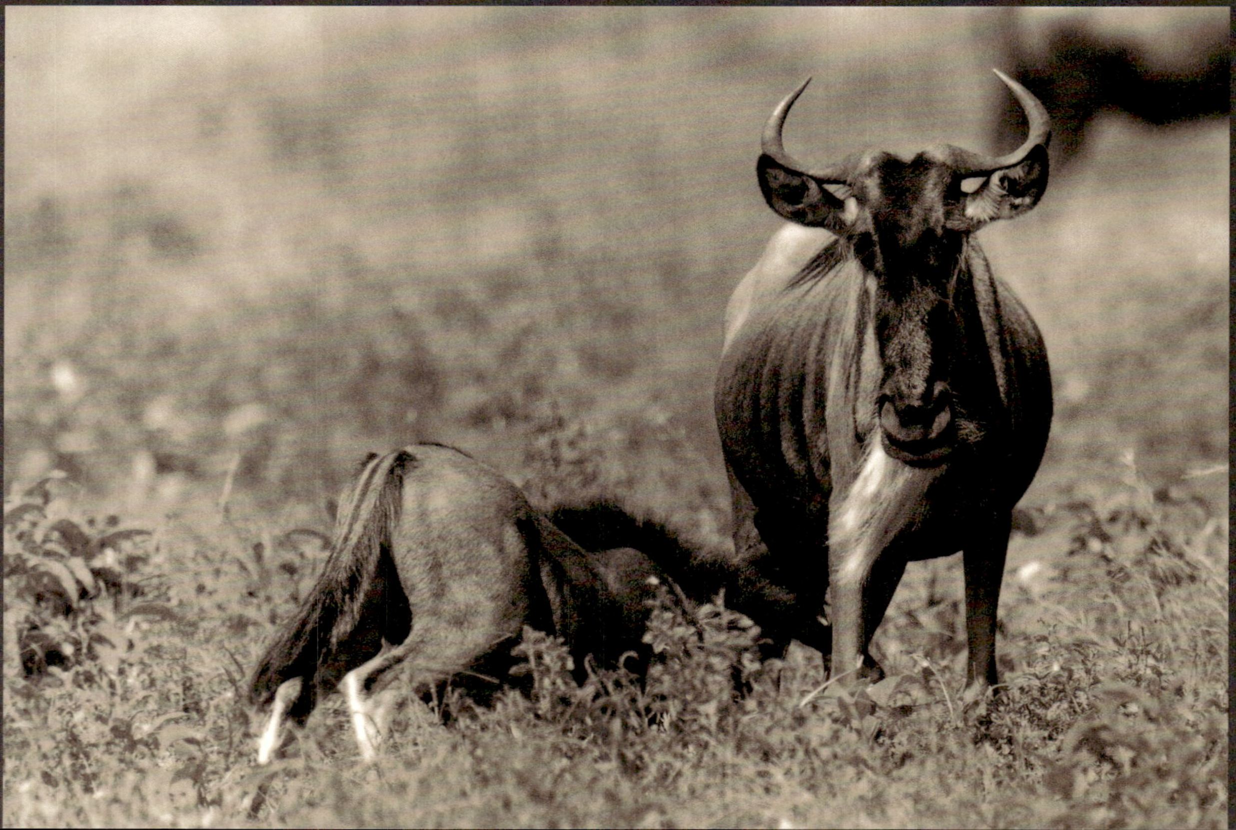

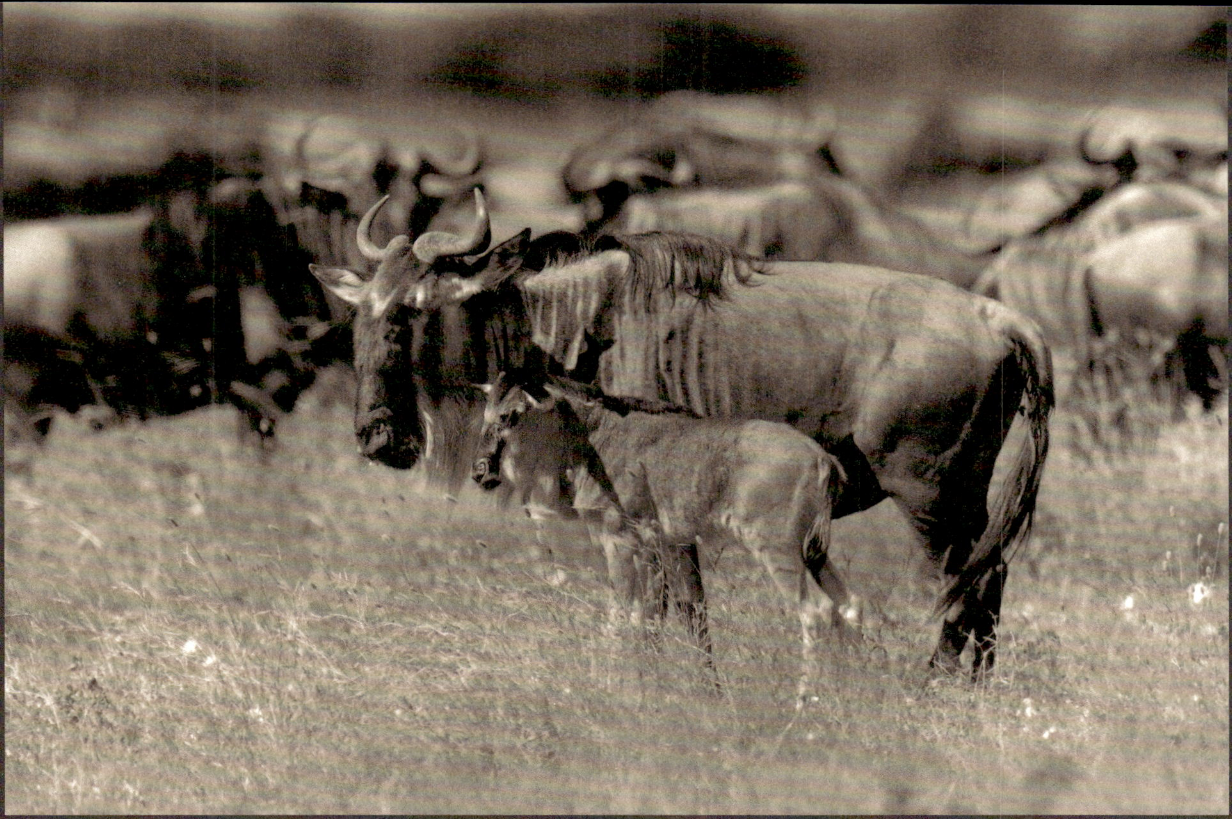

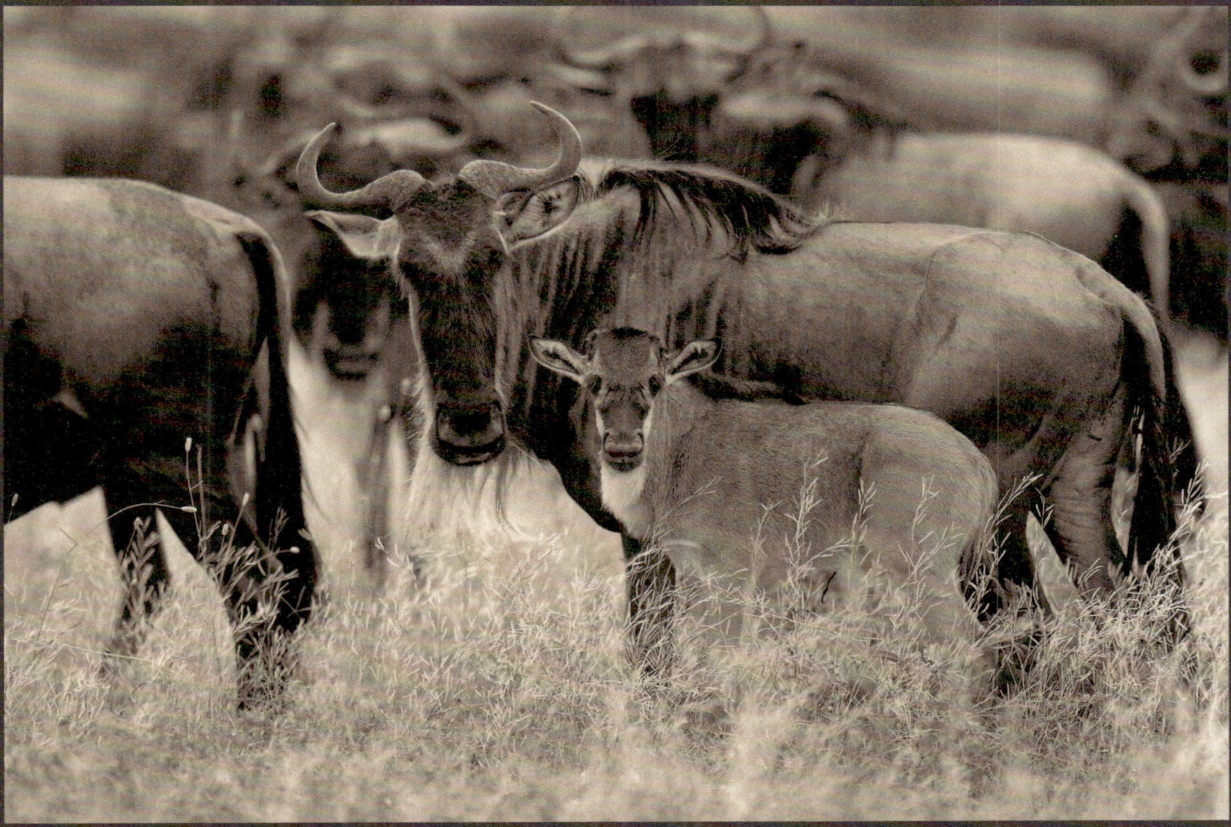

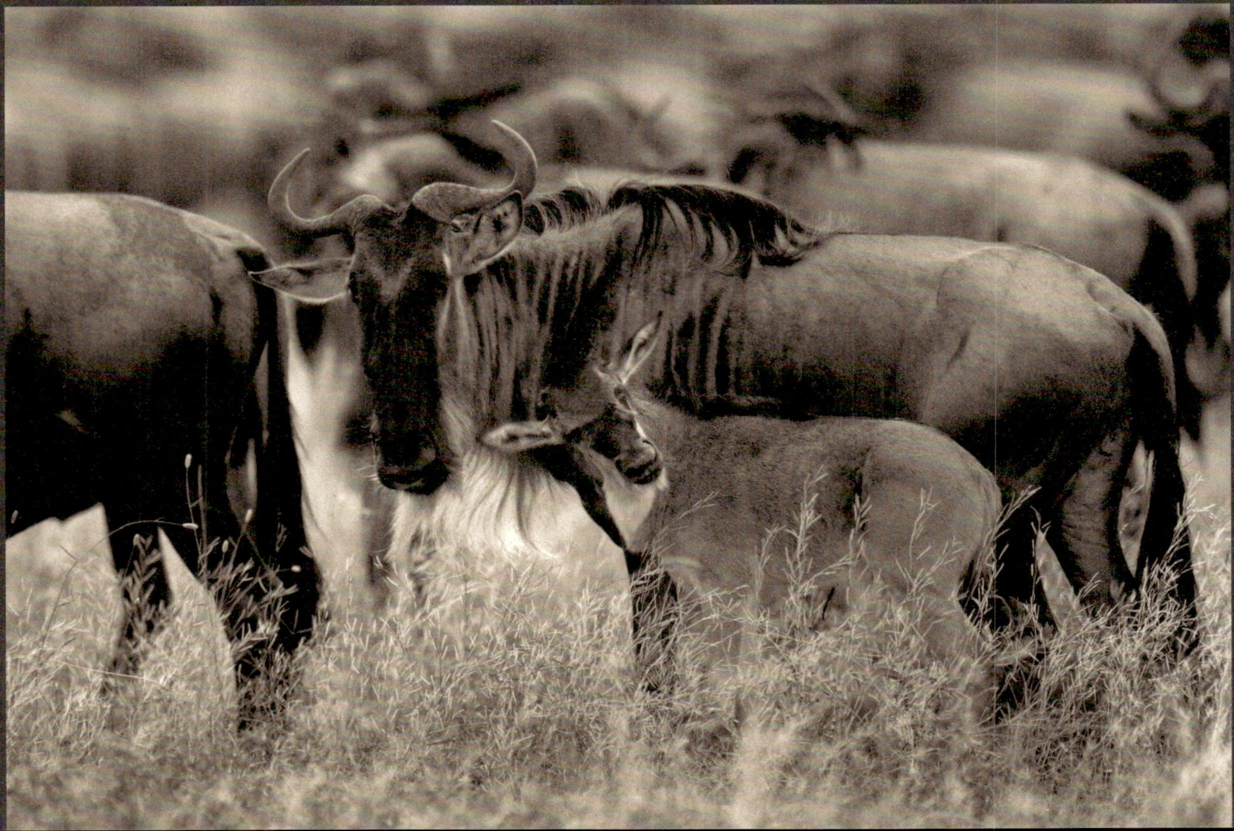

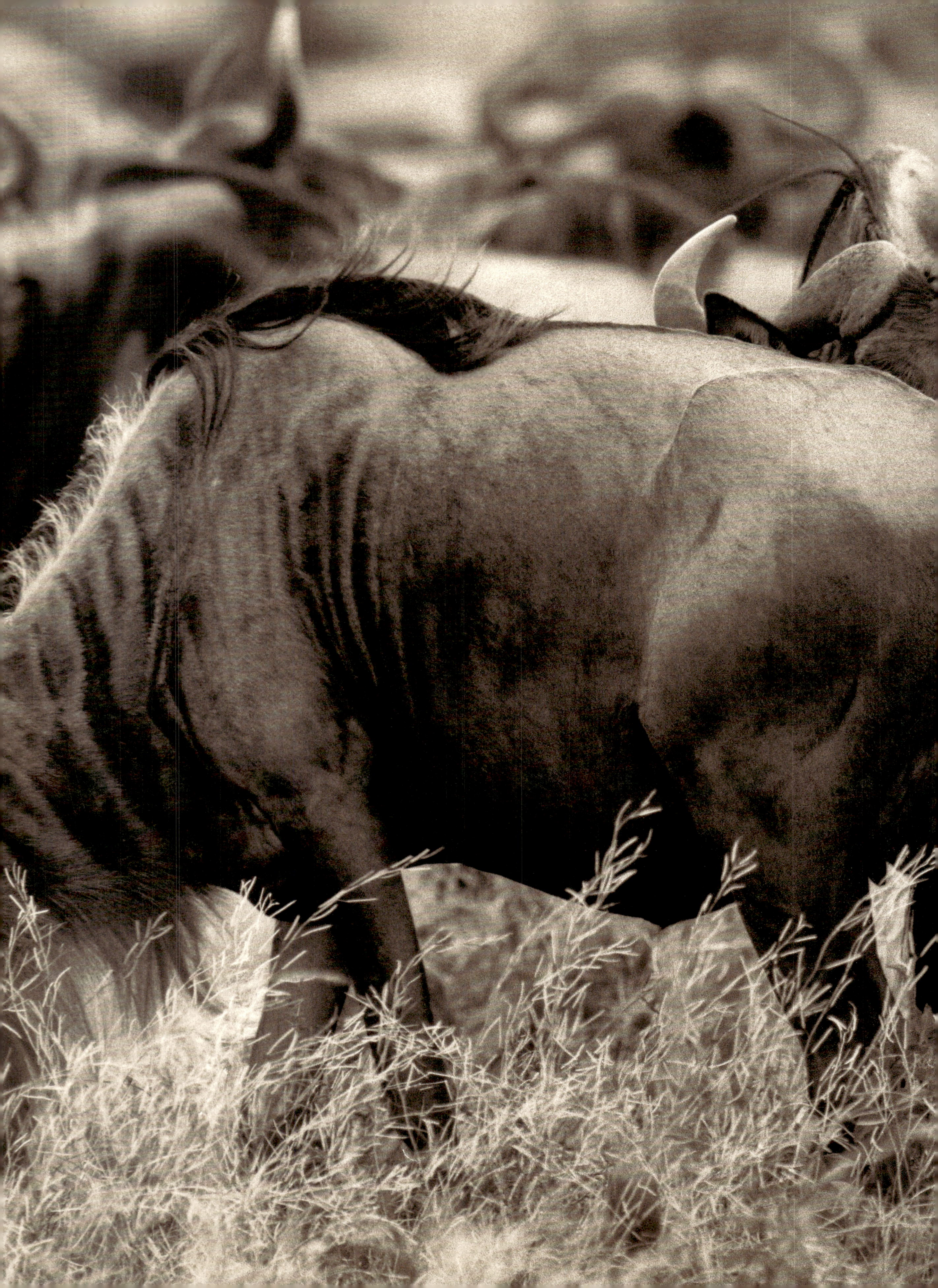

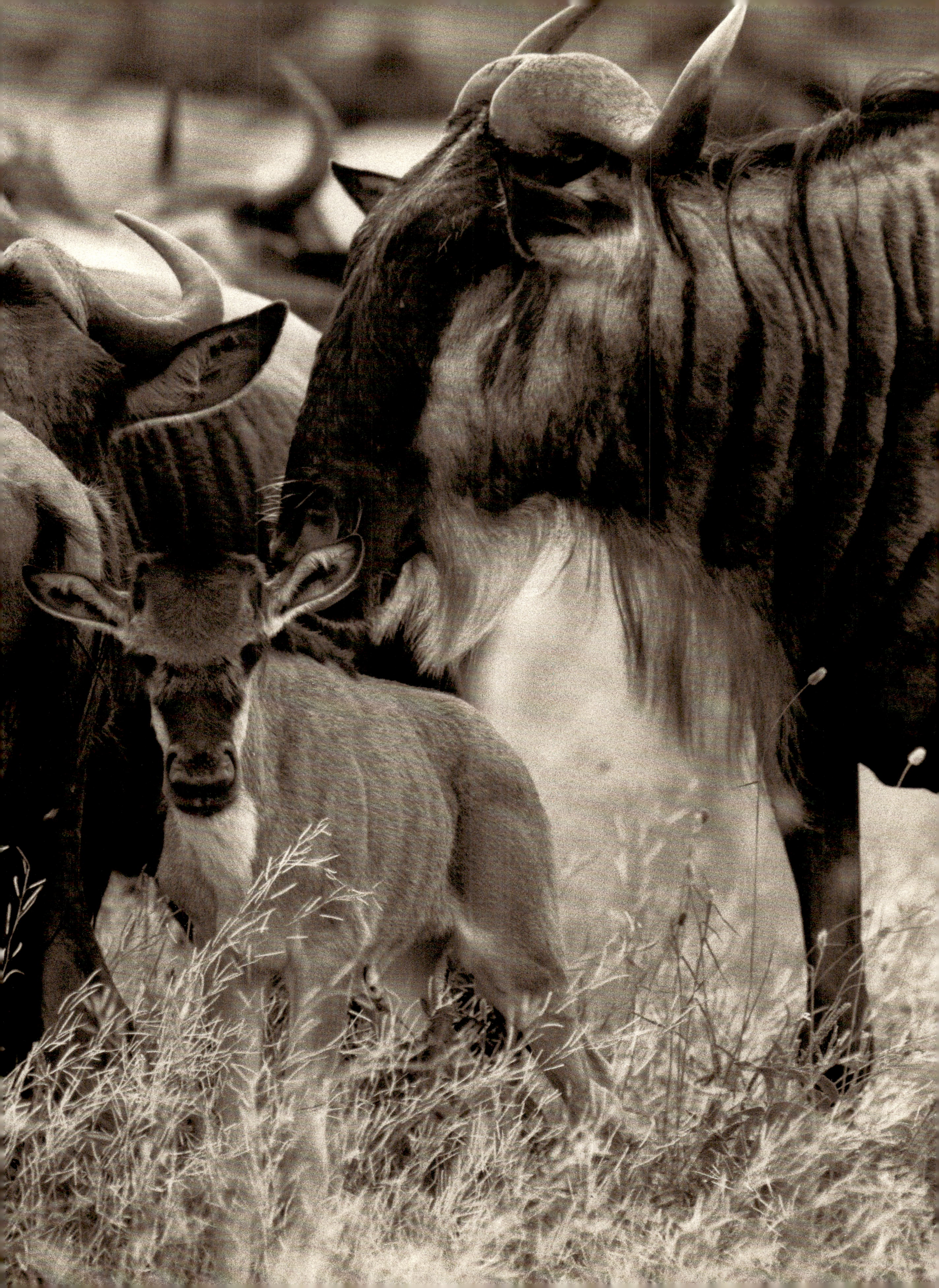

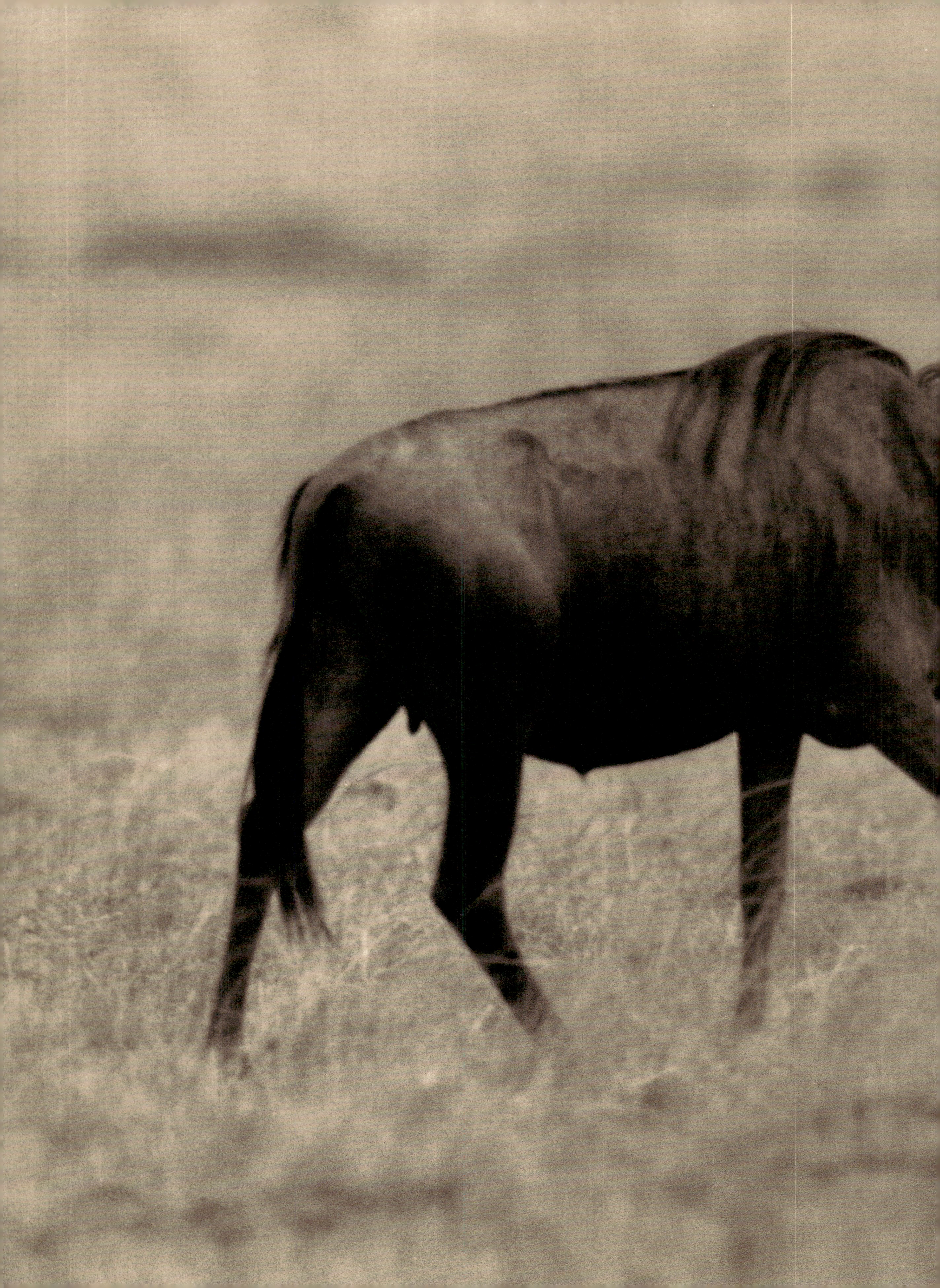

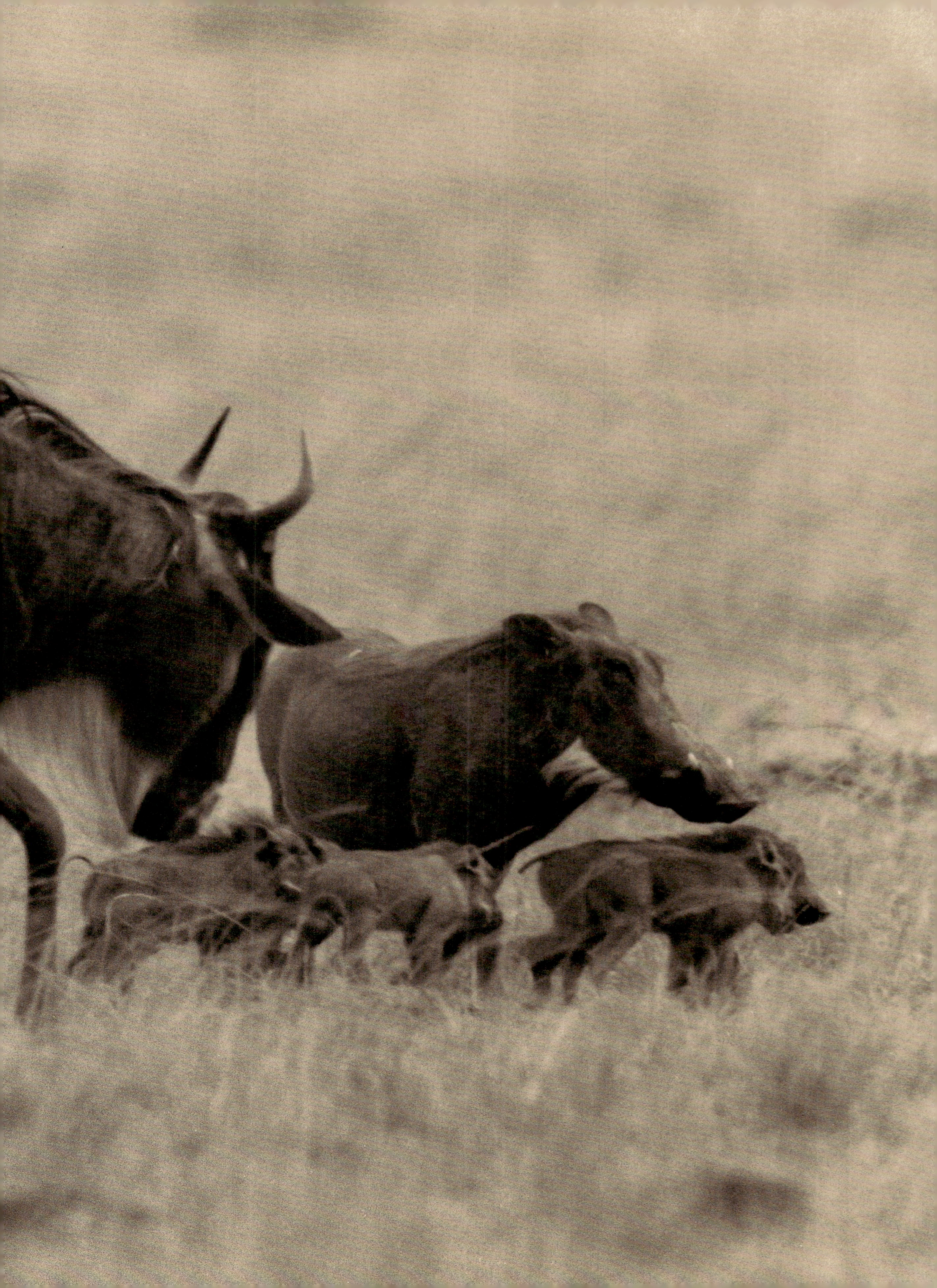

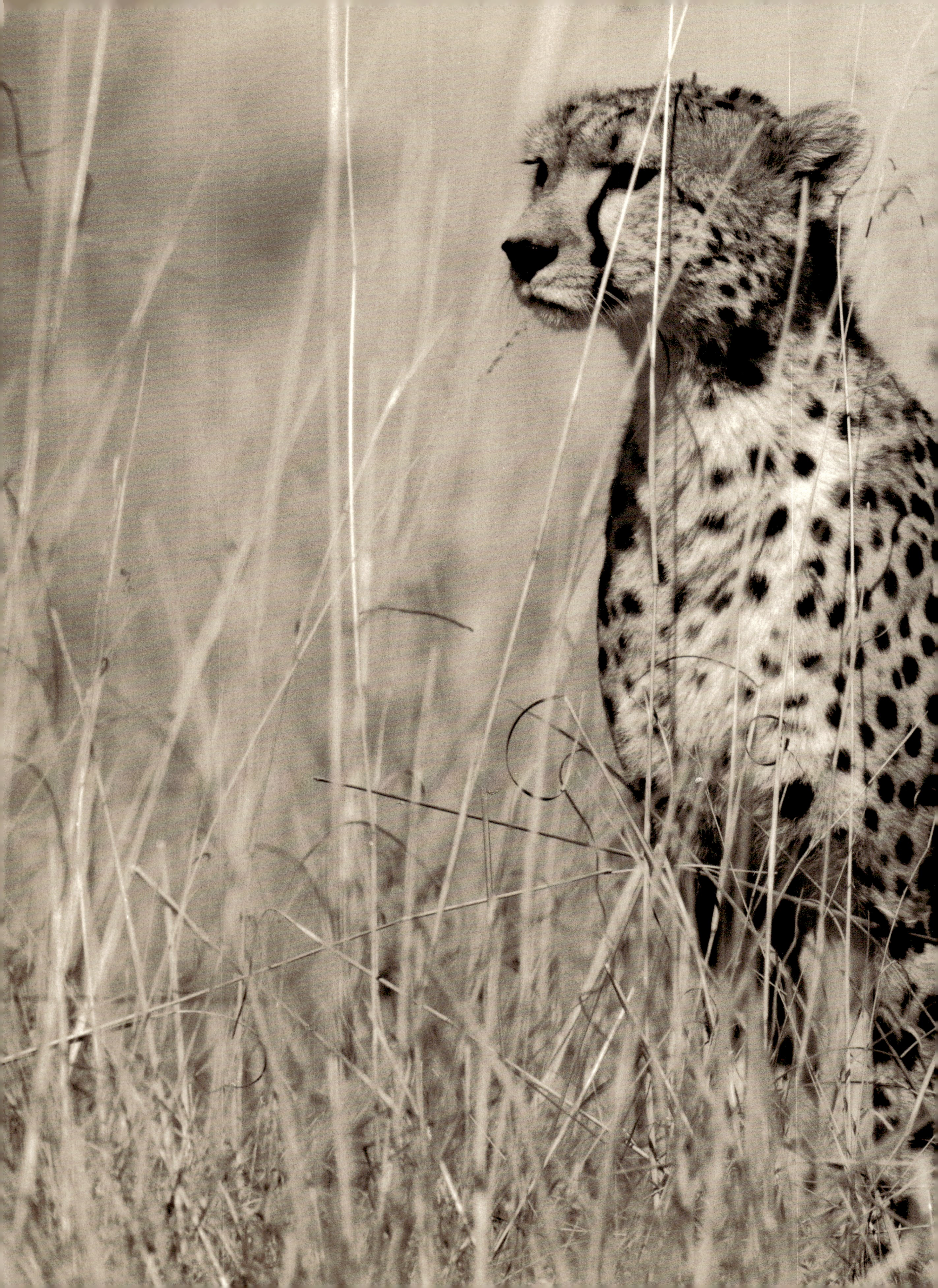

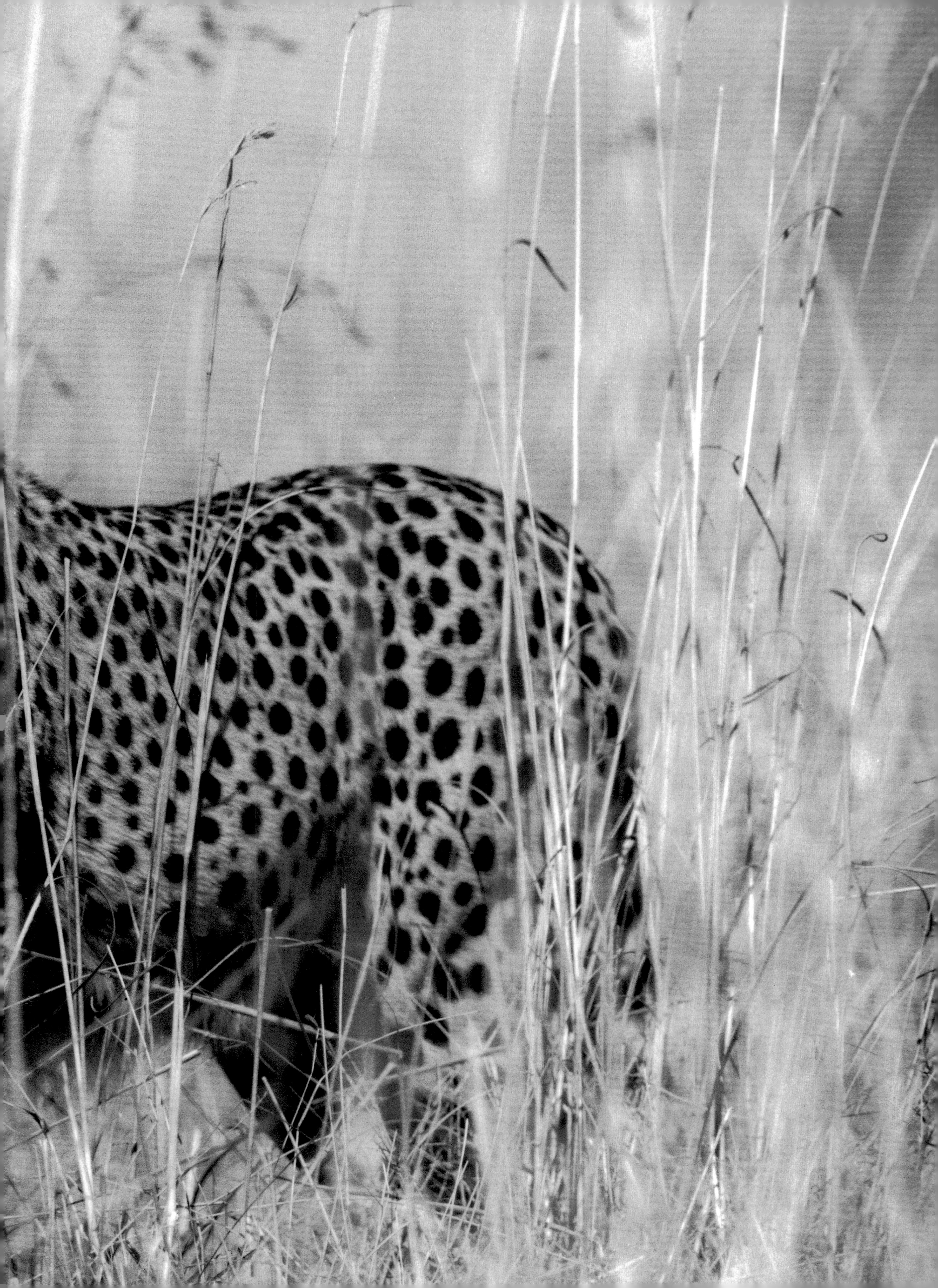

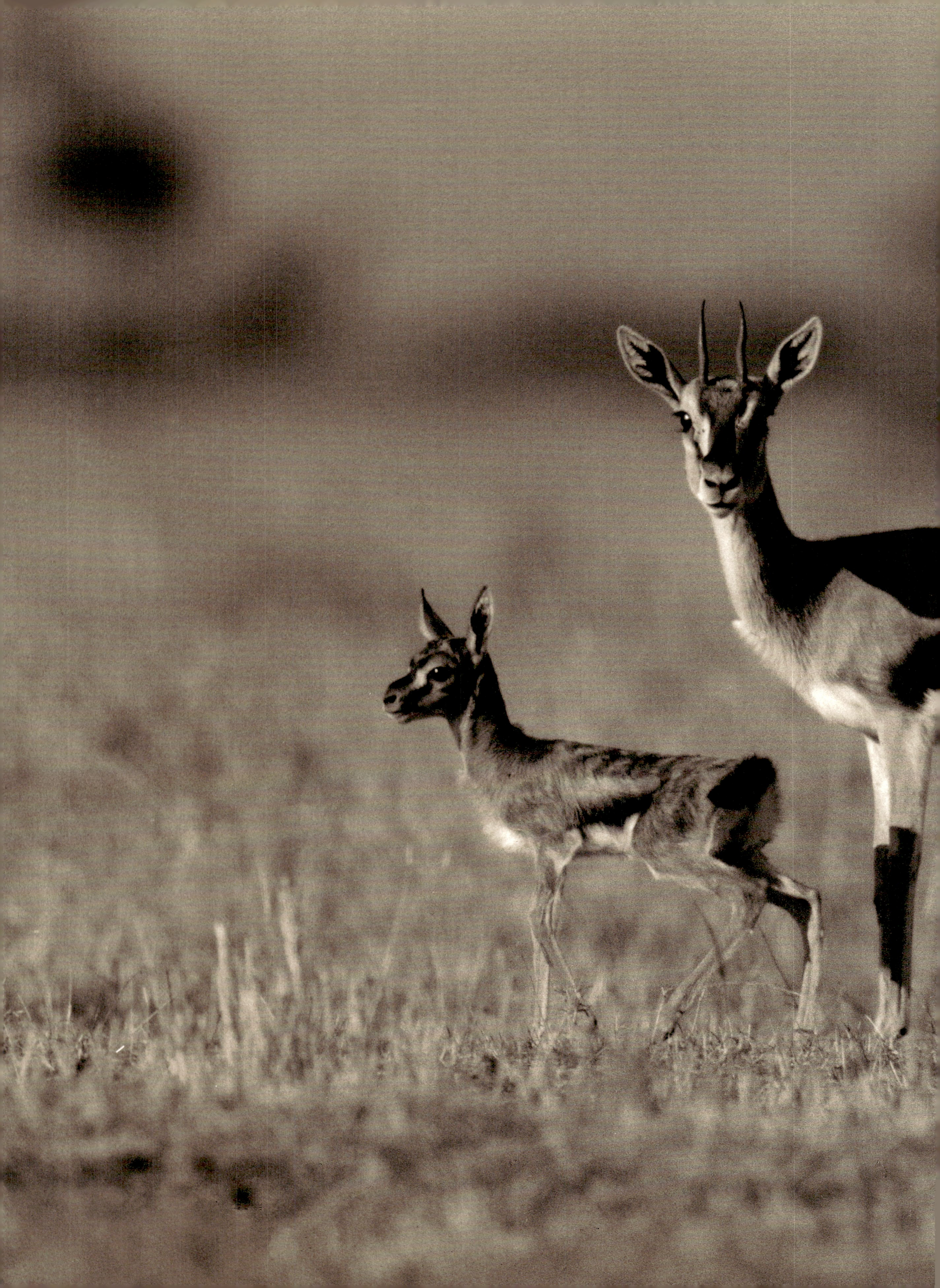

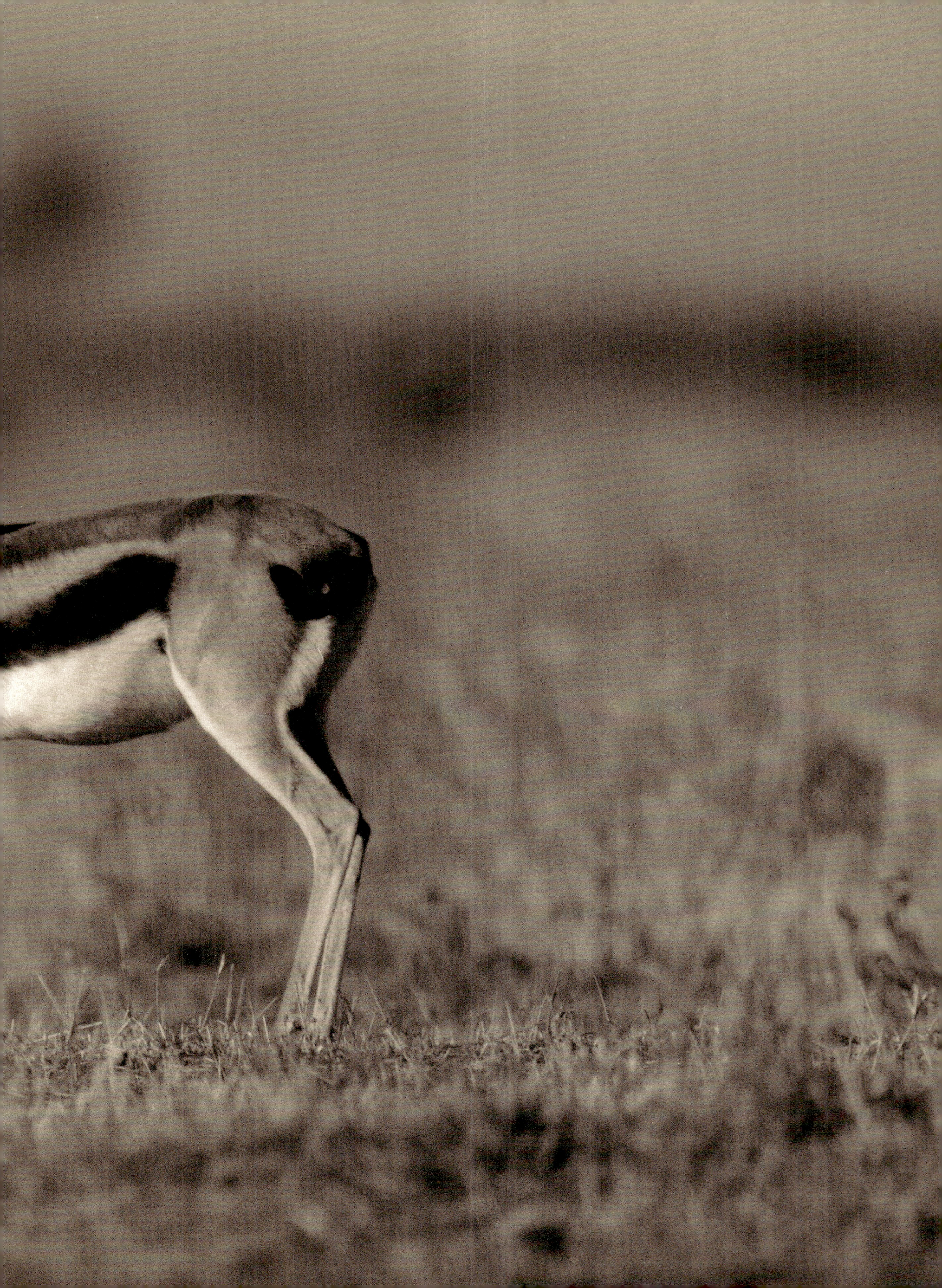

THE WHOLE POINT OF THE GAME IS TO MAKE MORE OF THE SAME, AND THIS IS JUST WHAT THEY ALL DO. EVERYONE IS IN THE REPRODUCTION BUSINESS AT THIS TIME OF YEAR; IT IS GOOD TIMES ALL ROUND. BOTH HERBIVORES AND CARNIVORES BENEFIT FROM THE WILDEBEEST GLUT OF THE PREDATOR MARKET: APART FROM ABUNDANT FOOD, THE CHANCES OF SURVIVAL OF THE YOUNG OF OTHER HERBIVORE SPECIES SUCH AS THOMMIES (PREVIOUS PAGE) AND WARTHOGS (PP.234–235) INCREASE THROUGH "SAFETY IN NUMBERS". THE ADULT LIONS, OF COURSE, HAVE MORE WILDEBEESTE VEAL THAN THEY CAN EAT, SO THERE IS PLENTY LEFT FOR THE CUBS. AND ALL THE YOUNGSTERS, CAT AND ANTELOPE ALIKE, HAVE ACCESS TO MOTHERS' MILK IN PLENTY BECAUSE THE MOTHERS HAVE ACCESS TO A SUPERABUNDANCE OF FOOD.

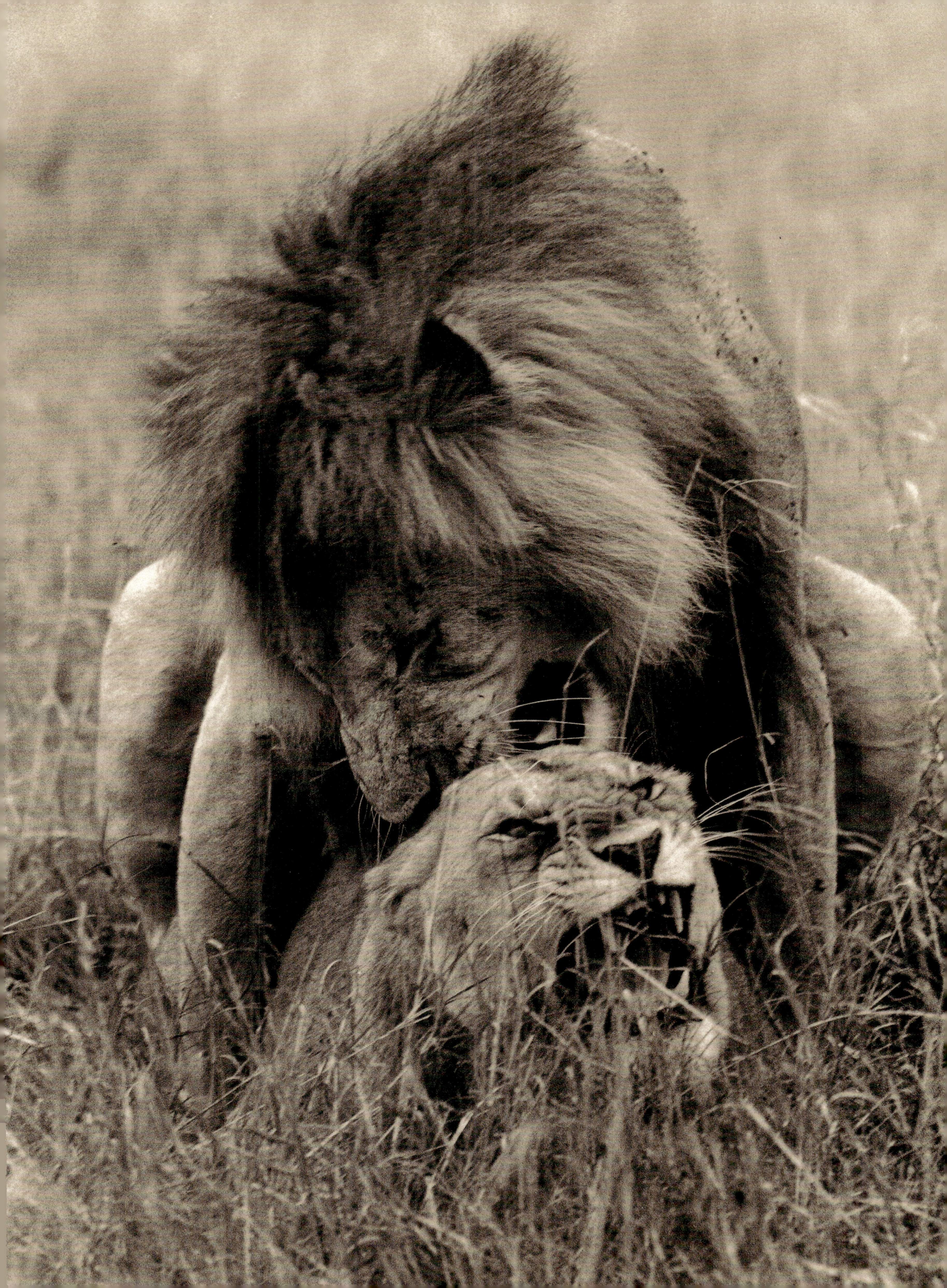

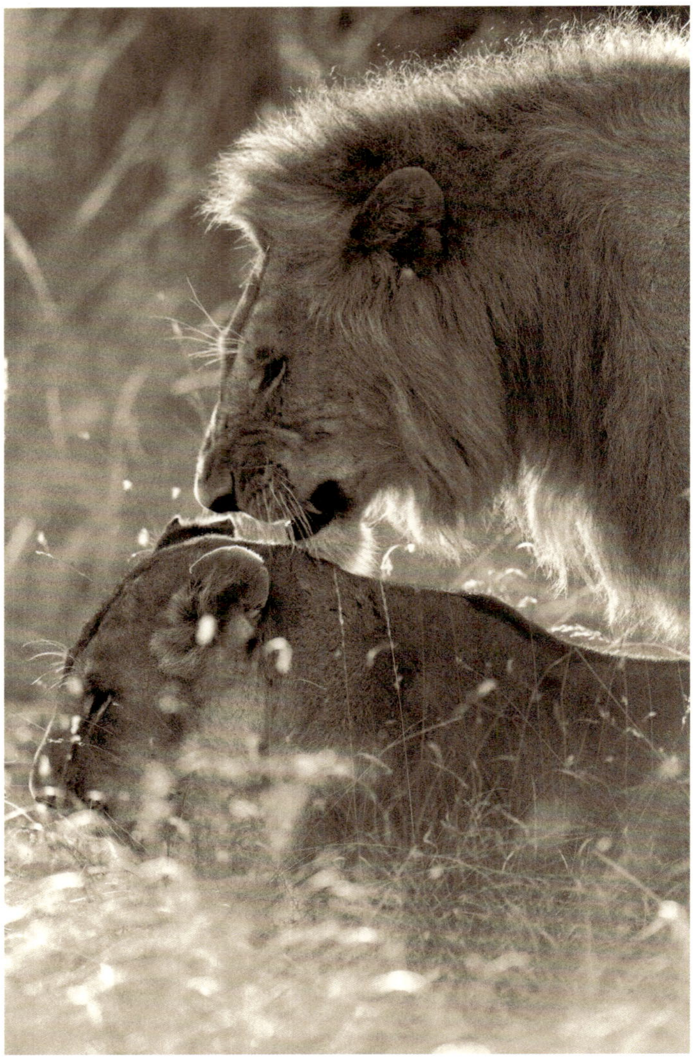

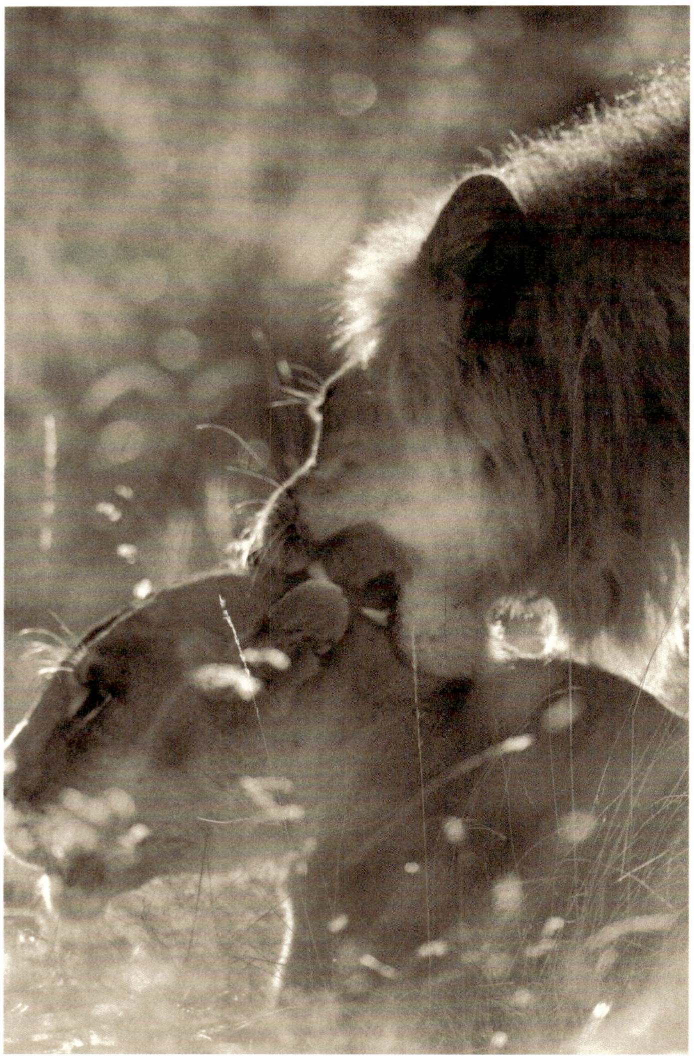

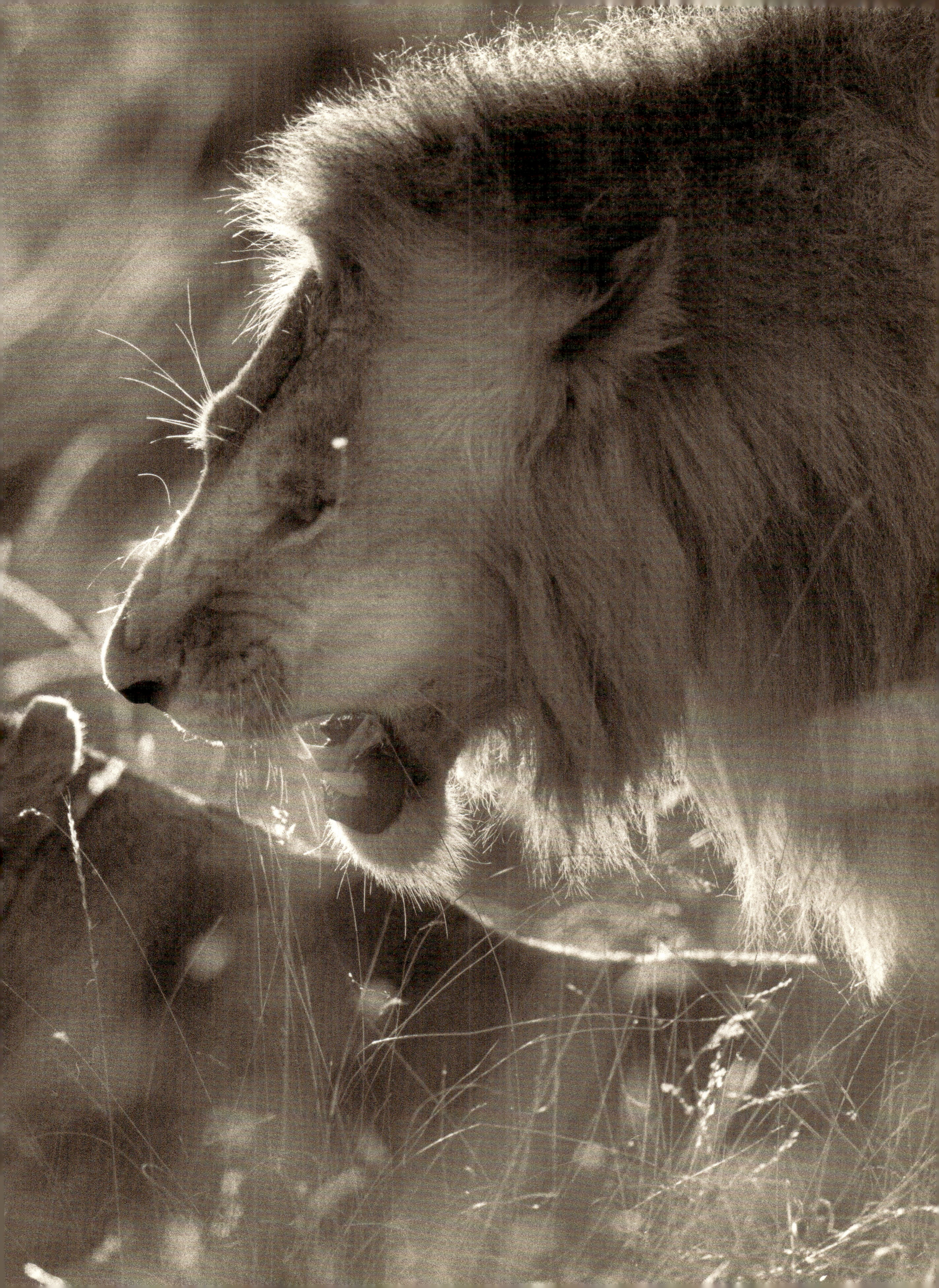

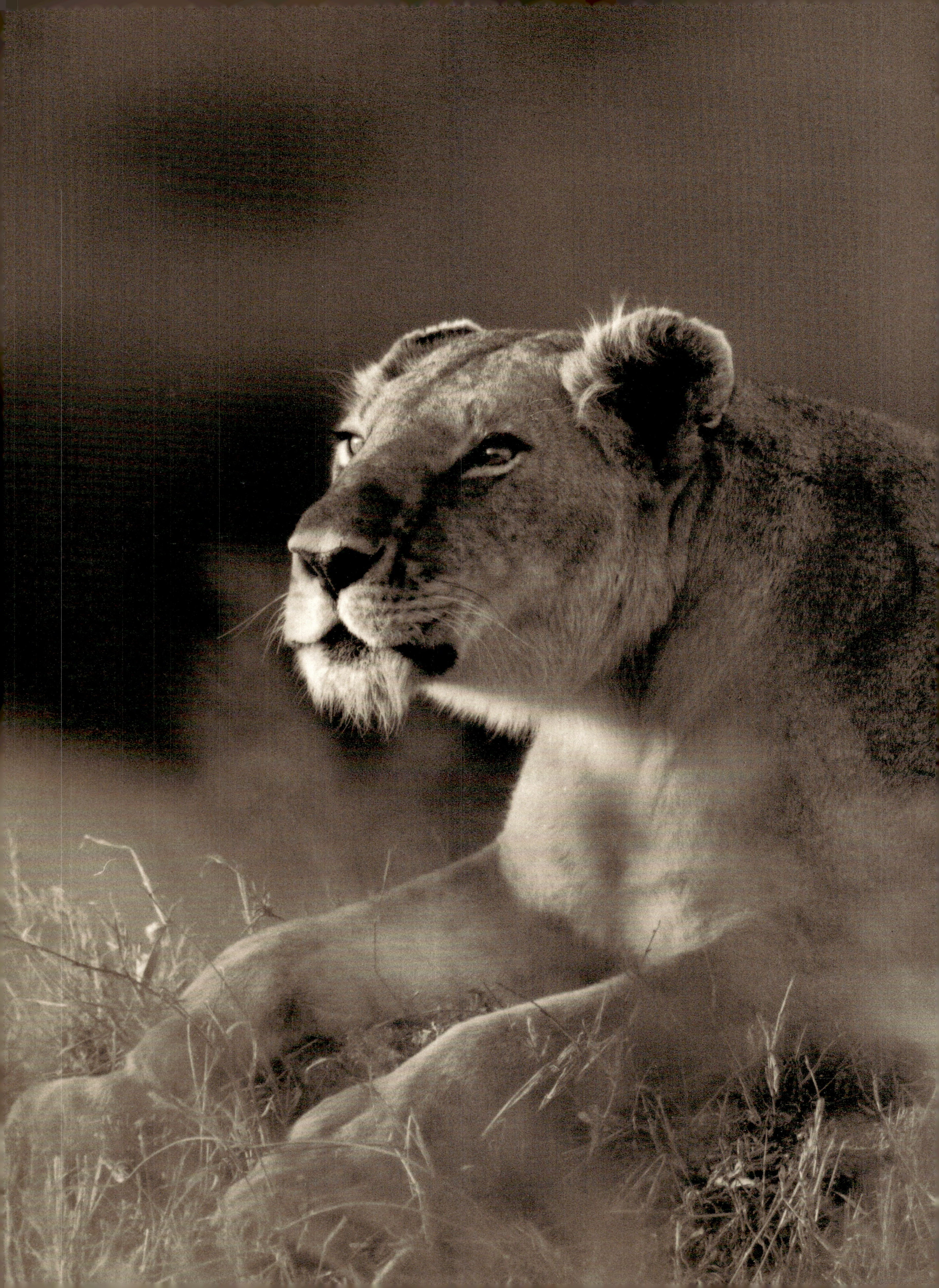

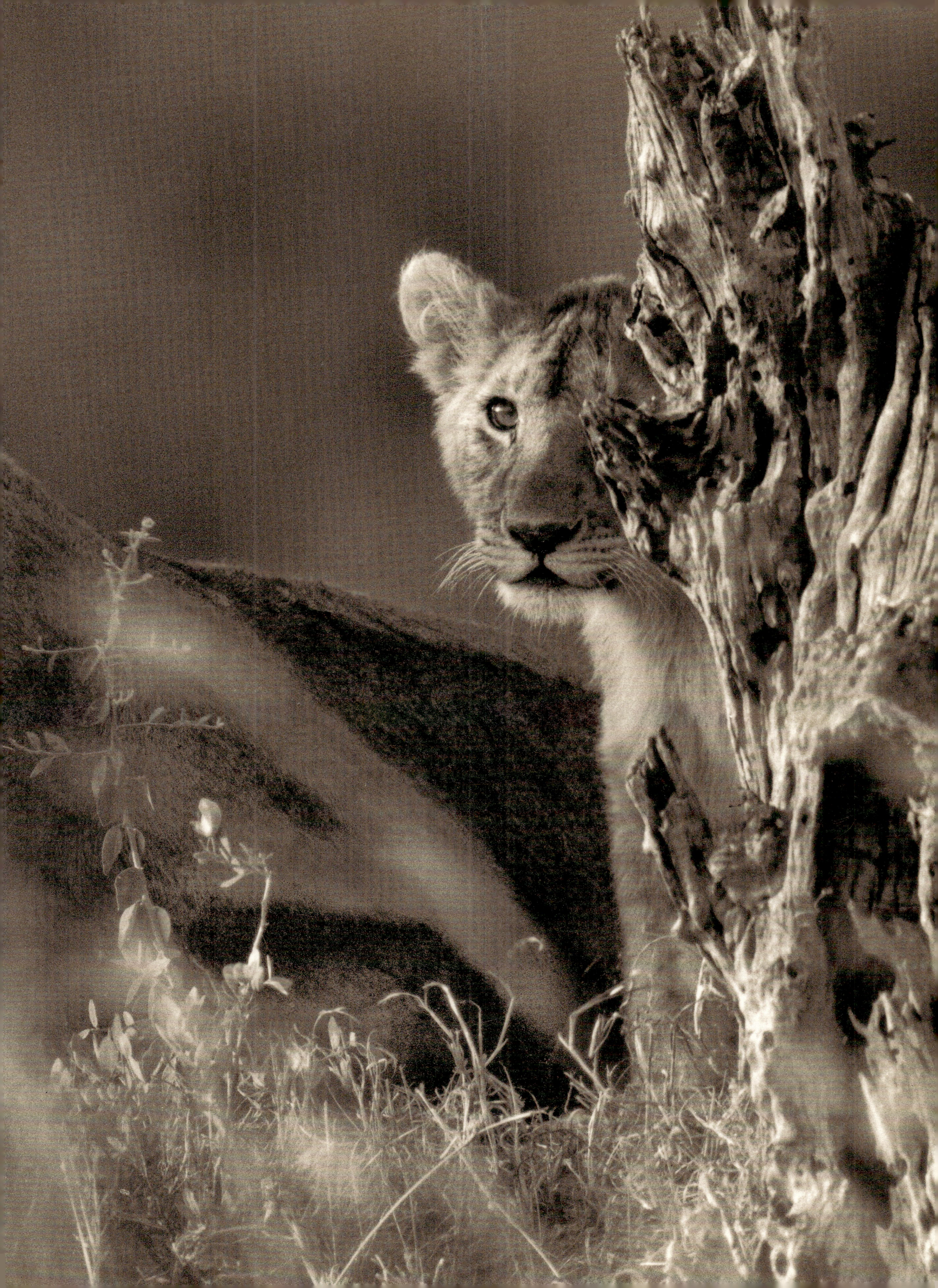

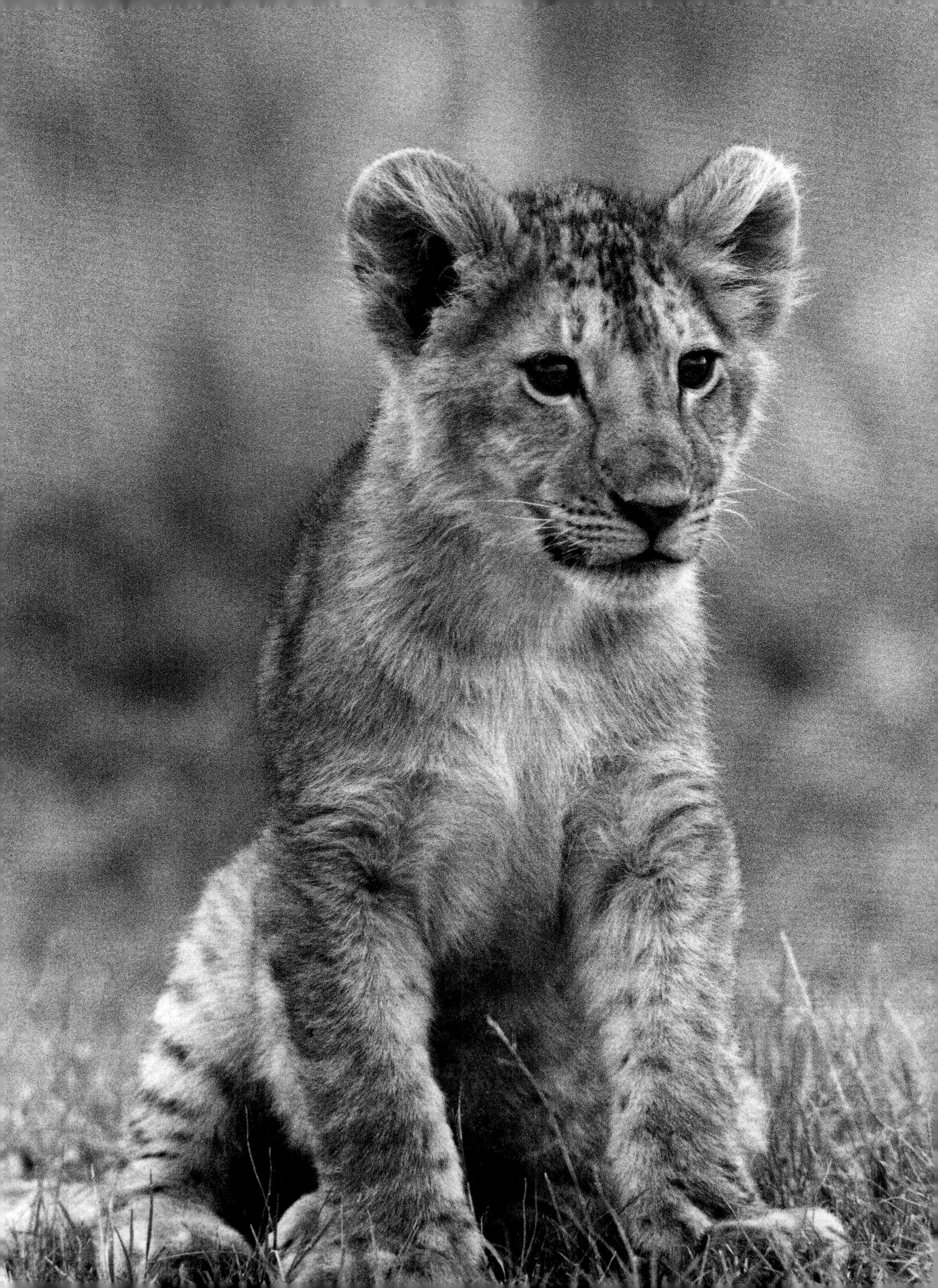

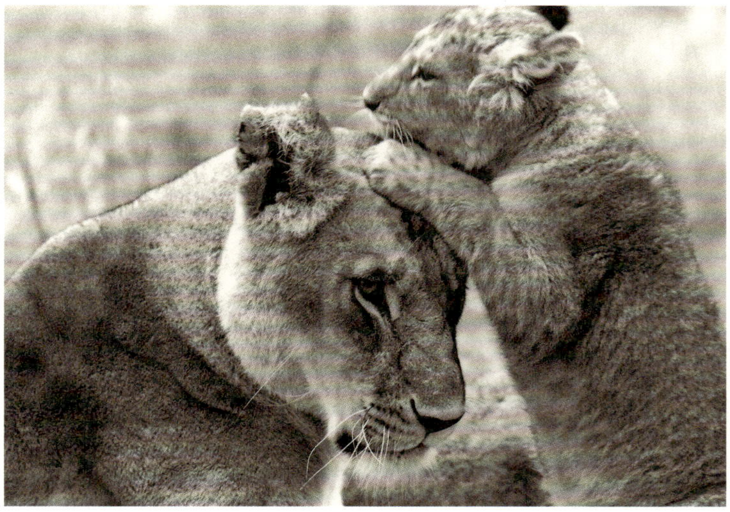
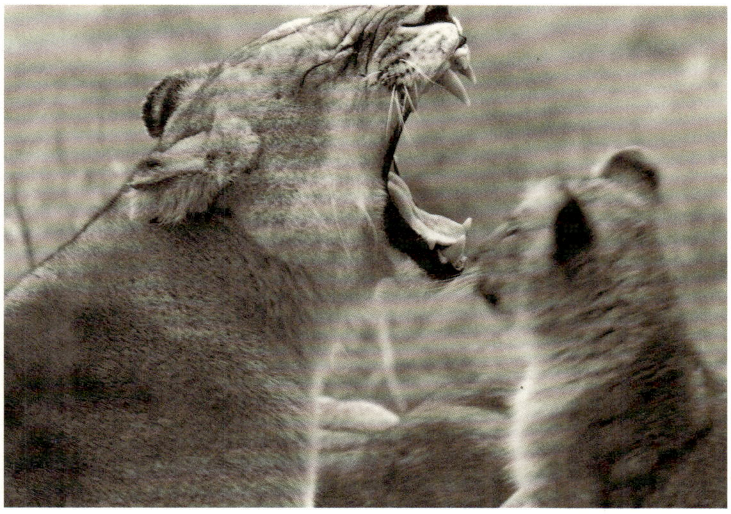
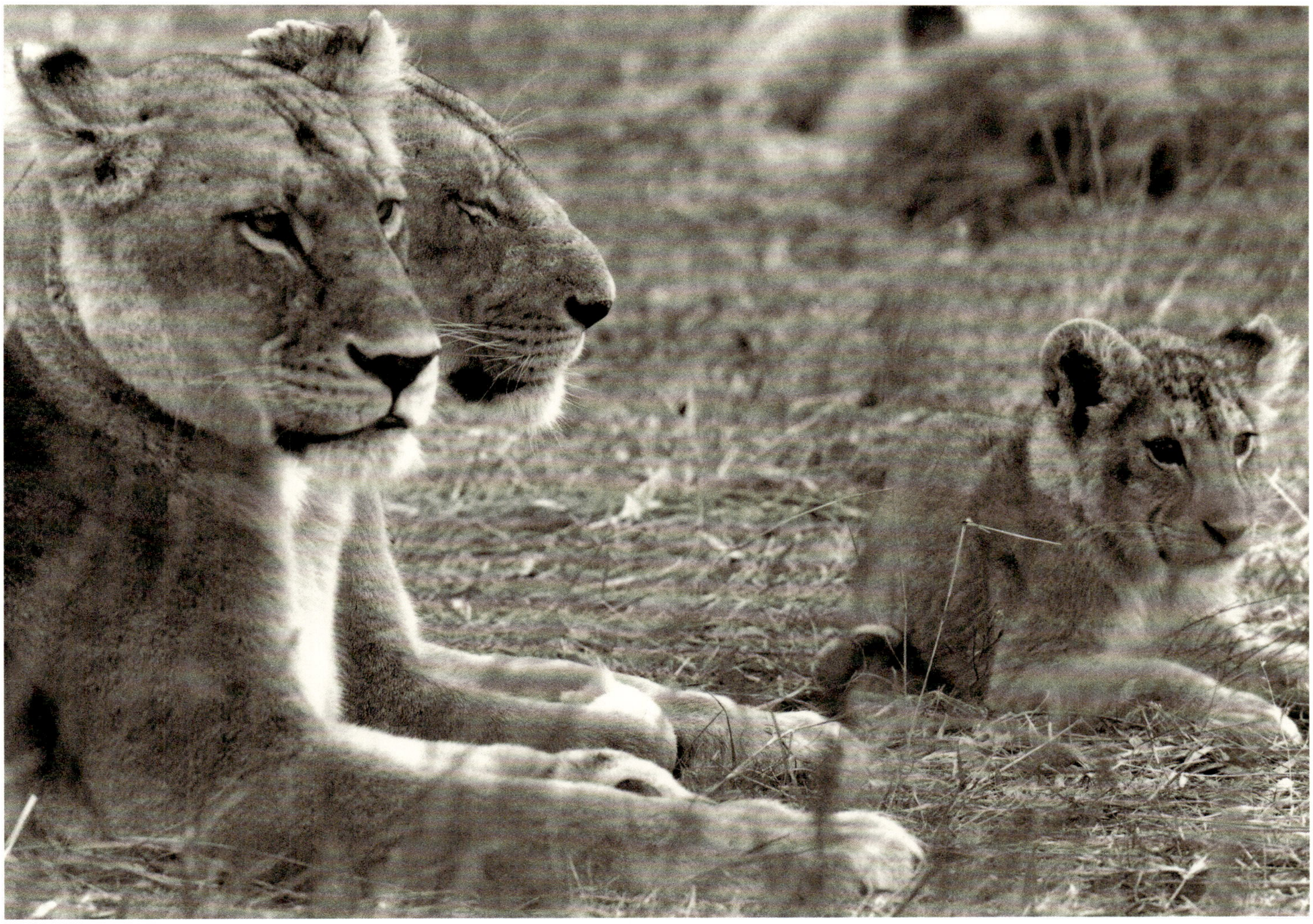

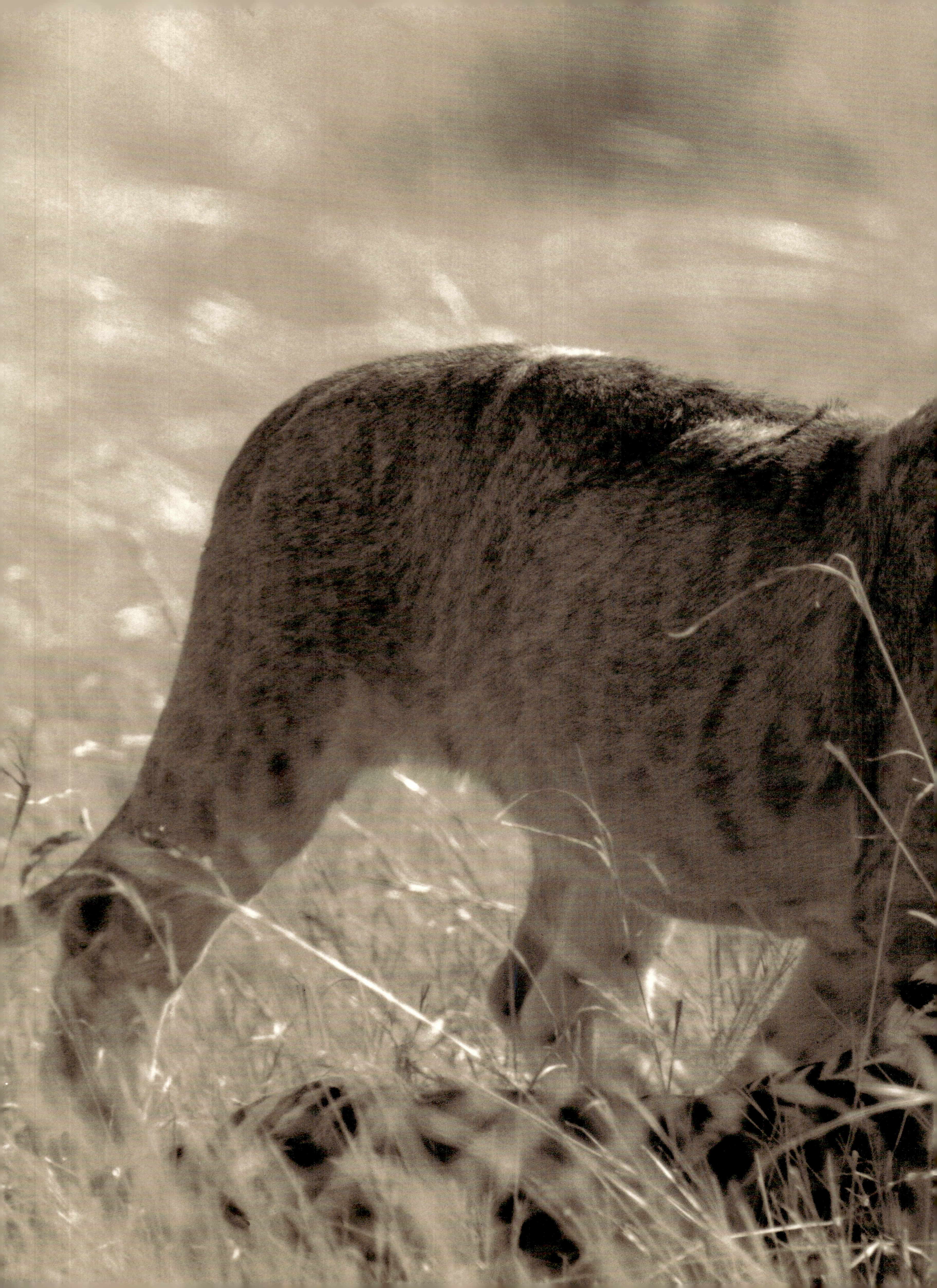

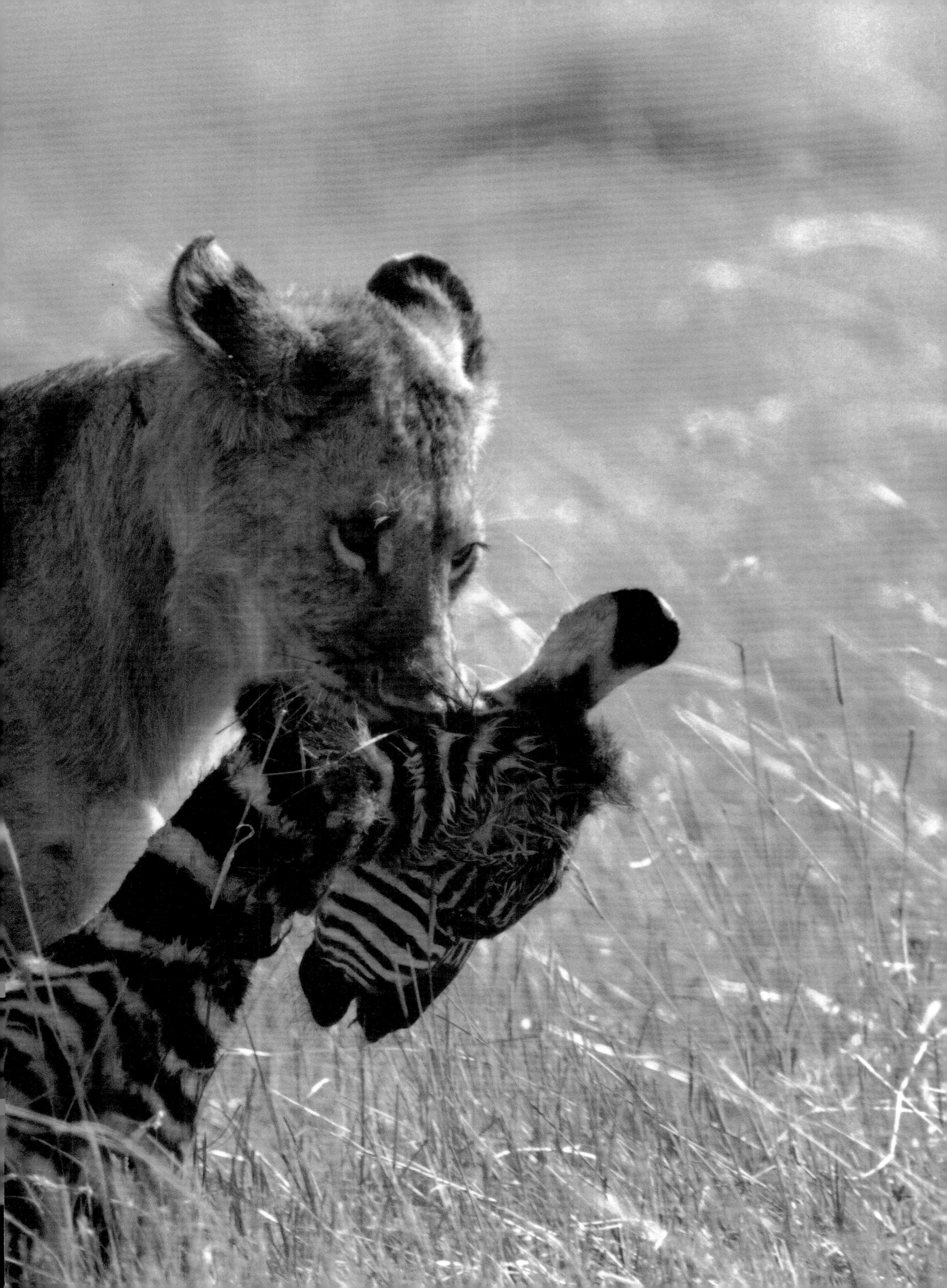

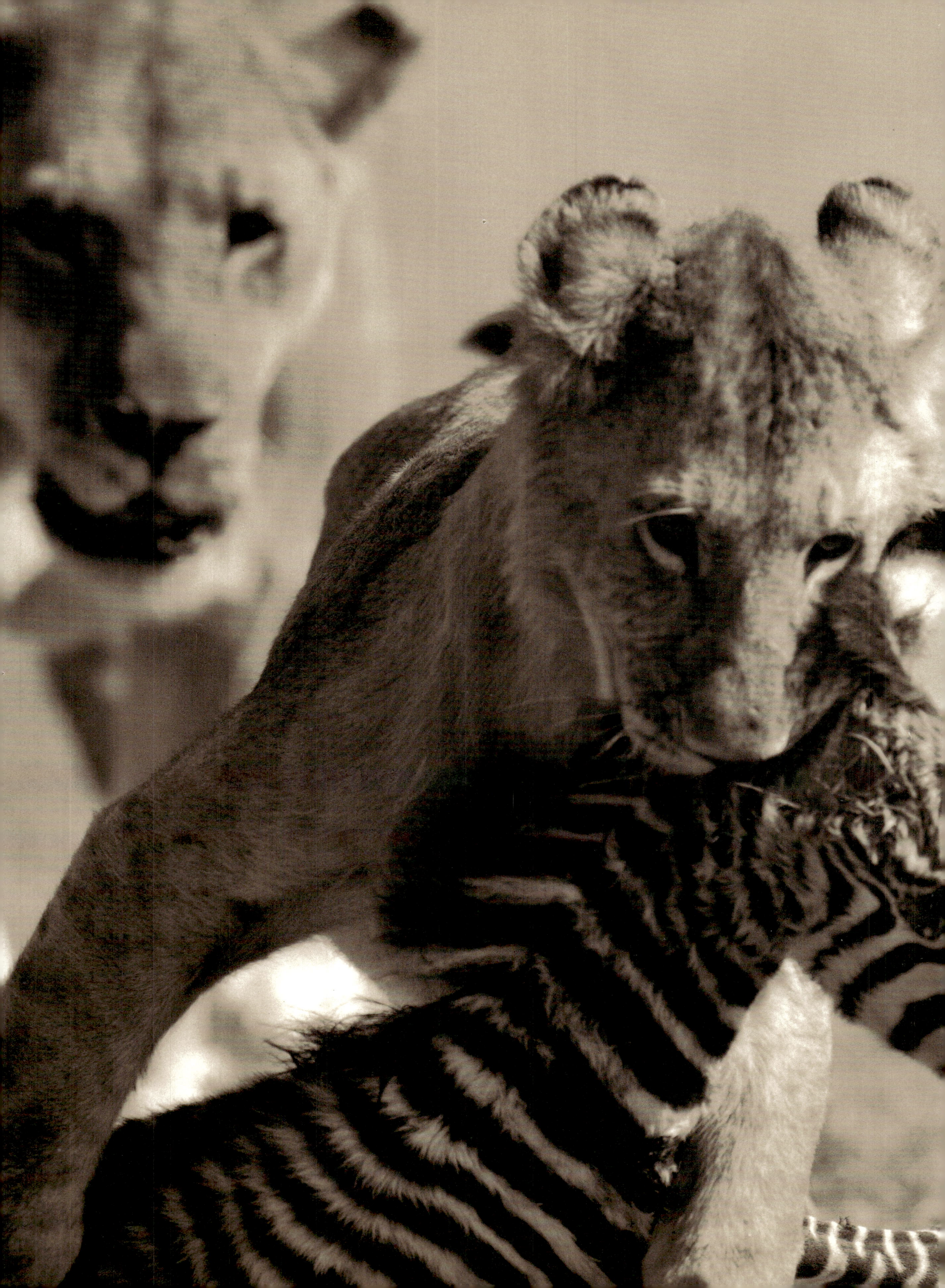

WITH THE YEARLY PASSAGE OF THE WILDEBEESTE AND ZEBRA THROUGH THEIR TERRITORIES, THERE IS PLENTY OF FOOD AROUND FOR LIONS TO CHASE AND HUNT DOWN, AND EVEN THE CUBS CAN OFTEN SNATCH A MORSEL FROM UNDER THE NOSES OF THE ADULTS. BEING CATS, THE POTENTIAL FOR INCREASE AMONG LIONS IS ENORMOUS AND EVERY YEAR THERE ARE SURPLUS CUBS THAT DIE OF STARVATION, USUALLY AFTER THE MOVABLE FEAST OF THE MIGRATION HAS MOVED ON. CONTRARY TO THE CONVENTIONAL PICTURE OF THE FEARSOME PREDATOR CONTROLLING THE NUMBERS OF GENTLE HERBIVORES, IN THE SERENGETI IT IS THE OTHER WAY ROUND.

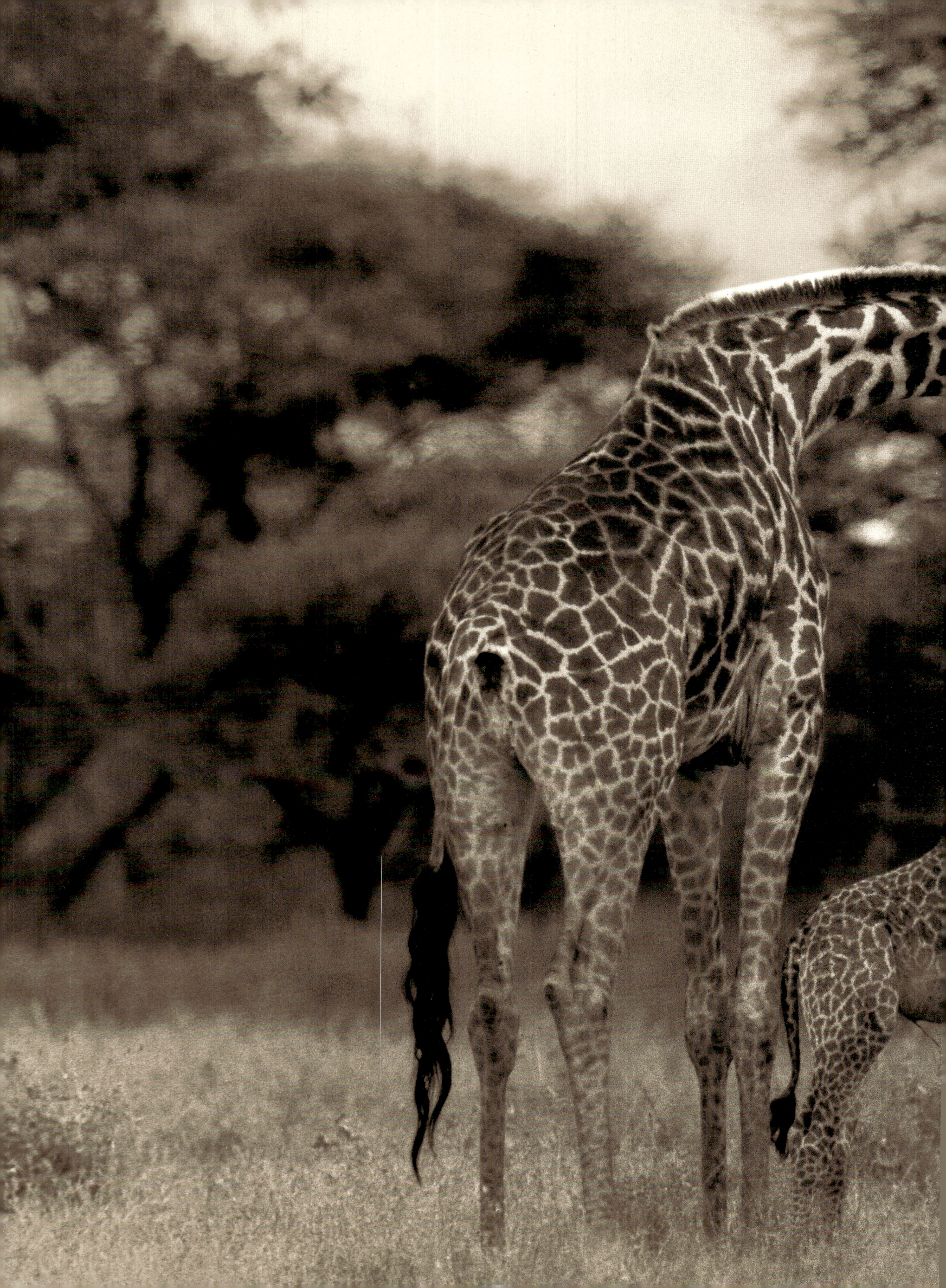

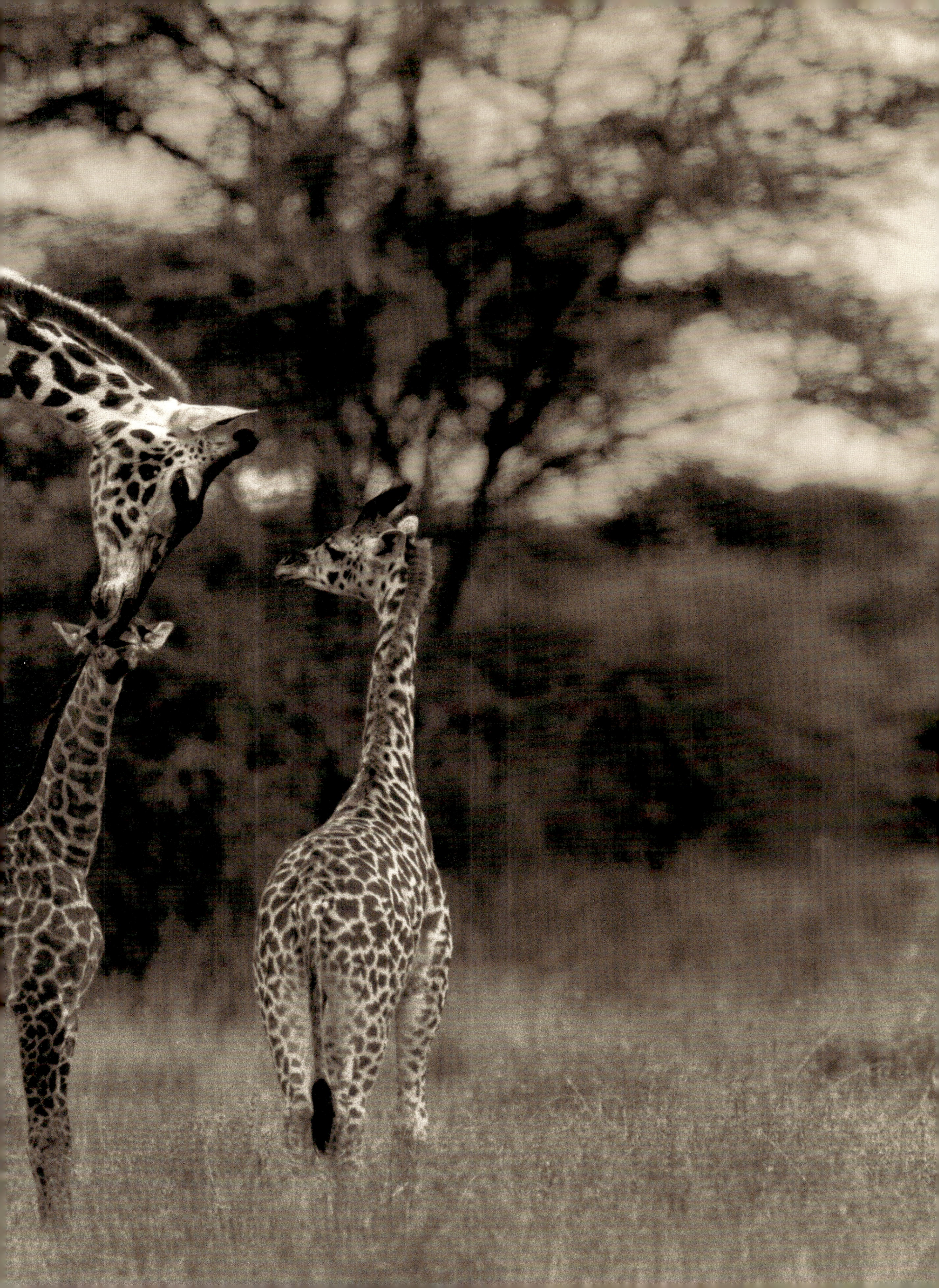

"THERE IS A PLACE WHERE THE GRASS MEETS THE SKY, AND THAT IS THE END"

MAASAI SAYING

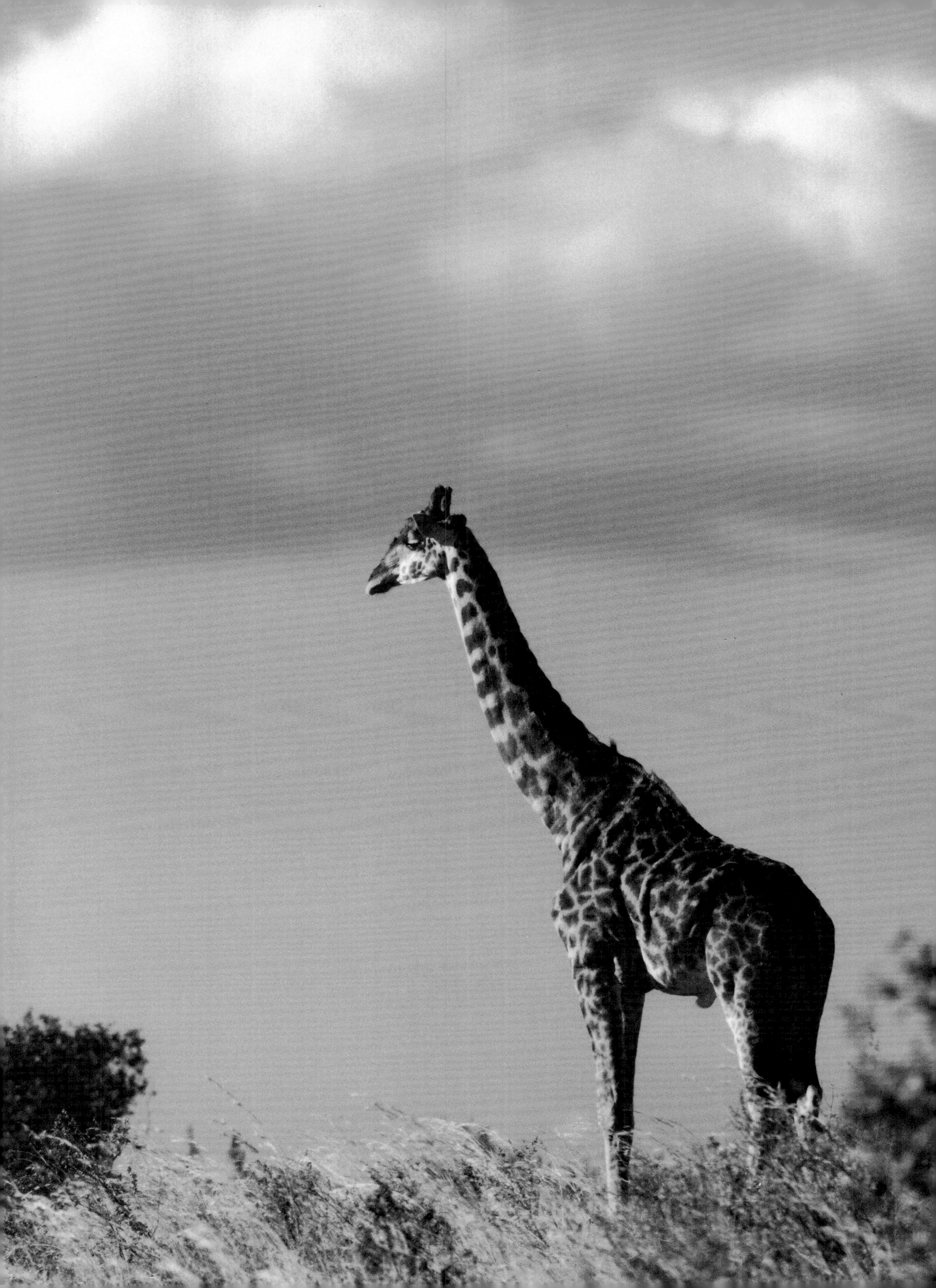

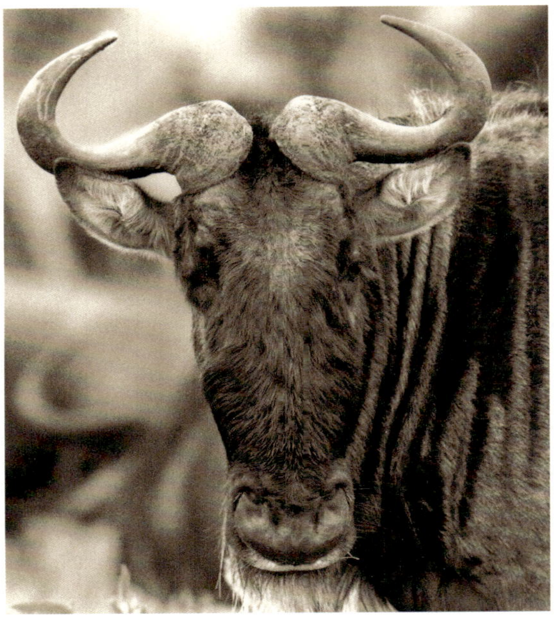

I would like to thank all those who collaborated on, and believed in, the creation of this book. My special thanks to: Elisabetta Pedrani who was the first to believe in my photography and who helped oversee every stage of the project; Massimo Bagnoli who assisted me in the Serengeti and Maasai Mara and whose experience of Africa was invaluable and an important psychological support; Massimo at De Stefani's photo lab in Milan who patiently helped me in the printing of my images; Nick Kariuki from Dallago's in Nairobi who organised my lodgings and helped make my trips possible; "Big Mama", the chief; George, our driver in the Maasai Mara; John for his kind assistance; all the rangers in the Maasai Mara and Serengeti National Parks who escorted me on my trips; Pratik Patel from Riescheux's travel company for his assistance in the Serengeti National Park; Mrs Serah Mutshia from the Kenyan embassy in Rome for her kind co-operation; Dr Guido Tosi and Dr.ssa Rossella, both Italian biologists working in Tarangire National Park in Tanzania, for their generous help and precious advice; Dr Harvey Croze and Dr Richard Estes who enriched my photographs with learned texts; and finally, to Harvill and in particular to Sophie Henley-Price, who believed in my work, for all the patience she showed me during the realisation of this volume.